Fundamental Ideas in Cosmology

Scientific, philosophical and sociological critical perspectives

Fundamental Ideas in Cosmology

Scientific, philosophical and sociological critical perspectives

Martín López-Corredoira
Instituto de Astrofísica de Canarias (IAC), Canary Islands, Spain

IOP Publishing, Bristol, UK

ISBN 978-0-7503-3775-5 (ebook)
ISBN 978-0-7503-3773-1 (print)
ISBN 978-0-7503-3776-2 (myPrint)
ISBN 978-0-7503-3774-8 (mobi)

DOI 10.1088/978-0-7503-3775-5

Version: 20220501

IOP ebooks

British Library Cataloguing-in-Publication Data: A catalogue record for this book is available from the British Library.

Published by IOP Publishing, wholly owned by The Institute of Physics, London

IOP Publishing, Temple Circus, Temple Way, Bristol, BS1 6HG, UK

US Office: IOP Publishing, Inc., 190 North Independence Mall West, Suite 601, Philadelphia, PA 19106, USA

Cover image: Cosmology research, conceptual image. Image credit: Nicolle R Fuller / Science Photo Library.

Contents

Abstract

Cosmological hypotheses should be very cautiously proposed and even more cautiously received. This scepticism is well-founded. There are scientific, philosophical and sociological arguments to support this claim. Cosmology is not a science like others since it contains more speculative elements than is usual in other branches of physics, with the possible exception of particle physics. The goal of cosmology is also more ambitious than routine theories in physics: cosmology aims to understand everything in our Universe without limit.

Physical observations (redshifts, cosmic microwave background radiation, abundance of light elements, formation and evolution of galaxies, large-scale structure) find explanations within the standard model, although many times after a number of ad hoc corrections. Nevertheless, the expression 'crisis in cosmology' stubbornly reverberates in the scientific literature: the higher the precision with which the standard cosmological model tries to fit the data, the greater the number of tensions that arise. Moreover, there are alternative explanations for most of the observations.

Only the standard model is considered by most professional cosmologists, while the challenges of the most fundamental ideas of modern cosmology are usually neglected, owing mainly to sociological factors. Funding, research positions, prestige, telescope time, publication in top journals, citations, conferences, and other resources are dedicated almost exclusively to standard cosmology. Moreover, religious, philosophical, economic, and political ideologies in a world dominated by anglophone culture also influence the contents of cosmological ideas. Nonetheless, the Universe is no mere social construct (a typical postmodern notion). Quite the contrary: the Universe exists independently of our human affairs. Although its global description may be misrepresented by our models, some of its properties and partial truths are derivable through scientific analysis.

Acknowledgement

I thank Terry Mahoney for the language revision of this book. Thanks are given to Terry Mahoney, Louis Marmet, Sébastien Comerón, Bjørn Ekeberg, Juan E Betancort Rijo, Francesco Sylos Labini, Eric J Lerner, Riccardo Scarpa and Bruno Binggeli for helpful comments and suggestions.

Author biography

Martín López-Corredoira

Martín López-Corredoira (1970–; Spain) obtained doctorates in Physics at the University of La Laguna (Tenerife, Canary Islands, Spain) in 1997 and Philosophy at the University of Seville (Spain) in 2003. He is staff researcher at the 'Instituto de Astrofísica de Canarias' (IAC, Tenerife) working in the fields of galaxies and cosmology, and has published more than 100 papers in major astrophysical journals (ApJ, AJ, A&A, MNRAS, IJMPD mainly), about half of them as first author. His contributions to cosmology include tests of different models, cosmic microwave background radiation anisotropies, large-scale structure of the Universe, ages of galaxies at high redshift, variations of constants and alternative gravity scenarios, historical evolution of measurements of cosmological parameters, and anomalous redshifts. His philosophical work includes the philosophy, sociology and history of science, the philosophy of nature and metaphysics (better termed anti-metaphysics), and other themes relating to political and ethical topics. The author is, in a broad sense, a philosopher-scientist, within a realist, materialist and sceptical tradition of continental European philosophy, but steadfastly eschewing from postmodern approaches. Previous published books include *Against the Tide. A Critical Review by Scientists of How Physics and Astronomy Get Done* (2008; as editor) and *The Twilight of the Scientific Age* (2013); and other titles in Spanish.

Chapter 1

Historical and conceptual introduction to the standard cosmological model

Some basic details, fundamental tenets, and observational pillars of the present-day version of the standard ΛCDM cosmological model are offered. This includes the historical development of the hypothesis, from Einstein to the present, and even farther back in time, to search for the origin of the fundamental ideas, along with the equations that relate the concepts of cosmology with the mathematical expression of the metric derived from general relativity. Reasons for scepticism concerning the theory are pointed out, to be further developed in later chapters.

1.1 Fundamental ideas in cosmology

Physical cosmology is a relatively recent area of scientific research, with roughly one century of life counting from the first cosmological speculations based on Einstein's general relativity. However, the interest of humanity in understanding the origin or eternity of the whole known Universe is as old as human culture, and is implicit or explicit in many cosmogonies contained in religious and philosophical ideas.

Among human societies there have always been some individuals with too ambitious megalomaniacal thoughts or delusions of grandeur[1] who considered that the whole of existence could be grasped in their hands, and they were pretty sure of their representations of the cosmos, obliging the society of their epochs to believe that they have reached absolute truths concerning the order of the Universe. In contrast, most people need to believe in something, they need to have the sensation that the forces that move the world are well identified, even if mysterious or

[1] The terms 'delusions of grandeur' and 'megalomaniacal thoughts' do not refer here to any kind of psychological judgement, but aim to characterise individuals whose ambitions regarding their creations or conquests exceed reality. For instance, these terms have been applied to personalities such as Napoléon Bonaparte (1769–1821) for his ambitious dream of conquering all of Europe. In the field of knowledge, however, we are clearly not dealing with any such ambition for political power, but more with a certain lack of recognition of our limits in achieving a complete knowledge of the entire Universe.

doi:10.1088/978-0-7503-3775-5ch1

embedded in dark elements. Only a few sceptical thinkers dared to make such claims as, 'I know only that I know nothing' (Socrates (470–399 BC)). The huge empty spaces of the cosmos and the darkness of night produce fear. Blaise Pascal (1623–1662) wrote, 'when I consider the short duration of my life, swallowed up in an eternity before and after, the little space I fill engulfed in the infinite immensity of spaces whereof I know nothing, and which know nothing of me, I am terrified. The eternal silence of these infinite spaces frightens me'. This fear makes people yearn for certainties, either gods or the solved mysteries of the Universe; hence, the creators of fantasies can achieve success in their epochs with reference to the global comprehension of nature. One may even imagine prehistoric magicians or priests looking at the sky with a mixed feeling of wonder and fear in their quest to understand how it all works, thus conceiving mainly religious ideas. Humanity has evolved, but some psychological motivations have changed little. Although our science is very different from religion or metaphysics and can claim much higher credibility, some humans—modern priests of science—still think they can get a complete explanation of the vast Universe. In that sense science has certain common characteristics with the distant past.

Even from its earliest stages, the rapidity of the development of physical cosmology did not prevent its creators from having complete faith in their speculative models, and—less than a century later—scientists today claim to have reached a fairly solid model, so much so that they no longer deem it necessary to discuss the fundamental ideas underlying cosmology. Rather, they claim, it is now time to concentrate on 'precision cosmology' (e.g., Primack 2005), in which only the minutiae of the theory are fine-tuned. In this current paradigm of the standard cosmological model, the fundamental properties of the Universe are believed to be well understood, with only minor adjustments to the basic model remaining to be made. The global picture of an expanding Universe originating during a singularity during the *Big Bang* is now taken for granted, and certain basic properties of the Universe are now considered to be known with amazing accuracy. For instance, the age of the Universe is calculated to be 13.787 ± 0.020 Gyr (Planck Collaboration 2020). It is difficult to believe that the precision with which we can determine the age of the Universe (20 million years) is much higher than that of the age of our planet. Does it not sound like the same kind of ambition that has accompanied the story of many previous representations of the cosmos throughout history? Indeed, this is reminiscent of the calculation by James Ussher (1581–1653) of the time and date of the creation within Christian religion as 'the entrance of the night preceding the 23rd day of October (…) the year before Christ 4004'.

The present book follows a different direction with respect to the one that is fashionable nowadays, which asserts that further discussion of the basic tenets of cosmology is ruled out. I shall attempt to show that 'fundamental ideas in cosmology' still warrant discussion, as there are many doubts concerning their validity, and there is a dearth of discussion of possibly erroneous statements concerning the foundations of standard cosmology. There is a significant number of results in isolated and disconnected papers that are usually ignored by leading cosmologists, and which are challenging and critical of the standard model. There are several observations that do not fit well into the current model. These observations, while posing a problem, are dismissed as peculiarities that will soon be fixed within the framework of an otherwise correct model. My intention here is to bring together many of these 'heretical' papers in order to help the more open-minded cosmologist to search the bibliography for tests of, and problems with, the standard model. The purpose of this is to give voice and visibility to those investigators who present and discuss observations that are unexplained, or apparently at odds with, the current standard cosmological model.

Most cosmologists are quite sure that they have the correct theory, and that they do not need to think about possible major flaws in the basic notions of their standard theory. They do not usually work within the framework of truly alternative cosmologies with different fundamentals[2] because they feel that these do not at present seriously compete with the standard model. These alternative models are certainly less developed because cosmologists do not work on them. It is a vicious circle. I consider this restrictive view to be unfair and I attempt here to open the door to further discussion of the fundamental observations of cosmology.

I critically review the most important assumptions of the standard cosmological scenarios. Some observations are discussed in order to show that the facts have not been strictly proven in some cases. Elsewhere, I show the durability of the standard theory against certain tests. I have chosen to review the general aspects of the foundations of cosmology as a whole rather than concentrate on certain branches of it because I am interested in expressing the caveats and open questions as a whole in order to extract global conclusions on cosmology. The goal is to bring together the work of many researchers who are not yet fully convinced of the standard view, thus allowing them to present their innovative ideas on theoretical or observational cosmology. I admit that some of the caveats presented may no longer hold, and that some of the observational measurements may be incorrect. Nonetheless, establishing who is right or wrong is not my mission here and I take no responsibility for the contents of the critical papers or those of the defenders of the standard model. I am conscious that many critical papers may need further analysis with regard to the problems they posit before reaching a firm conclusion on whether the standard model is correct or not. But my role here is not to defend any of the ideas in the citations listed in the bibliography, but to remain neutral and show that the debate is alive, and that the game of creating a definitive cosmological theory is not over yet.

The aim of this book is not to criticise one theory in order to promote another, substituting one too ambitious enterprise by another, but rather to describe a socio-cultural state of the art in cooking cosmological recipes, letting different voices be heard within a healthy tradition of pluralism. I can already advance the general conclusion, whose details will be given throughout this book: a sceptical view, a

[2] I mean with important variations in the fundamentals of the model, such as questioning the expansion of the Universe, the alternative origin of the cosmic microwave background radiation or light elements, etc. Instead, they merely investigate the same cosmological model with the same fundamentals and play with different dark energy models or other small variations.

perspective of considering that we are still very far from that complete view of the Universe, and a recognition that, although there are many elements in theories that bring knowledge of some partial reality of the cosmos, we are far from having a complete picture. There are too many elements of our culture that contaminate our research to permit us to consider the global cosmological enterprise as a purely scientific and objective one.

This book is divided into three parts. The first part (chapters 1 and 2) gives some basic descriptions of the theoretical ideas underlying cosmological theories; the second part (chapters 3–7) presents the recent debate concerning the comparison of observational data with the fundamental ideas of the standard cosmological model; the third part (chapters 8–10) offers philosophical and sociological perspectives. While the first two parts present scientific contents in the literature, the third is more a personal interpretation of the topics covered in this book. Nothing lasts forever, neither cosmological theories and their arguments in favour or against, nor any other perspectives. Nonetheless, the reader needs to bear in mind that the technical discussions in the scientific part will most probably become outdated within a few decades. If this book were to be read after, say, 2100, when new problems and discussions would presumably have arisen in cosmology, the text of the first two parts would most probably be considered obsolete. There are indeed some topics discussed in this book that are already considered obsolete, but which are included in order to provide a broad overview of debates in recent decades. Nevertheless, the viewpoints in the third part might still remain valid, since they are general reflections on science and humanity and the mentality of cosmologists at any given epoch.

1.2 Cosmology in western culture before the twentieth century

Every civilisation has its own cosmology that attempts to explain the order of our visible cosmos, although the observable Universe was of course much smaller in the past than it is today, and almost restricted to the Earth, Moon, Sun and planets of the solar system and the sphere of fixed stars. The most primitive societies have only proposed magical–animistic or mythological–religious cosmological views. Advanced civilisations have developed more rational perspectives. There are also many metaphysical ideas related to our vision of the totality of the existence (*Weltanschauung*), instead of cosmology. Only astronomical views concerning cosmology are expounded here, leaving aside all ethnoastronomical aspects. I do not offer a full description of every cosmology in the history of astronomy—not even of astronomical cosmologies generated in western culture, since a proper treatment of these would require several books. That is not my purpose here, but I briefly mention some of them to illustrate the paths humanity has taken before reaching our present-day scientific outlook.

Within western culture, it was ancient Greece that witnessed the dawn of reason and an empirical approach that gave rise to philosophy and the natural sciences, thus setting a new horizon of intellectual ideas that tried to explain our cosmos. Indeed, cosmological speculation and natural philosophy were born together, and both disciplines flourished together at different periods throughout history

(Gale 1993). Thales of Miletus (c. 623–c. 545 BC) devised a cosmology based on water as the essence of all matter, with the Earth as a flat disc floating on a vast sea, ideas that were indeed contained in previous Mesopotamian or Babylonian cosmologies. Anaximander (c. 610–c. 546 BC) conceived a Universe with the Earth at its centre. Among his many other achievements, he was already aware that the Moon reflected the light of the Sun and described the Earth as spherical—so already at that time the idea of a flat Earth was obsolete—and knew the origin of equinoxes and solstices. Also, Pythagoras (c. 570–c. 490 BC) and his disciples knew that the Earth was spherical and had decomposed solar motion into two components: a yearly and a daily one. Philolaus (c. 470–c. 385 BC), a follower of Pythagoras, proposed a model in which the Earth, Moon, Sun and planets all moved around a central fire. Since the Earth was much closer to this central fire than the rest of the heaven bodies, it would be almost at the centre of the Universe. More fully developed mathematical models or philosophical considerations applied to the planetary motions would come later, as conceived in the mind of Plato (c. 427–347 BC), Eudoxus of Cnidus (c. 390–c. 337 BC), Aristotle (385–322 BC), and Heraclides Ponticus (c. 390–c. 310 BC), all from a geocentric point of view. Heliocentric models would start with Aristarchus of Samos (c. 310–c. 230 BC) and Seleucus of Seleucia (c. 190–c. 150 BC). As is well known, however, the Aristotelian geocentric view, maintained by Apollonius of Perga (c. 262–c. 190 BC) and Ptolemy (AD c. 100-c. 170), would prevail until the end of the Middle Ages in western civilisation. The soul of the cosmos in Plato's view was its principle of eternal and recurring circular and uniform motions, and this doctrine prevailed in Aristotle's writings and generated later models, even the heliocentric ones. In Ptolemaic astronomy, there was a complex machinery of epicycles, equants and deferents, devised to save the idea of uniform circular motion with the Earth at the centre.

Not only were the motions of the planets, together with those of the Earth, Moon and Sun, given explanations, but other astronomical considerations extended the cosmological horizon. The Milky Way was a very evident visible structure in the sky. Although ancient philosophers had no idea of its dimensions or the distances involved, speculation concerning its nature flourished very early on. For instance, Anaxagoras (500–428 BC) posited that the Milky Way was a reflection of light emitted by stars different from the Sun. Theoretical speculations about the size of the Universe were also in the mind of Greek philosophers. Archelaus (nicknamed 'the physicist', fl. 5th century BC), a disciple of Anaxagoras and master of Socrates, claimed that the Universe has no limits. Archimedes of Syracuse (c. 287–c. 212 BC) would also make the Universe vastly larger than was then believed, because no stellar parallaxes were measurable at that time, and a moving Earth in a heliocentric model should produce a parallax in the apparent positions of the fixed stars, unless the stars were too far away for their parallaxes to be measured. However, Archimedes proposed a maximum size of 0.5 parsec by assuming that the ratio of the diameter of the Universe to the diameter of the orbit of the Earth around the Sun was equal to the ratio of the diameter of the orbit of the Earth around the Sun to the diameter of the Earth.

The medieval schoolmen absorbed parts of the cosmologies of Plato and Aristotle, although adding other elements connected to religion. For instance, in 1225, the bishop Robert Grosseteste (1175–1253) described a cosmological model reminiscent of the present one, in which the Universe is created in an explosion and subsequent condensation. God issued His first fiat, 'Let there be light', and at that precise instant light issued from the divine and entered matter. The power of this luminosity was not so great as to produce a further expansion of the outermost parts of this mass to the highest degree. Grosseteste speculated that there was some form of coupling between light and matter, consequently giving rise to the material body of the entire cosmos. Expansion takes place when matter reaches a minimum density and subsequent emission of light from the outer region leads to the creation of the inner bodily mass so as to create nine celestial spheres (Sparavigna 2014, Bower *et al* 2018).

In *The Divine Comedy*, an allegorical vision of the afterlife and Christian world-view, Dante Alighieri (*c.* 1265–1321) offers a typical example of cosmovision. He puts the Earth at the centre of the Universe in accordance with the Aristotelian model. Inside the Earth is found Hell, divided into nine circles, representing increasing levels of sin. Between the surface of the Earth and sphere of the Moon lies Purgatory. Outwards, the Earth is surrounded by whirling spheres made of transparent solid matter. Added to the eight Aristotelian spheres corresponding to the planets, Moon, Sun, and the fixed stars, there is a ninth sphere, the *primum mobile*, the source of the motion of all the inner planetary spheres. Beyond the *primum mobile* lies the spiritual Universe, the mind of God, or Empyrean heaven; this sphere thus marks the boundary between the natural and supernatural worlds (see figure 1.1). This is certainly not a purely scientific approach, but a religious interpretation mixed with previous scientific ideas. Some of its elements might still be present in some degree in the construction of putatively *scientific* modern cosmology (Binggeli 2006, 2017) (see section 9.3). The cosmologies of much older religions, such as Vedism, developed in 1700–1100 BC, or Jainism (c. 500 AD), also provided fanciful explanations of the entire Universe. In the Middle Age, other cultures gave place to more sophisticated scientific models of the solar system such as that of Aryabhata (476–550), Nilakantha Somayaji (1444–1544) in India, and those of the Maragha school in Muslim countries (13th–16th centuries) based on the Aristotelian model.

The new change of paradigm, or recovery of an old one proposed by Aristarchus of Samos, arose in Europe with Nicolaus Copernicus (1473–1543), Giordano Bruno (1548–1600), Galileo Galilei (1564–1642) and Johannes Kepler (1571–1630), who reintroduced and finally established the heliocentric model. With Kepler's discovery of his first two laws of planetary motion, introducing elliptical orbits and abandoning uniform circular motion in the heavens, the stage was set for the development by Isaac Newton (1643–1727) of a new physics and theory of universal gravitation. Moreover, beyond the solar system, Bruno defended an infinite Universe with an infinite number of suns (stars), and Galilei was the first to observe the Milky Way Galaxy as individual stars through the telescope, instead of a continuous cloud or nebulosity as previously thought. The new ideas and observations expanded the limits of the Universe far beyond the solar system, and the new cosmologies would

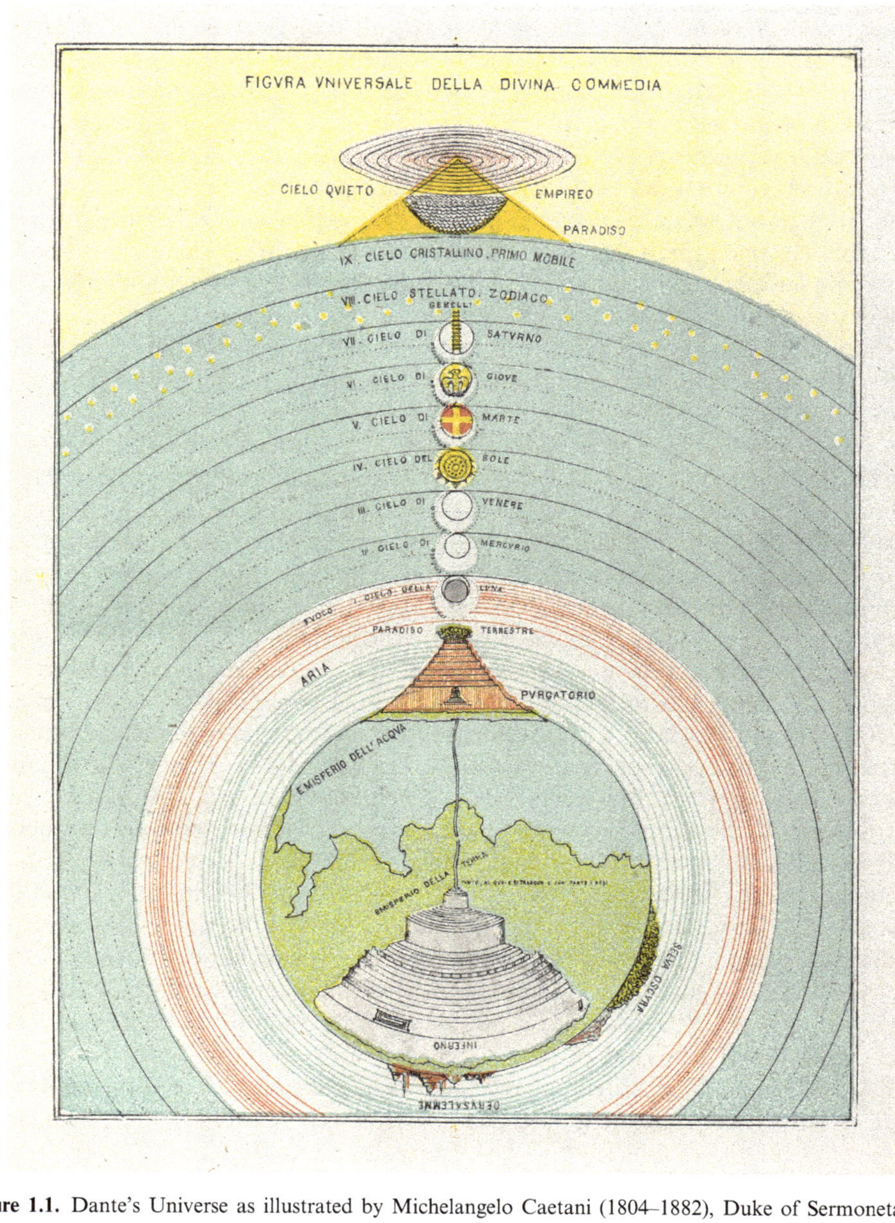

Figure 1.1. Dante's Universe as illustrated by Michelangelo Caetani (1804–1882), Duke of Sermoneta and Prince of Teano (Reproduced from https://commons.wikimedia.org/wiki/File:Michelangelo_Caetani,_Overview_of_the_Divine_Comedy,_1855_Cornell_CUL_PJM_1071_01.jpg. Original Caetani 1821. Image stated to be in the public domain.).

have to give explanations for this new, much wider, scenario, which would be either infinite or at least include all of the visible stars in the Milky Way.

Newton's physical cosmology—leaving aside his religious beliefs—was the scenario in which to refer a scientific vision of the Universe in 18th and 19th centuries, thanks in particular to his theory of gravitation, which imposed some order in the dynamics of the stars and planets, and also the asteroids, comets, Moon and Sun, which constituted all the astronomical objects then known. The cosmological question in Newton's physics, however, posed a great problem that needed to be solved: Bentley's paradox concerning the permanence and stability of the system. Why has not the whole Universe collapsed through the action of gravity? According to Newton, each star in a finite Universe should be attracted towards every other star, such that they all fall together at some central point. Newton acknowledged the problem in a letter to Richard Bentley (1662–1742), a leading Cambridge philosopher at the time (Kerzsberg 1986). Newton also thought about solving this issue with an infinite Universe containing a totally homogeneous distribution of mass, but he realised that this was an unstable solution: even a very small nudge exerted on a star would cause a slight deviation from uniformity in the mass distribution that would produce a cascade causing all matter to collapse. In the end, Newton skipped the paradox by claiming that God prevented the collapse by making 'constant minute corrections', so this stability and permanence must be an action of God (Kerzsberg 1986). The impression after the end of 17th century was that there were no cosmological models that could be fully understood in scientific terms, but that key to understanding was through a thorough understanding of the gravitational interaction among the different bodies of the Universe. This idea has remained embedded in the mentality of all cosmologists, who still think that a gravity theory immediately gives us a cosmological theory.

Western intellectuals of the 18th century were impressed by the power of Newtonian physics + gravitation in comprehending all astronomical phenomena; they were in awe of the genius of the mathematician who could unlock the secrets of the Universe. Newton's laws served not only to explain the observed motions of the planets, but also to make predictions of other events. Society has always been fascinated by predictions, such as the supposed first ever prediction of a solar eclipse by Thales of Miletus in 585 BC, or the confirmation by Eddington and his collaborators in 1919 of Einstein's theory of general relativity with the measurement of the Sun's deflection of the light from stars observed during an eclipse. All these events drew people towards reason and science, as happened, for example, when Edmund Halley (1656–1742) predicted the return of the comet later named after him in late 1758–early 1759, which earned him much acclaim when his successful prediction gloriously vindicated Newton's theory of gravitation (Wallis 1984). We might perhaps compare the physicists who could manipulate gravity equations to foretell how the heavens moved to the magicians and prophets of primitive societies. There is the suggestion of a sense of immensely powerful mini-gods who alone understand the Universe as a whole, and this sense of wonder is the essence of cosmology as an intellectual movement. Consequently, when the next towering genius of gravitation arrived in the form of Einstein, the ground was laid for further claims to possessing keys to understanding the entire Universe, as we shall see in the next section.

The word *Cosmology* applied to the study of the Universe as a whole was first used in a work on metaphysics *Cosmologia Generalis* (1731) by Christian Wolff (1679–1754), a scientific study of the Universe which involved physics, astronomy and philosophy, but also including esotericism and religion. Nonetheless, 18th and 19th science would be more devoted to a Universe without metaphysics, without God, a materialistic science, aimed at understanding not only the structure and dynamics of matter, but also its origin and evolution.

In 1750, Thomas Wright (1711–1786) published *An Original Theory or New Hypothesis of the Universe*, in which he suggested that the stars were located in spherical shells or rings around a centre by which they were attracted and gave it some metaphysical interpretation. He envisaged the Milky Way to be a transversal section of the Universe when viewed from the great centre. The philosopher Immanuel Kant (1724–1804) read the book by Wright and did not accept the supernatural claims but, inspired by it, proposed in his work *Allgemeine Naturgeschichte und Theories des Himmels* (1755) that the Universe to comprise a number of 'island universes', and gravity to create a hierarchical structure of planets around stars, the stars accumulating to form island universes, which would also be clustered into groups of island universes. At the beginning of the 20th century, this scenario would be confirmed, the galaxies being these island universes and the Milky Way itself a galaxy. Johann Heinrich Lambert (1728–1777) also developed a theory of the generation of the Universe that was similar to the nebular hypothesis that Wright and Kant had proposed. Lambert published his own version of the nebular hypothesis of the origin of the solar system in *Cosmologische Briefe über die Einrichtung des Weltbaues* later, in 1761, although started independently in 1749, before Wright's and Kant's publications. Lambert hypothesised that the Sun was part of a group of stars which travelled together through the Milky Way, and that there were many such groupings (star systems) throughout the Galaxy. The appearance of the Milky Way could be accounted for by assuming it to be made of a ring of stars all about equally distant (Gray 1978).

The nineteenth century would be more conservative in its speculations concerning cosmological scenarios and more focused on the development of a serious science, such as astrophysics, which tried to explain certain phenomena in the Universe separately. Nonetheless, there are remarkable examples of the interest of different scientists and thinkers on the mysteries of the extension of the cosmos in space and time. Heinrich Wilhelm Olbers (1758–1840) put forth his famous paradox, according to which the darkness of the night sky conflicts with the assumption of an infinite and eternal static Universe, the only possibility in terms of Newtonian physics. Indeed, the idea had already been proposed much earlier, for instance by Thomas Digges (c. 1546–1595).

The poet and writer Edgar Allan Poe (1809–1849) would speculate a solution to this paradox and would also suggest the expansion and collapse of the Universe in his literary text *Eureka: a Prose Poem* (1848): he rejected the idea of an infinite Universe to solve Olber's paradox and reasoned that a Universe governed by gravitation would collapse into a heap if not kept apart by some form of repulsion. Poe postulated that God had, in an enormous explosion at the creation, thrust all the stars apart which would first expand and then contract into a final catastrophe, the end of the world. The similarity of this scenario with the present one may us lead to think that, although scientifically unexplained, a Universe in expansion was already in the air before the Newtonian gravity was substituted to make possible this fantasy of Poe. But in Poe's idea of a finite Universe we have the problem of the borders of the Universe, which could not be explained by the poet in a logical way.

Another key element in this century would be the mathematical development of non-Euclidean geometries with more than two dimensions. The Russian mathematician Nicolái Lobachevsky (1792–1856) stated that if space is either Euclidean or negatively curved, like the surface of a saddle, it must be infinite. However, there are flat non-Euclidean universes that may be finite. Using a two-dimensional analogy, such a space could have the topology of a giant torus (Silk 2001). There are indeed 18 distinct types of flat spaces, with only ten of them being compact, the others being infinite in one or more directions (Silk 2001). This gives a solution to Olbers' paradox, by making possible a finite Universe without borders. The non-Euclidean geometries would also be an element to be included in the metrics of the cosmological models to be developed in the next century. In particular, Bernhard Riemann (1826–1866) founded the field of Riemannian geometry, a tool that would later prove to be necessary for the mathematical formulation of general relativity.

1.3 Origin and evolution of the standard cosmological model

There remains much to discuss concerning the origin and evolution of the Universe. It may still not be clear whether the Universe had a beginning or not, or whether its evolution followed the present models or not. There are aspects of nature that still defy explanation, especially where great distances in space and time are concerned and for which the history of the Universe that cannot be probed at present. Nevertheless, the scientific community is much closer to hand, and it is easier to track the origin and evolution of human ideas, together with their causes and motivations.

Since the beginning of the 20th century, a continuous evolution and perfection of what we today call the standard cosmological model has been produced, although some authors like to distinguish separate periods within this evolution. For instance, Lerner (1991) distinguished four periods prior to 1991: (1) before the end of World War II; (2) between 1945 and 1965, from Gamow to the official discovery of the cosmic microwave background radiation; (3) between 1965 and 1980; (4) from 1980 onwards, with the introduction of inflation. Another possible historical division of the development of cosmology into six periods was proposed by Luminet (2008): (1) initial period (1917–1927); (2) the period of development (1927–1945); (3) the period of consolidation (1945–1965); (4) the period of acceptance (1965–1980); (5) the period of enlargement (1980–1998), and (6) the period of high precision experimental cosmology (1998–); which is equivalent to Lerner's proposal with the exception that Lerner's first period is subdivided into two blocks, and after 1998, the year of the establishment of the dark energy hypothesis, is added a new period, which of course was not contained in Lerner's division because it had been stated much sooner, in 1991. Certainly, there are phase transitions—small revolutions—within the smooth evolution. Locating them in time is somewhat subjective; it is easier to see history as a continuous quest for the confirmation of an idea conceived *a priori*.

At the beginning of 20th century, two great achievements in physics and astronomy initiated the journey towards the standard cosmological model as we know it today.

First, the observational evidence for the existence of many galaxies separated by very large distances—much larger than the usual distances managed by astronomers previously—the Milky Way thus being only one galaxy among many. It was definitively established after a period of discussion that finished with the Great Debate in 1920 (Hetherington 1993) between the American astronomers Heber D Curtis (1872–1942), who defended the hypothesis that some nebulae (now called galaxies) were not part of the Milky Way but were located at very large distances from it, and Harlow Shapley (1885–1972), who claimed that these nebulae were part of the Milky Way. At first, Shapley was more convincing among astronomers[3]. Curtis was later demonstrated to be the clear winner, although Shapley was at least right in his statement that the Sun was not at the centre of the Milky Way. In my opinion, this was the most important revolution in astronomy after Copernicus and Galilei. The Sun was now no longer at the centre of our Galaxy, and the Milky Way no longer occupied a privileged position in the Universe, but was merely one galaxy among many other galaxies. A fresh blow to the belief that our civilisation and our planet occupy an important position in the Universe. This achievement gave rise to the subsequent development of extragalactic astronomy and, implicitly, a new cosmological vision was emerging out of this scenario: a vision of a Universe of vast spaces, impossible to imagine, where galaxies are the fundamental components in a larger-scale structure.

The other great achievement came from physics in the form of Albert Einstein's (1879–1955) general relativity. Certainly, his earlier discovery of special relativity was also very important, but for astronomical, and particularly from the perspective of cosmology, general relativity was the long-awaited breakthrough. Newton's magnificent achievements had blocked the free expansion of cosmological ideas because of the problems in solving the stability of systems without an eventual collapse and having recourse to godly intervention[4]. Einstein was like a Messiah of Gravity, resurrecting fervour for a global comprehension of the Universe. The manifestations of the new paradigm would come immediately after among many of the brightest minds in physics and astronomy (plus the many thousands of amateurs who try to imitate him or challenge relativity today). The father of general relativity himself produced the first steps towards a cosmology in an early proposal (Einstein 1917) when he posited a static model that included a cosmological constant to guarantee stability. He would later recognise this proposal as his 'biggest blunder'. This hurry to produce a cosmology, only two years after the publication of general relativity, may be the reason why this first approach did not last long. It evidences

[3] Indeed, Shapley was more interested in defending the hypothesis that the size of the Milky Way was very large (diameter of 100 kpc) and that the question of spiral nebulae was secondary, but he was wrong on both counts. Another concern of Shapley's in the debate was his candidacy for the directorship of the Harvard college Observatory.

[4] Indeed, a model of an expanding Universe could be obtained even within a Newtonian cosmology, as was shown by Milne (1933, 1934), by maintaining an infinite Euclidean space, with Newtonian gravity and regarding expansion as a pure Doppler effect in the recession of the galaxies. Many facts and equations that were explained by the standard model with general relativity could also be explained with Newtonian cosmology. There remained some problems (stability, Olbers' paradox), but there were also proposals to solve them without general relativity (Baryshev and Teerikorpi 2012 [section 7.1.3]).

how eager Einstein and his contemporaries were to probe the deep truths of the wider Universe.

The models that would constitute the basis of our present standard cosmology came a little later. The basic idea assumed is that the current Universe is homogeneous on a large scale, and that the distances among all the different objects are currently growing owing to the expansion of the Universe, a recession of objects with respect to one another on a large scale. On small scales, different objects could cluster together because their gravitational attraction overcomes the expansion. The Russian physicist Alexander Friedmann (1888–1925) developed the basic aspects of the application of general relativity to a cosmological model (Friedmann 1922, 1924).

The German astronomer Carl Wirtz (1876–1939) noted in 1924 a correlation between the faintness of a galaxy and its redshift. Edwin P Hubble (1889–1953) and Milton Humason (1891–1972) measured the distance of a number of galaxies during the same year and would later find the famous Hubble–Lemaître law of the linear relationship between radial velocities and distances. The redshifts used by Hubble had been measured by Vesto Slipher (1875–1969), published by Arthur S Eddington (1882–1944) (Eddington 1923) and added to by Humason (1929) before Hubble (1929)'s famous announcement, interpreting the law as a proof of the expansion of the Universe, whose theoretical models were known to Hubble (Sandage 1995). It is curious that Hubble derived his law using a sample of only 24 nearby galaxies. At such small distances, peculiar motions dominate over recession. Therefore, his famous discovery was based on a coincidence that all the galaxies with dominant peculiar motions happened by chance to follow a linear law of velocity with distance. This shows us how often the theoretical preconceptions guide research (Bonometto 2001).

Prior to Hubble's publication in 1927, the Belgian Catholic priest, physicist and astronomer Georges Lemaître (1894–1966) developed a theoretical model of an expanding Universe in an extension of the work of Friedmann. The work by Lemaître (1927) was published in French in a small Belgian journal, and also tells us about the recession of galaxies and the recession rate in the linear velocity–distance relationship, including an analysis of observational data, as rediscovered later by Hubble in 1929. Hubble did not cite the paper by Lemaître and took all credit for the discovery of the expansion of the Universe. His collaborator Humason told once in an interview offered in 1965 that Hubble knew about the velocity–distance relationship of Lemaître from a talk offered at an IAU meeting in Holland in 1928 (Llallena Rojo 2017 [p 90]), although it is not entirely clear whether this was true or whether Hubble was really unaware of Lemaître's discovery.

In 1931, Lemaître's paper of 1927 was translated into English for the journal *Monthly Notices of the Royal Astronomical Society* (MNRAS) and this translation completely omitted some paragraphs and formulae from the original paper that referred to the analysis of data showing a recession of galaxies and the Hubble–Lemaître law (Nussbaumer and Bieri 2009, Bergh 2011, Block 2012). The reasons for this omission have been widely discussed (Block 2012). It was later clarified (Livio 2011, Luminet 2013) that Lemaître himself did the translation of his paper and deliberately omitted the relevant paragraphs. Lemaître wrote in a letter to the editor of MNRAS: 'I did not find advisable to reprint the provisional discussion of radial velocities which is clearly of no actual interest, and also the geometrical note, which could be replaced by a small bibliography of ancient and new papers on the subject. I join a French text with indication of the passages omitted in the translation.' What an awkward action!

Another line of development of the cosmological model was suggested by the Japanese physicist Seitaro Suzuki, who suggested that the observed helium–hydrogen ratio might be explained 'if the cosmos had, at the creation, the temperature higher than 10^9 degrees' (Suzuki 1928). Lemaître in 1931, with the expansion and the arrow of time from the second law of thermodynamics in mind, developed his concept of the 'primeval atom' (Lemaître 1931, 1946), the first version of what later would be called the 'Big Bang'. According to him, the initial state of matter in the Universe might be thought of as a sea of neutrons. Lemaître thought that cosmic rays were relics of primordial decays of atoms, which was demonstrated later to be wrong. Moreover, his ideas on stellar evolution were also demonstrated to be wrong during the 1930s so, by the end of the decade, the primeval-atom hypothesis had been generally rejected by the scientific community.

Another important protagonist during the gestation of modern cosmology was the above-mentioned British astronomer Arthur S Eddington (1882–1944), who was one of the main scientists responsible for spreading and publicising general relativity and its implications for cosmology, apart from giving the most remarkable impulse to the theory by his observations of the solar eclipse in 1919. Eddington (1929, 1931) was also a defender of a finite Universe and the extrapolation of the second law of thermodynamics to the whole Universe. Another British physicist supporting this approach was James Jeans (1877–1946), in a work titled 'The Physics of the Universe' (Jeans 1928). Lemaître and Eddington proposed philosophical arguments that excluded infinite universes that were also compatible with general relativity.

Some philosophers or historians of science (e.g., Gale 1993) consider the birth of modern cosmology to have occurred on 29 September 1931, when the British Association convened a special session devoted solely to the topic 'The Evolution of the Universe' (Dingle 1931). At the meeting, all the major cosmological workers had reached consensus on two essential points: (1) they had a science; (2) this science was deployed about the general theory of relativity as its central model. Hubble, however, refrained from accepting the consensus of a relativistic model of an expanding Universe until at least 1937.

After World War II, George Gamow (1904–1968), a Russian physicist who emigrated to US in 1934 where he would develop his cosmological ideas, compared the detonation of an atomic bomb with the origin of the Universe and popularised the 'Big Bang'[5] theory (Gamow 1947). He and one of his students, Ralph Alpher

[5] In fact, the name 'Big Bang' was not given by Gamow, but by one of the opponents of his theory, Fred Hoyle (1915–2001), who dubbed Gamow's primaeval atom theory as the 'Big Bang' in order to ridicule it. However, the name caught on. Several decades later, in 1993, the journal *Sky & Telescope* set a competition for a suitable alternative name for the standard theory. After receiving many proposals, they could not find anything to beat 'Big Bang'. Hoyle would say 'Words are like harpoons, once they go in, they are very hard to pull out' (Horgan 1996 [chapter 4]).

(1921–2007), published a paper in 1948. Gamow, who had certain sense of humour, decided to put the reputed physicist Hans Bethe (1906–2005) as second author, even though he had not participated in the development of the paper, so the result was a paper by Alpher, Bethe and Gamow (Alpher *et al* 1948), to rhyme with 'alpha, beta and gamma'. Later, Robert Herman (1914–1997) joined the research team, but—according to Gamow—he stubbornly refused to change his name to 'Delter'.

Rather than searching for an explanation of cosmic rays, as Lemaître did, Gamow and his collaborators attempted to explain the abundance of the elements, assuming that there was no process that could explain the present-day abundances. They contended that the heavier elements must have been formed during an early hot initial stage of expansion, since they thought that stars could not achieve temperatures high enough to produce them. This would be rejected by Hoyle (1946, 1947), who showed that heavy elements can indeed be formed in the stars. A detailed theory by Margaret Burbidge (1919–2020) and collaborators would later show how stars could produce elements in proportions very close to those observed to exist (Burbidge *et al* 1957). However, the theory by Burbidge *et al* could not explain the abundances of helium, a quarter of all matter, and it was hard to see how certain light elements (deuterium, lithium, beryllium and boron) could survive at all. Another attempt was made at the beginning of the 1960s by the Soviet physicist Yákov Zel'dovich (1914–1987), who proposed a cold Universe scenario that predicted the conversion of all matter not into helium, as in the former version, but into pure hydrogen (Zel'dovich 1963).

Alpher and Herman (1949) and Gamow (1953) also predicted an early stage of the Universe that would produce a relic radiation that could be observed at present as a background in microwave wavelengths, corresponding to the epoch of decoupling of matter and radiation. Alpher and Herman calculated the necessary mass density of neutrons and protons to make the helium abundance agree with the observed value (Burbidge 2006). Nevertheless, Gamow and his coworkers were of the opinion that the detection of that microwave radiation was completely unfeasible (Novikov 2001). The first published recognition of the relic radiation as a detectable microwave phenomenon was in 1964 by the Russian cosmologists Andrei Doroshkevich (1937–) and Igor Dmitriyevich Novikov (1935–) (Doroshkevich and Novikov 1964). Then came the official discovery of the cosmic microwave background radiation by Arno Allan Penzias (1933–) and Robert Woodrow Wilson (1936-) (Penzias and Wilson 1965), although this same radiation had been previously directly or indirectly observed by other researchers[6].

Another piece of evidence supporting the standard model of the expanding Universe came from Malcolm Longair (1941–) and Martin Ryle (1918–1984), who

[6] Shmaonov (1957) from the former Soviet Union was measuring radio waves coming from space at a wavelength of 3.2 cm and concluded that the absolute effective temperature of the background radiation appeared to be 4 ± 3 K, independent of the direction of the sky. It is also possible that a team of Japanese radio astronomers measured this radiation at the beginning of 1950s (Novikov 2001). It was also found by Andrew MacKellar (1910–1960) in 1941 as the radiation necessary to excite rotating cyanide molecules (Novikov 2001). Herzberg (1950) also mentions that there is a strange excitation of molecular spectra, as if a 2.3 K radiation existed.

argued that the data indicate that the Universe must be evolving (Longair 1966, Ryle 1968). The galaxies at high redshift—that is, at great distance—showed distributions and properties different from those at low redshift. Since at larger distances we are observing the past Universe, given the limited speed of light, this implies that the distant galaxies belong to an epoch of the Universe that was much earlier than the present one. This would be a strong argument against alternative models, particularly the steady state model (see section 2.2), which assumed that the Universe never changes. The story of the understanding of galaxy evolution also involves Philip James Edwin Peebles' (1935–; Nobel Prize for Physics in 2019) initial suggestion of a galaxy formation model that starts in the early Universe with baryon only perturbations to get around the smoothness of the Cosmic Background Radiation (Peebles and Yu 1970).

This confirmation of the predicted microwave radiation, even if the predictions did not completely fit the observations (see section 5.1), and evolution of the Universe gave confidence to those cosmologists who supported the standard model. Many hitherto sceptical physicists and astronomers became convinced that they now had a solid theory. By the mid-seventies, cosmologists' confidence was such that they felt able to describe in intimate detail events of the first minutes of the Universe (Weinberg 1977). Nonetheless, there were problems that remained to be solved, such as why the Universe appeared to be the same in all directions (isotropic), why the cosmic microwave background radiation was evenly distributed, and why its anisotropies were so small. Why was the Universe flat and the geometry nearly Euclidean? How did the large-scale structure of the cosmos originate? Clearly, work on the fundamental pillars of the cosmological edifice remained to be done.

In the 1980s, proposals were brought forth to solve these pending problems, with inflation as the leading idea in the solution of cosmological problems at the beginning of the Universe, and the idea of non-baryonic dark matter as a new paradigm that allows the theory to fit the numbers of some observations. Grand Unified Theories of particle physics would also support the existence of non-baryonic dark matter. In chapter 3, I give details of the motivation and evolution of these two ideas. Also, the joining of cosmology and particle physics and scenarios containing baby universes, wormholes, superstrings and other exotic ideas were born. This excess of theoretical speculation not based on observations has led some authors to call this epoch the era of post-modern cosmology (Bonometto 2001). This union between cosmology and particle physics is due in part to the halting of particle physics experiments because of their escalating cost, a situation that led many particle physicists to move over into cosmology, wishfully contemplating the Universe as the great accelerator in the sky (Disney 2000, White 2007). Alas, particle physicists lack the necessary astronomical background—complained Disney—to appreciate how soft an observational, as opposed to an experimental science, of necessity has to be.

In the 1990s, a third patch was applied to the theory in an effort to solve new inconsistencies with the data in the form of dark energy, which supposedly produced acceleration in the cosmic expansion. The problems to be solved were basically the new Hubble–Lemaître diagrams with type Ia supernovae as putative standard candles, the numbers obtained from cosmic microwave background radiation anisotropies, and especially estimates of the age of the Universe, which were inconsistent with the calculated ages of the oldest stars. I will offer further details about the emergence of the dark energy idea in chapter 3.

The renovated standard model including these new elements added ad hoc would come to be called the ΛCDM cosmological model, where Λ stands for dark energy and CDM stands for cold dark matter, the favoured subgroup of models of non-baryonic dark matter[7]. A graphical scheme like the one given in figure 1.2 represents

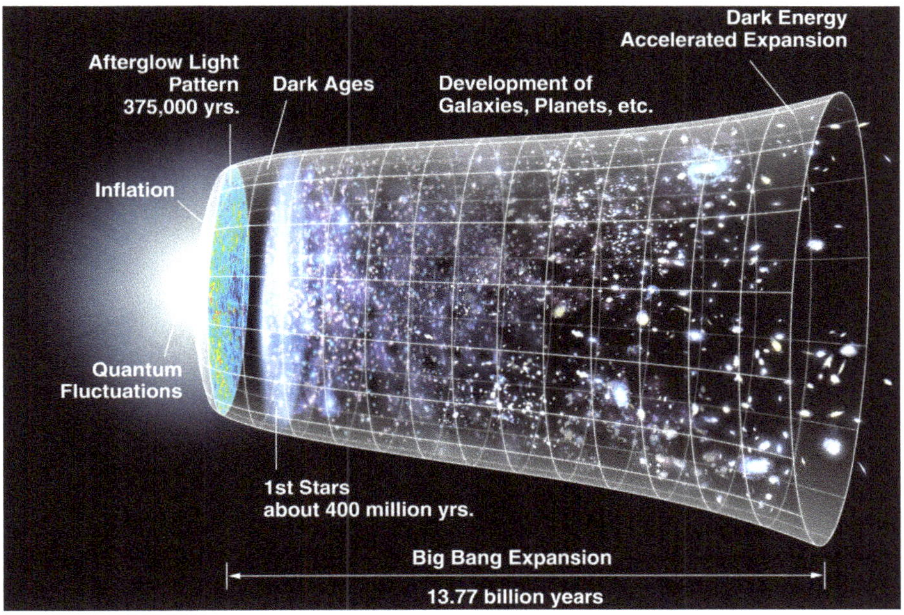

Figure 1.2. Graphical representation of the evolution of the Universe over 13.77 billion years, including a period of 'inflation' that produced a burst of exponential growth in the Universe. Later, the expansion of the Universe gradually slowed down as the matter gravitationally pulled in on itself. More recently, the expansion has begun to speed up again as the repulsive effects of dark energy have come to dominate the expansion of the Universe. The Cosmic Microwave Background Radiation was formed 375 thousand years after the beginning of the Universe. (Reproduced from https://map.gsfc.nasa.gov/media/060915/index.html by NASA/WMAP Science Team, WMAP # 60915, uploaded 30 May 2018. Image stated to be in the public domain).

[7] The definition of cold or hot dark matter refers to the velocity of the particles that constitute it, which grows when the mass of the particles is lower, so cold matter means massive particles. In the cold dark matter (CDM) theory, the structure of the large-scale distribution of galaxies grows hierarchically, with small objects collapsing under self-gravity first and then merging in a continuous hierarchy to form larger and more massive objects. It is the opposite of the hot dark matter (HDM) paradigm, which was more commonly used in the early 1980s, where structure does not form hierarchically (bottom-up), but forms by fragmentation (top-down), with the largest superclusters forming first in flat pancake-like sheets and subsequently fragmenting into smaller pieces that constitute the galaxies. There are also models that are a mixture of cold and hot dark matter, called warm dark matter (WDM), or interacting dark matter (iDM) that have been competing with CDM in recent years (e.g., Bose *et al* 2019).

the Universe according to the standard model. Some cosmologists referred to it as 'concordance cosmology', to emphasise that this model is in agreement with all the known observations. As said by Merritt (2017), when cosmologists speak of 'concordance', they mean that it is possible to find a single set of parameters that provides an acceptable fit to the conjunction of observational data sets, but not that there is an independent confirmation of the value of any single parameter.

Some authors, critical of the standard model (e.g., Stubss 2007, Hartnett 2008), prefer to call it 'consensus cosmology' rather than 'concordance cosmology', wishing to emphasise that this new cosmology is above all a sociological question of agreement among powerful scientific teams in order to establish the orthodoxy of a fundamental dogma. This agreement would be mainly between two powerful cosmological groups, the teams dedicated to the analysis of supernovae and the cosmic microwave background, who found a rough coincidence in the necessary amount of dark energy, although with large error bars, that reinforced their belief that they had discovered an absolute truth, thus compelling the rest of the community to accept this truth as a solid standard, while at the same time discarding the results of other less powerful cosmological groups that presented different values of the parameters. Talking about consensus cosmology, Rudolph ('Rudy') Schild (1940–) once queried, 'Which consensus? Do you know who consented? A bunch of guys at Princeton who drink too much tea together' (Unzicker and Jones 2013 [chapter 3]).

With this, we reach 1998 to cover the period from the beginning of the period of precision cosmology until today (e.g., Primack 2005, Luminet 2008). Rather than major discoveries or proposals, I would emphasise the lack of discussion on the fundamental ideas in cosmology dating from this epoch, when it becomes a tenet of belief that all the major problems have been solved. This state of complacency has resulted in an excess of confidence in the robustness and superiority of the standard model with respect to any alternative model. Certainly, some minor topics are being debated, such as the equation of state of dark energy, the types of inflation or the coldness or hotness of dark matter, but these are subtleties within the major fundamental scheme. This is the epoch in which the main enterprise of cosmology consists in spending big money on megaprojects that will achieve accurate measurements of the values of the cosmological parameters and solve any small problems that remain to be explained. This is also the epoch of highest social recognition of cosmology: Not only do schools, museums, and popular science journals talk about the Big Bang as well established, to be compared to Darwin's evolution and natural selection theory, but cosmology now occupies a privileged ranking among the most prestigious natural sciences. For instance, cosmology and its dark elements have been awarded with Nobel Prizes in Physics in 2011 and 2019, respectively for the putative discovery of the dark energy that produces the acceleration of the expansion, and for the inclusion of the dark components in our understanding of the Universe. One may wonder whether unconfirmed quasi-metaphysical speculations should properly form part of the body of recognised knowledge of physics, leaving behind the conservative tradition of Nobel committees not awarding prizes for speculative proposals. Einstein did not receive either of his Nobel Prizes for his discovery of special and general relativity; neither did Curtis for his definitive recognition of the true nature of galaxies in the Great Debate of 1920. Neither Lemaître nor Hubble received the Nobel Prize for their discovery of the expansion of the Universe, but we now have committees that give maximum awards for the highly speculative proposal of the acceleration of the expansion, whose reality has yet to be confirmed. We certainly do live in very a special time for cosmology. However, as we will see in later chapters, this brand of epistemological optimism has declined with time, and the expression 'crisis in cosmology' is stubbornly reverberating in the media. The initial expectation of removing the pending minor problems arising from increased accuracy of measurements has backfired: the higher the precision with which the standard cosmological model tries to fit the data, the greater the number of tensions that arise, the problems proliferating rather than diminishing. There will be much discussion of these tensions throughout chapters 3–7.

1.4 Pillars of the standard model

1.4.1 General relativity and basic equations of the standard cosmological model

Students and professional researchers in theoretical cosmology are used to thinking that they have understood everything about the topic after having learned mathematically to formalise a set of simplistic ideas, but merely knowing how to solve some equations does not mean that we can lay any claim to understanding the Universe[8] and, even if we have understood all of the physical or metaphysical backgrounds of a cosmological model, who says that this is the true representation of the Cosmos? Understanding the Universe is too ambitious an enterprise, and I would be satisfied if I could understand the humans that produced ideas about the Universe. It is the precise purpose of the present book to try to understand where these ideas stem from, and the astronomical observations and social influences that have motivated them, by reviewing hundreds of references related to the subject. The reader may find the technical details in the bibliography, of which I present a summary description in the coming chapters. I also encourage the lay reader of these pages to undertake the reading of other technical books on cosmology, among the vast number of texts available in technical libraries. I will not recommend any text in particular, because it is difficult to choose among the excellent books and review papers that have been written.

Presenting the mathematical developments or technicalities of observational techniques as a subject for undergraduate or postgraduate students is not the aim of the present pages, whose purpose is rather to offer a critical discussion oriented towards professionals who have already learnt or have been working on cosmological research. Also, specialists in the history, philosophy and sociology of science will find in these pages the fundamental ideas of cosmology and the debates

[8] As expressed by the philosopher Friedrich Nietzsche (1844–1900): 'The calculation of the world, the possibility of expressing with formulae all the things that are happening, is it understanding? What would we understand about a musical composition if we calculated all of the things in it that are calculable and reducible to formulae' (*The Will to Power*).

concerning it without having to lose themselves in a skein of equations whose disentangling consumes most of the efforts of the reader. Nonetheless, even though mathematical developments are not an essential part of this work, I would like to show in this section some basic equations for the benefit of those readers with knowledge of physics but without a previous knowledge of cosmology (the rest of this section may be skipped by the reader with insufficient knowledge of physics).

In the usual metric notation in gravitation, ds denotes a certain infinitesimal interval within a particular mathematical space with N dimensions and is related to the interval in each dimension x_μ ($\mu = 1, \ldots, N$) by

$$ds^2 = \sum_{\mu=1}^{N} \sum_{\nu=1}^{N} g_{\mu\nu} dx_\mu dx_\nu. \tag{1.1}$$

In tensor notation, the right-hand side of this equation is usually written as $g_{\mu\nu} dx^\mu dx^\nu$, where the multiplication of variables with subindex and superindex denotes a sum over the index; $g_{\mu\nu}$ are the components of the metric tensor. In the application to our 4-dimensional spacetime (three dimensions for space and one for time), the simplest example in a Euclidean static space (Lorentzian–Minkowski manifold) is, in Cartesian coordinates:

$$ds^2 = dt^2 - \frac{1}{c^2}(dx^2 + dy^2 + dz^2), \tag{1.2}$$

or, equivalently, expressed in spherical coordinates (r, θ and ϕ are the three variables that define a position of a point in the space)

$$ds^2 = dt^2 - \frac{1}{c^2}(dr^2 + r^2 d\theta^2 + r^2 \sin^2 \theta d\phi^2). \tag{1.3}$$

One of the core ideas of general relativity is that the metric of spacetime is determined by the matter and energy content, and that the dynamics of particles is determined by the geodesic in the geometry that minimises ds between two points in that spacetime. The coefficients $g_{\mu\nu}$ of the metric follow

$$R_{\mu\nu} - \frac{1}{2} R g_{\mu\nu} + \Lambda g_{\mu\nu} = \frac{8\pi G}{c^4} T_{\mu\nu}, \tag{1.4}$$

where $R_{\mu\nu}$ are the components of the Ricci curvature tensor, which is a function of $g_{\mu\nu}$ and its first and second derivatives with respect to the variables $x_{\mu\nu}$, and reflects the degree to which the geometry of a given metric tensor differs from that of Euclidean space; $R \equiv \sum_{\mu=1}^{N} \sum_{\nu=1}^{N} g_{\mu\nu} R_{\mu\nu}$ is the scalar curvature; Λ is the cosmological constant (or quintessence if we allowed it to vary with time instead of being a constant); and $T_{\mu\nu}$ is the stress–energy tensor. In a perfect fluid, i.e. one with an isotropic pressure p and a unique density ρ and no viscosity, it follows that

$$T_{\mu\nu} = (\rho + p) u_\mu u_\nu - p g_{\mu\nu}, \tag{1.5}$$

where $u_\mu \equiv \frac{dx_\mu}{ds}$.

The starting point of the standard cosmological model is the cosmological principle, which decrees that the Universe on a large scale is homogeneous and isotropic, and is expanding. The standard model assumes a Friedmann–Lemaître-Robertson–Walker (FLRW) spacetime, which gives a metric

$$ds^2 = dt^2 - \frac{a^2(t)}{c^2}\left(\frac{dr^2}{1 - kr^2} + r^2 d\theta^2 + r^2 \sin^2\theta d\phi^2\right), \quad (1.6)$$

where k stands for the curvature and is constant with time, being equal to zero for a flat metric, positive for a closed Universe and negative for an open Universe; $a(t)$ is an adimensional scale factor that changes with time t due to the expansion, with $a(t_0) = 1$ and t_0 representing our present epoch. Again, r, θ, and ϕ are the three variables in spherical coordinates that define the position of a point in space, but here they stand for comoving coordinates; that is, they do not take into account the expansion. The physical coordinates are derived when we multiply the common scale factor $a(t)$ by the comoving coordinates. In order to know $a(t)$, one needs to know the field equations obtained through the use of general relativity, and the equation of state of the different components.

When the FLRW metric is used in conjunction with the Einstein field equations (1.4), we obtain the two equations that were studied by Friedmann (Wesson 2014). The assumption that the density ρ and pressure p of the cosmological fluid are isotropic and homogeneous renders the partial differential equations (1.4) as ordinary differential equations in the scale factor $a(t)$:

$$8\pi G\rho = 3\left(\frac{\dot{a}^2}{a^2} + \frac{kc^2}{a^2}\right) - \Lambda c^2, \quad (1.7)$$

$$\frac{8\pi Gp}{c^2} = -2\frac{\ddot{a}}{a} - \frac{\dot{a}^2}{a^2} - \frac{kc^2}{a^2} + \Lambda c^2. \quad (1.8)$$

Joining these two equations, we get

$$\ddot{a} = \frac{-4\pi G}{3}\left(\rho + \frac{3p}{c^2}\right)a + \frac{\Lambda c^2}{3}a, \quad (1.9)$$

$$\dot{\rho} = -\left(\rho + \frac{p}{c^2}\right)\left(\frac{3\dot{a}}{a}\right). \quad (1.10)$$

A positive cosmological constant Λ experiences repulsion, producing an acceleration of the expansion. If $\Lambda = 0$ we have the Friedmann models, which were the favoured ones before the reintroduction of the cosmological constant (dark energy) in the 1990s. Equation (1.10) indicates a stability relation for the Universe, in the sense that the density adjusts in proportion to the expansion rate and the combination density and pressure.

The cosmological redshift z (different from a Doppler effect) is a better parameter to use as a cosmological measure than either the distance or the time. For an object in which the spectral lines emitted at wavelength λ_{emitted} are observed at a wavelength $\lambda_{\text{observed}}$, the definition of redshift is $z \equiv \frac{\lambda_{\text{observed}} - \lambda_{\text{emitted}}}{\lambda_{\text{emitted}}}$ In terms of the scale factor of the FLRW metric at present (t_0) and at the emission epoch (t_e), it is given by $(1 + z) = \frac{a(t_0)}{a(t_e)}$. The Hubble–Lemaître constant H_0 relates to the previous equations as $H_0 \equiv \frac{\dot{a}(t_0)}{a(t_0)}$.

In equations (1.7), (1.8), we observe that the Λ term behaves mathematically as an extra substance, known as dark energy, in vacuum space with density ρ_Λ and pressure p_Λ such that

$$\rho_\Lambda = \frac{-p_\Lambda}{c^2} = \frac{\Lambda c^2}{8\pi G}. \tag{1.11}$$

Note that a positive Λ produces a positive vacuum density with negative pressure. We can define a total density $\rho_t \equiv \rho + \rho_\Lambda$ and total pressure $p_t \equiv p + p_\Lambda$. The term included with density ρ and pressure p can also be decomposed in two terms that carry some energy: matter (m) and radiation (r), so $\rho = \rho_m + \rho_r$, $p = p_m + p_r$. The Friedmann equations can be solved when we have the equation of state that governs the different components of the density, assumed to be perfect fluids $p_i = \omega_i \rho_i c^2$, where the index i denotes the components Λ, m, or r. Hence, we can rearrange equations (1.7), (1.9) into

$$8\pi G\rho_t = 3\left(\frac{\dot{a}^2}{a^2} + \frac{kc^2}{a^2}\right), \tag{1.12}$$

$$\ddot{a} = \frac{-4\pi G}{3}\left[\sum_i \rho_i(1 + 3\omega_i)\right]. \tag{1.13}$$

The equation of state of matter is $\omega_m = 0$ while for radiation $\omega_r = 1/3$, and for the Λ term is usually taken as $\omega_\Lambda = -1$ although it can be left as a free parameter too (Cappi 2001). Defining the ratios $\Omega_i \equiv \frac{\rho_i}{\rho_c}$, where $\rho_c \equiv =\frac{3H^2}{8\pi G}$, $H(t) = \frac{\dot{a}(t)}{a(t)}$; and $\Omega_k \equiv -\frac{k\,c^2}{H\,a}$, we can rewrite equation (1.12) as:

$$\Omega_t = \Omega_m + \Omega_r + \Omega_\Lambda = 1 - \Omega_k. \tag{1.14}$$

The comoving distance from us to a source with redshift z can be written as:

$$r(z) = \frac{c}{H_0}\int_0^z \frac{dx}{E(x)}, \tag{1.15}$$

where

$$E(z) = \sqrt{\Omega_k(1 + z)^2 + \Omega_m(1 + z)^3 + \Omega_r(1 + z)^4 + \Omega_\Lambda(1 + z)^{3(1+\omega_\Lambda)}}. \tag{1.16}$$

The look back time of a source redshift z is

$$t(z) = \frac{1}{H_0}\int_0^z \frac{dx}{(1 + x)E(x)}. \tag{1.17}$$

There are two other commonly used definitions of distance from us to a source of redshift z: the luminosity distance and the angular distance. The luminosity distance, d_L, follows the relationship in an equivalent Euclidean Universe of total flux F related to the total luminosity of an object by $F = \frac{L}{4\pi d_L^2}$. It is obtained from the comoving distance $r(z)$ as

$$d_L(z) = \begin{cases} \dfrac{c(1+z)}{H_0\sqrt{-\Omega_k}} \sin\left(\dfrac{H_0\sqrt{-\Omega_k}\, r(z)}{c}\right), & \Omega_k < 0 \\ (1+z)r(z), & \Omega_k = 0 \\ \dfrac{c(1+z)}{H_0\sqrt{\Omega_k}} \sinh\left(\dfrac{H_0\sqrt{\Omega_k}\, r(z)}{c}\right), & \Omega_k > 0 \end{cases}. \tag{1.18}$$

The angular distance d_A is defined such that it behaves as in an Euclidean space, making the linear size D of an object proportional to its angular size α and inversely proportional to that distance; i.e. $\alpha = \frac{D}{d_A}$. The way to calculate it within the standard cosmology is:

$$d_A(z) = \frac{d_L(z)}{(1+z)^2}. \tag{1.19}$$

The set of equations of this subsection define the geometry of the Universe and are used for all kinds of tests relating different quantities: the angular size test, the Hubble–Lemaître diagram relating fluxes with redshifts, etc. We shall see many examples in chapter 4. Nevertheless, the physics of cosmology does not end with these equations. There is much more, related to all the different phenomena that can be modelled within the standard scenario. First, we have not included in these equations inflation (see section 3.2), which modifies $a(t)$ for the early times of the Universe. Predictions of the abundances of the light elements, to be compared with the observed abundances (see chapter 6), comes from the specific nuclear physics developed to tackle that question. Also, the cosmic microwave background radiation (see chapter 5) can be explained independent of the relativistic geometrical description. Of course, the Friedmann cosmological equations are necessary to derive the conditions of density and pressure that give way to other events, but this physics is not implicit in general relativity alone. The large-scale structure and evolution of galaxies (see chapter 7) is pure gravitational in nature in the standard model, but it requires further mathematics, usually with the help of computer simulations, since the above expression accounts only for the average properties of the distribution of matter on very large scales, but not the fluctuations with respect to the average that form clusters of galaxies, filaments, and voids. In any case, the basic paradigm, the framework that supports the mathematical construction of the cosmological edifice is contained in the above equations.

This cosmology is built on the simplistic idea that the whole Universe is an expanding, homogeneous (on a large scale), and isotropic distribution of matter, radiation, and dark energy whose dynamics is governed by gravity and only by gravity, the same general relativity laws being assumed to hold everywhere and throughout time. These are assumptions that cannot be considered as irrefutably solid pillars. There are indeed alternative models based on different assumptions, some of which will be reviewed in chapter 2. In any case, even if we assume that the standard model has been built on the correct foundations, is it, perhaps, not a little pretentious that we give these equations the name of *cosmology*, which embodies knowledge about the entire existing physical world?

1.4.2 Observational pillars

It was already mentioned in the brief historical overview the observations that led cosmologists to believe they have a model that truly represents the Universe. I summarise them:

- The redshifts of all galaxies follow the Hubble–Lemaître law, by which they are related to the distance of the galaxy, plus the Doppler effect due to their peculiar motions. This was interpreted as a proof of the expansion of the Universe. In chapter 4, this fundamental pillar will be discussed.
- The cosmic microwave background radiation of 2.725 K coming from all directions with very small anisotropies is compatible with a high energy primordial Universe. In the standard model, fluctuations in the hot matter–radiation fluid at $z_{rec} \approx 1100$ oscillate like sound waves. The peaks (Sakharov oscillations) in the power spectrum are a consequence of these sound waves, which we see at the epoch of recombination z_{rec}. This pillar will be discussed in chapter 5.
- The abundance pattern of the light elements is to be explained in terms of primordial nucleosynthesis. This will be discussed in chapter 6.
- The defenders of the standard model think that the formation and evolution of galaxies can only be explained in terms of gravitation in the cold (or warm) dark matter theory of an expanding Universe. The present-day scenario is one of hierarchical formation, in which the galaxies are formed in continuous episodes of accretion and merging during their evolution. This will be examined in chapter 7.
- Olbers' paradox, the fact that the night is dark, is solved with a finite Universe, and general relativity provides the scenario for a finite although unlimited and borderless Universe. Even an infinite Universe is possible within the general relativistic context, but the limited speed of light limits the horizon within which galaxies are visible, thus solving Olbers' paradox. Indeed, the key question is the limited age of the Universe rather than its limited size, and that the Universe is not eternal but had a beginning 13.8 Gyr ago forms part of the basic assumptions of the standard models. The question of whether there are objects in the Universe older than this proposed age will be discussed in chapter 7.

Some observations will be discussed or rediscussed in order to show that these supposedly established facts have not been strictly proven in some cases, but also, in other cases to show the solidity of the standard theory against certain tests.

1.5 Towards a sceptical position on cosmology

Scepticism has played an important role in the history of ideas. Reasonable doubts have led the human intellect towards new horizons, it has enabled humanity to avoid dogmas and be open-minded. In philosophy, there is a great tradition of epistemological scepticism, even to the extreme of denying the possibility of knowledge. One of the most remarkable thinkers with this view was the pre-Socratic philosopher Gorgias (c. 483 BC–375 BC), who argued that nothing exists, that even if there were something we could not know it, and that even if we could know it we could not communicate it to others (Jones 1952 [p 60]); the school of Academic Scepticism during 3rd and 2nd century BC also denied that knowledge is possible, although they held that some beliefs are more reasonable or probable than others (Popkin 1967). There is indeed some degree of scepticism in most philosophers, although some of them, such as the modern philosophers René Descartes (1596–1650) and David Hume (1711–1776), regard doubt as the most important element of their philosophies.

Science cannot adopt such an extreme scepticism as that of Gorgias, otherwise there would not be science since there would be no motivation to obtain or communicate knowledge on the grounds that nature would be deemed not to exist. Science is based on the assumption that nature certainly does exist, and that we can obtain knowledge about it. But scientific method is cautious and does not accept as truth the first idea that enters the researcher's head. 'Therefore, the most rational stance as a scientist is to doubt all or almost all explanations, interpretations, and evidence in science, as they are all tentative' (Brewer 2020 [chapter 61]). There is a tradition among good scientists to cast doubt upon many important discoveries, and this has been good for science, because it obliged it to consolidate and corroborate theories until they could no longer be reasonably doubted. For instance, in the development of modern atomic theory since John Dalton (1766–1844), much doubt was cast on the real existence of atoms, even as late as the beginning of the 20th century. Scientists today no longer doubt the existence of atoms; if any did, they would lose all credibility among their peers. This solidity of science has proved beneficial and has been gained through the painstaking construction of hypotheses, thanks to the prudent attitude of many scientists who thought like natural philosophers rather than theologians. Hence, waiting a few generations before treating the existence of dark matter and dark energy as an incontrovertible truth would be a wise move.

Cosmological hypotheses in particular should be very cautiously proposed and even more cautiously received. Whether modern physical cosmology is a science at all is topic to be discussed elsewhere (see chapter 10). In any case, even if accepted as scientific enterprise, cosmology would not be a science like others since it contains more speculative elements than is usual in other branches of physics, except, perhaps, particle physics. The goal of cosmology is also more ambitious than the usual theories in physics. For example, astrophysics contains certain branches, such as stellar physics, galactic structure, etc, that have a limited number of topics to understand, but cosmology aims to understand everything without limit. Given that it is a mere one century of existence as a science, in only one century since the beginning of its construction, its claims must be taken with a grain of salt.

To put it bluntly, if we compare figures 1.1 and 1.2, can we not see that we are playing the same game now as many centuries ago? If we forget about the scientific explanations, I see in both figures a similar state of mind that tries to represent the cosmos in several layers with the elements that we have developed in our fantasies. The astronomer Bruno Binggeli (1953–) indeed thinks that this similarity is not a coincidence and compares modern cosmology to a symbolic expression of the states of our mind (Binggeli 2006) (see section 9.3). Examining the history of cosmological ideas, as we have done in this chapter, we can see humanity stumbling over the same stone again and again. Like a *matryoshka* doll with infinite layers although open from the inside out, our western culture since ancient Greece has been opening the different layers of our Universe hoping to find the last one, and every time it finds a new element (galaxies, dark matter, dark energy, …), naively thinking that the last piece that completes the understanding of the vast Universe has finally been found. Other astrophysicists have also noted this historical similarity: Jayant V Narlikar (1938–) says, 'There is one trait which the cosmologists of old seem to share with their modern counterparts, viz. their fond wish that the mystery of the nature of the Universe would be solved in their lifetime' (Narlikar 2001). The astrophysicist Michael J ('Mike') Disney (1937–) calls it the 'fortunate epoch' assumption, the idea that we live in the first human epoch that possesses the technical means to tease out the crucial observations (Disney 2000).

Mike Disney is indeed a remarkable example of the modern sceptical astrophysicist, with a long career of contributing significant advances to extragalactic astrophysics. In his bold paper 'The case against cosmology' (Disney 2000), he identifies present-day cosmology as a dogma with a series of gratuitous or quasi-gratuitous assumptions: apart from the above-mentioned 'fortunate epoch', there is: the 'non-theological' assumption, according to which speculations are not made which cannot, at least in principle, be tested against observational or experimental data; the 'good-luck' assumption, under which the portion of the Universe susceptible to observation is supposed to be representative of the cosmos as a whole; the 'simplicity' assumption that the Universe was constructed using a significantly lower number of free parameters than the number of clean and independent observations we can make; and the 'non-circularity' assumption that the laws of physics that have significantly controlled the Universe since its beginning are, or can be, known to us from considerations outside cosmology itself. He concludes:

> We believe the most charitable thing that can be said of such statements is that they are naive in the extreme and betray a complete lack of understanding of history, of the huge difference between an observational and an experimental science, and of the peculiar limitations of cosmology as a scientific discipline (Disney 2000).
>
> Only beasts could remain indifferent to questions about the origin, structure and fate of the cosmos in which they live. Only saints could resign themselves to never knowing the answers. The upshot has been that every civilization

> known to anthropology has put together such meagre observations as it possesses, has interpreted them in the light of currently fashionable ideas, and then manufactured as plausible a cosmological story as it can to tell its students and its children. The trouble is that none of those cosmologies have stood the test of time. Have we any reason to be more confident in the Big Bang Cosmology (BBC) which is fashionable today? (Disney 2011)

We may interpret this as too daring, too exaggerated a parody that is out of place in the present cosmological scene. Ćirković (2002) criticises Disney (2000), saying that his claims are rhetorical with no new ideas about the sociology and philosophy of science, and that his critique is unfair, biased, and constrained within an extreme inductivism. Other disciplines operate in a similar way to cosmology and they are sciences, says Ćirković. But we could also pay attention to some of Disney's sentences and see that there is some background of truth in what he claims, in spite of the exaggeration.

In any case, Disney's position is an exception within the exception of anti-Big Bang cosmologists. Most (professional or amateur) researchers who are critical with the standard model still live in the delusion of grandeur that accompanied the creators of the Big Bang theory by believing that this theory can be substituted by another, an alternative model (see chapter 2 for many examples), in most cases created by themselves. A sceptical attitude is common among philosophers of science and academics with a humanistic education, an attitude that is usually accompanied by an anti-scientific stance and a vision that 'man is the measure of all things' (Protagoras, c. 485 BC–c. 411 BC), putting everything in the same sack, and reducing any natural truth to a cultural relativistic standpoint so they do not distinguish between the truths of cosmology and the truths of other particular sciences. This is no better than the misguided attitude of those tavern philosophers who, with no training in either philosophy or science, declare their disbelief in the Big Bang while presenting no justification for their claim.

Not all attitudes and ideas have the same value, regardless of their number of followers. As a matter of fact, if we remove from cosmology those opinions of dogmatic individuals, both orthodox and heterodox, those with no idea of modern cosmology, and those who adopt clearly anti-scientific positions, the number of individuals is very small. *Pulchrum est paucorum hominum*[9]. Here, 'beauty' takes the form of the brilliance of intellect, which observes with rigour and reason transcends mere delusion. But, of course, being part of a very small group—that group often consisting of just one person—with some ideas in common does not necessarily give us a clearer perspective. A belief that one is an unacknowledged genius or that one's views happen to coincide with the majority are insufficient qualification in the quest for knowledge. Chapter 8 will be dedicated to the sociological aspects of cosmology and to an analysis of why individuals orientate their research towards either orthodox or heterodox positions.

[9] 'Beauty is for the few', an expression from the original German edition of Nietzsche's *Twilight of the Gods*.

The view that will be developed in this book is certainly sceptical vis-à-vis the standard cosmological model, but this scepticism is not mere pose. It must be argued with the same kinds of scientific arguments that are used to defend the standard model, with due reference to the observational pillars that support the fundamental ideas of modern cosmology. This may leave us open to an accusation of adopting an instrumentalist stance[10] with regard to cosmology (Soler Gil 2012), on the grounds that we say it works insofar as it explains the observations because many elements were introduced ad hoc to make it work. However, I am not saying that no model contains any truth or reality, since such an attitude would reduce scepticism in cosmology to another dogma in itself. As a matter of fact, I will show in chapters 3–7 that there are some partial realities in the standard model that look quite robust. The overall picture of a Universe completely understood by the theory is somewhat hazy, but at least some partial truths can be observed in quite a clear-cut way.

References

Alpher R A, Bethe H and Gamow G 1948 The origin of chemical elements *Phys. Rev.* **73** 803–4

Alpher R A and Herman R 1949 Remarks on the evolution of the expanding universe *Phys. Rev.* **75** 1089–95

Baryshev Y V and Teerikorpi P 2012 *Fundamental Questions of Practical Cosmology* (Dordrecht: Springer)

Bergh S 2011 The curious case of Lemaître's equation no. 24 *J. R. Astron. Soc. Can.* **105** 151–2

Binggeli B 2006 *Primum Mobile. Dantes Jenseitsreise und die moderne Kosmologie* (Zurich: Ammann Verlag & Co)

Binggeli B 2017 Primum mobile. Dante's journey and modern cosmology *Paper presented at Società Dante Alighieri, Rome, 2017, Nov. 16th* https://www.brunobinggeli.ch/pdf/DanteRoma-en.pdf)

Block D L 2012 Georges Lemaître and Stigler's Law of Eponymy *Georges Lemaître: Life, Science and Legacy (Astrophys. Space Sci. Lib. 395)* ed R D Holder and S Mitton (Berlin: Springer) pp 89–96

Bonometto S A 2001 Modern and post-modern cosmology *Historical Development of Modern Cosmology (ASP Conf. Ser. 252)* ed V J Martínez, V Trimble and M J Pons-Bordería (San Francisco, CA: Astronomical Society of the Pacific) pp 219–36

Bose S, Vogelsberger M, Zavala J, Pfrommer C, Cyr-Racine F-Y, Bohr S and Bringmann T 2019 Ethos—an effective theory of structure formation: detecting dark matter interactions through the Lyman-α forest *Mon. Not. R. Astron. Soc.* **487** 522–36

Bower R G, McLeish T C B, Tanner B K, Smithson H E, Panti C, Lewis N and Gasper G E M 2018 A medieval multiverse?: Mathematical modelling of the thirteenth century universe of Robert Grosseteste *Proc. R. Soc.* A **470** 20140025

[10] In the philosophy of science, instrumentalism is the opposite of realism in the interpretation of a scientific theory. From an instrumentalist viewpoint, a successful scientific theory reveals no truth about nature's unobservable objects, but is merely a set of mathematical formulae that work successfully to predict observations that state or summarise regularities. The classical example is the Ptolemaic geocentric model, which, even being a false representation of the solar system, was able to predict, albeit far from accurately, the positions of the planets.

Brewer D D 2020 Essentials of scientific research *A Practical Guide* (Seattle, WA: Evidence Guides)

Burbidge E M, Burbidge G R, Fowler W A and Hoyle F 1957 Synthesis of the elements in stars *Rev. Mod. Phys.* **29** 547–50

Burbidge G R 2006 The state of cosmology *Current Issues in Cosmology* ed J-C Pecker and J V Narlikar (Cambridge: Cambridge University Press) pp 3–16

Caetani M 1821 *La materia della Divina commedia di Dante Alighieri dichiarata in VI tavole* (Firenze: G C Sansoni)

Cappi A 2001 Testing cosmological models with negative pressure *Astrophys. Lett. Commun.* **40** 161–76

Ćirković M M 2002 Laudatores temporis acti, or why cosmology is alive and well–a reply to Disney *Gen. Relativ. Gravit.* **34** 119–30

Dingle H 1931 The evolution of the universe *Nature* **128** 699–722

Disney M J 2000 The case against cosmology *Gen. Relativ. Gravit.* **32** 1125–34

Disney M J 2011 Doubts about big bang cosmology *Aspects of Today's Cosmology* ed A Alfonso-Faus (Rijeka: InTech) pp 123–32

Doroshkevich A G and Novikov I D 1964 Mean density of radiation in the metagalaxy and certain problems in relativistic cosmology *Sov. Phys. - Dokl.* **9** 111–3

Eddington A S 1923 *Mathematical Theory of Relativity* (Cambridge: Cambridge University Press)

Eddington A S 1929 *The Nature of the Physical World* (New York: Macmillan)

Eddington A S 1931 The end of the world: from the standpoint of mathematical physics *Nature* **127** 447–53

Einstein A 1917 Kosmologische betrachtungen zur allgemeinen relativitätstheorie *Sitzungsberichte der Königlich Preussischen Akademie der Wissenschaften* (Berlin: Akademie der Wissenschaften) pp 142–52

Friedmann A 1922 über die krümmung des raumes *Z. Phys.* **10** 377–86

Friedmann A 1924 über die möglichkeit einer welt mit konstanter negativer krümmung des raumes' *Z. Phys.* **21** 326–32

Gale G 1993 Philosophical aspects of the origin of modern cosmology *Encyclopedia of Cosmology* ed N S Hetherington (New York: Garland Publishing) pp 481–95

Gamow G 1947 *One, Two, Three, Infinity* (New York: Mentor)

Gamow G 1953 The expanding universe and the origin of galaxies *Kgl. Dan. Vidensk. Selsk. Mat. Fys. Medd* **27** 3–15 (http://gymarkiv.sdu.dk/MFM/kdvs/mfm%2020-29/mfm-27-10.pdf)

Gray J J 1978 Johann Heinrich Lambert, mathematician and scientist, 1728–1777 *Historia Math.* **5** 13–41

Hartnett J G 2008 Crisis in cosmology continues with conference of big-bang dissidents creation.com (19th September)

Herzberg G 1950 *Spectra of Diatomic Molecules* (New York: Van Nostrand)

Hetherington N S 1993 Great debate *Encyclopedia of Cosmology* ed N S Hetherington (New York: Garland Publishing) pp 260–2

Horgan J 1996 The End of Science *Facing the limits of Knowledge in the Twilight of the Scientific Age* (Cambridge: Addison-Wesley)

Hoyle F 1946 The synthesis of the elements from hydrogen *Mon. Not. R. Astron. Soc.* **106** 343–83

Hoyle F 1947 On the formation of heavy elements in stars *Proc. Phys. Soc. London* **59** 972–8

Hubble E P 1929 A relation between distance and radial velocity among extra-galactic nebulae *Proc. Natl Acad. Sci. USA* **15** 168–73

Humason M L 1929 The large radial velocity of N.G.C. 7619 *Proc. Natl Acad. Sci. USA* **15** 167–8
Jeans J 1928 The physics of the Universe *Nature* **122** 689–700
Jones W T 1952 *A History of Western Philosophy* (New York: Harcourt Brace)
Kerzsberg P 1986 The cosmological question in Newton's science *Osiris* **2** 69–106
Lemaître G 1927 Un univers homogène de masse constante et de rayon croissant rendant compte de la vitesse radiale des nébuleuses extra-galactiques *Ann. Soc. Sci. Brux.* A **47** 49–59
Lemaître G 1931 The beginning of the world from the point of view of quantum theory *Nature* **127** 706
Lemaître G 1946 L'Hypothèse de l'atome primitif *Essai de Cosmogonie* (Neuchâtel: Editions du Griffon)
Lerner E J 1991 *The Big Bang Never Happened: A Startling Refutation of the Dominant Theory of the Origin of the Universe* (Toronto: Random House)
Livio M 2011 Lost in translation: mystery of the missing text solved *Nature* **479** 171–3
Llallena Rojo A M 2017 *El Big Bang y el origen del Universo* (Barcelona: RBA)
Longair M S 1966 On the interpretation of radio source counts *Mon. Not. R. Astron. Soc* **133** 421–36
Luminet J-P 2008 The rise of Big Bang models, from myth to theory and observation *Antropogenesi, dall' energia al fenomeno umano* ed A Pavan and E Magno (Bologna: Il Mulino) p 5
Luminet J-P 2013 Editorial note to: Georges Lemaître. A homogeneous universe of constant mass and increasing radius accounting for the radial velocity of extra-galactic nebulae *Gen. Relativ. Gravit.* **45** 1619–33
Merritt D 2017 Cosmology and convention *Studies in History and Philosophy of Modern Physics* **57** 41–52
Milne E A 1933 World-structure and the expansion of the universe *Z. Astrophys* **6** 1–95
Milne E A 1934 A Newtonian expanding universe *Q. J. Math.* **5** 64–72
Narlikar J V 2001 The evolution of alternative cosmologies *Historical Development of Modern Cosmology (ASP Conf. Ser. 252)* ed V J Martínez, V Trimble and M J Pons-Bordería (San Francisco, CA: Astronomical Society of the Pacific) pp 175–99
Novikov I 2001 Discovery of CMB, Sakharov oscillations and polarization of the CMB anisotropy *Historical Development of Modern Cosmology (ASP Conf. Ser. 252)* ed V J Martínez, V Trimble and M J Pons-Bordería (San Francisco, CA: Astronomical Society of the Pacific) pp 43–53
Nussbaumer H and Bieri L 2009 *Discovering the Expanding Universe* (Cambridge: Cambridge University Press)
Peebles P J E and Yu J T 1970 Primeval adiabatic perturbation in an expanding universe *Astrophys. J.* **162** 815–36
Penzias A and Wilson R 1965 A measurement of excess antenna temperature at 4080 mc/s *Astrophys. J.* **142** 419–21
Planck Collaboration 2020 Planck 2018 results. VI. Cosmological parameters *Astron. Astrophys.* **641** A6
Popkin R H 1967 Skepticism *The Encyclopedia of Philosophy* ed P Edwards (New York: Macmillan) p 455
Primack J R 2005 Precision cosmology *New Astron. Rev.* **49** 25–34
Ryle M 1968 The counts of radio sources *Annu. Rev. Astron. Astrophys.* **6** 249–66

Sandage A 1995 Practical cosmology: inventing the past *The Deep Universe, 23rd Sass Fee Advanced Course in Astrophysics* ed B Binggeli and R Buser (Berlin: Springer) pp 1–232
Shmaonov T 1957 Commentary *Pribori Tekh. Exp. (Russia)* **1** 83
Silk J 2001 How big is the universe? *Historical Development of Modern Cosmology (ASP Conf. Ser. 252)* ed V J Martínez, V Trimble and M J Pons-Bordería (San Francisco, CA: Astronomical Society of the Pacific) pp 109–20
Soler Gil F J 2012 *Discovery or Construction? Astroparticle Physics and the Search for Physical Reality* (Frankfurt am Main: Peter Lang)
Sparavigna A C 2014 Robert Grosseteste's thought on light and form of the world *Int. J. Sci.* **3** 54–62
Stubss C W 2007 Addressing the crisis in fundamental physics *Int. J. Mod. Phys.* **D16** 1947–52
Suzuki S 1928 On the thermal equilibrium of dissociation of atom-nuclei *Proc. Phys.-Math. Soc. Japan* **10** 166–9
Unzicker A and Jones S 2013 *Bankrupting Physics. How Today's Top Scientists Are Gambling Away Their Credibility* (New York: Palgrave)
Wallis R 1984 The glory of gravity–Halley's Comet 1759 *Ann. Sci.* **41** 279–86
Weinberg S 1977 The First Three Minutes *A Modern View of the Origin of the Universe* (New York: Basic Books)
Wesson P S 2014 Einstein's equations, cosmology and astrophysics *Springer Handbook of Spacetime* ed A Ashtekar and V Petkov (Berlin: Springer) pp 617–27
White S D M 2007 Fundamentalist physics: why dark energy is bad for astronomy *Rep. Prog. Phys.* **70** 883–97
Zel'dovich Y B 1963 Prestellar state of matter *Zh. Eksp. Teoret. Fiz* **43** 1561

Chapter 2

Some examples of alternative cosmologies

Here I show a few remarkable examples of alternative cosmological theories, ranging from a compilation of variations on the Standard Model (inhomogeneous Universe, Cold Big Bang, varying physical constants or gravity law, zero-active mass, Milne, cyclical models), through the more distant quasi-steady-state cosmology, plasma cosmology, or Universe models as a hypersphere, to the most exotic cases including static models with non-cosmological redshifts of galaxies. A global cosmological theory that provides a full explanation of everything in the Universe does not yet exist, in either standard or alternative views, but some alternative models may explain certain aspects of some observational data better than the ΛCDM model.

With few exceptions (e.g., Peebles 1993 [chapter 7]), academic books on cosmology usually describe only the standard model and do not mention alternative theories. There is however a rich variety of alternative ideas with intelligent proposals that merit consideration. Here I will offer a sample of these alternative models; not all of them, since their number is huge and it is impossible to mention all of them in a single chapter of this book. I will not give a complete list of models[1], but this sample is large enough to give an idea of which theoretical approaches are being discussed in cosmology from heterodox points of view that are dissident either with respect to the standard model or with respect to the dominant discrepant theories. Defending any particular theory against standard cosmology or globally criticising it is not my purpose here, although I may mention some aspects that are being debated concerning them. Here, I just review the literature and offer a classification of the different models, much as a botanist or zoologist classifying different species might do.

[1] Indeed, nobody can offer a complete list of alternative cosmological models, because they proliferate like mushrooms in forests during autumn, but there are more complete lists than mine. For instance, see the webpage of the *Alternative Cosmology Group (ACG)* by Louis Marmet: http://cosmology.info/essays/models_marmet.html.

doi:10.1088/978-0-7503-3775-5ch2

The wild fauna and flora of cosmologists indeed encompass a wide spectrum of colourful suggestions.

I will mention some of the most remarkable models in the scientific literature of recent decades and also some contributions of minor impact, mostly coming from professional physicists or astronomers. There is also a vast literature produced by non-professional and amateurs who try to open new routes within the golden odyssey of cosmological model creation, but very few of them will be mentioned here. They are examples of curious ideas which, although not fully developed, could be the seeds of competitive models when they are further elaborated. To a greater or lesser extent, all of these alternative models suffer from a lack of development in comparison with the standard ΛCDM, so the first thing we must take into account when reading this chapter is that they cannot of course compete with the standard model in all aspects because many of these alternative ideas are in the hands of very few individuals—occasionally only a single individual—who cannot produce hypotheses and ad hoc refinement of them to fit the ever increasing deluge of observational data at the same speed as the thousands of researchers working on the standard model. I will go on to talk about sociological matters in chapter 8. In any case, it may happen that an alternative theory might explain certain aspects of some observational data better than the ΛCDM model; moreover, even if it fails to explain other types of observations, it might be only a matter of time and ad hoc speculation of the existence of new unknown/dark elements for it to be made to fit those data too. One might say that, in a certain sense, cosmology is a game whose purpose is to invent new fantastic creatures—not gods as in a religion, but mathematical abstractions that have a similar wildcard function—which can adapt our inherently prejudiced worldview to the reality.

2.1 Variations on the standard model

The present-day standard model of cosmology gives us a representation of a cosmos whose dynamics is dominated by gravity (Friedmann equations derived from general relativity) with a finite lifetime, large scales homogeneity, expansion and a hot initial state, together with other elements necessary to avoid certain inconsistencies with observations (inflation, non-baryonic dark matter, dark energy, etc). There are also other models that are closer to the main characteristics of the standard model, but they differ in some minor aspects. Many of these models are indeed investigated by some mainstream cosmologists. They are alternative models that stem from the variations on the standard model. Here are some examples:

- Rather than new cosmological models, there are multiple proposals for keeping the fundamentals of the standard model as stated before the 1980s and looking for variations in the type of dark matter, the different equations of state of dark energy or even without dark energy, or the hundreds of variations on the type of inflation or alternative proposals such as cosmic strings, walls and other textures. These topics will be further treated in chapter 3. There are also variations in the number of neutrino families for nucleosynthesis; the formation of structures in a monolithic way (galaxies all forming at once) rather than the standard hierarchical scenario (galaxies being formed in continuous episodes of accretion and merging), etc. I do not think these are truly 'alternative' cosmological models; rather, they can mostly be embedded in the same classical Big Bang paradigm with the same fundamental pillars except for the final developments for the specific ΛCDM model.

- Inhomogeneous Universe. The density distribution of the Universe is not homogeneous on very large scales. It may obey a fractal distribution (e.g., Baryshev *et al* 1994, Gabrielli *et al* 2005), when the mass within a sphere of radius R is not proportional to R^3 for large enough R (in the regime in which there should be homogeneity), but proportional to R^D with a fractal dimension $D < 3$. There is little theoretical background to support a cosmology of these characteristics, but some observations may point in this direction. Another idea stems from timescape cosmology (Wiltshire 2009), in which apparent cosmic acceleration is an effect related to the calibration of clocks and rods of observers in bound systems relative to volume-average observers in an inhomogeneous geometry in ordinary general relativity. Back reactions have caused time to run more slowly or, in voids, more quickly, thus producing the illusion that supernovae are farther away than they are and also implying that the expansion of the Universe is in fact slowing down. An inhomogeneous isotropic Universe described by a Lemaître–Tolman–Bondi solution of Einstein's field equations can also provide a positive acceleration of the expansion without dark energy (Romano 2007).
- Cold Big Bang. This theory evolved from the 1960s (Layzer 1990). Rather than a very high temperature at the beginning of the Universe with later progressive cooling, the Universe starts with $T = 0$ K. Explanations are offered for the origin of the light elements in primordial and/or stellar nucleosynthesis (Aguirre 1999), the cosmic microwave background radiation (Aguirre 2000) in terms of thermalisation by intergalactic particles—a mixture of carbon/silicate dust and iron or carbon whiskers—of stellar radiation originating in Population III (further details of this idea are given in section 2.2), and other phenomena explained by the standard Hot Big Bang.
- Variations or oscillations of physical constants (c, G, h, the fine structure constant, or others) with time or distance. For instance, Goldman *et al* (1992) propose variable G with distance, with which they explain the apparent variation of Ω with the scale with no need of dark matter; or different cosmological models with a variable speed of light c over time (Albrecht and Magueijo 1999, Shojaie and Farhoudi 2004, Ellis 2007, Miller 2021) or other variables (Unzicker 2015), or several physical constants varying at the same time (Gupta 2020). The consequences differ according to the models. They preserve the basic aspects of the standard model while keeping expansion and finite time since the beginning of the Universe, but they change with regard to a number of characteristics (Gohar 2017).
- Modifications of aspects of the gravity law. These theories not only change G, but also the gravity force equation. There are several tens of these alternative

theories which I will not mention here. They can be found in other reviews (e.g., Clifton *et al* 2012).

The most popular alternative gravity theory is the modification of gravity law proposed in 'Modified Newtonian Dynamics' (MOND) (Sanders and McGaugh 2002, Famaey and McGaugh 2012, Sanders 2015, Merritt 2020), which modifies the Newtonian law for accelerations lower than $\sim a_0 \approx 1 \times 10^{-10}$ m s^{-2}. This was in principle a phenomenological approach. Its proponents attempted to incorporate elements that make it compatible with more general gravitation theories; for example, the AQUAdratic Lagrangian theory (AQUAL) (Bekenstein and Milgrom 1984) or the Quasi linear approximation of MOND (QMOND) (Milgrom 2010), which expanded MOND to preserve the conservation of momentum, angular momentum, and energy, and follow the weak equivalence principle. A relativistic gravitation theory of MOND was developed under the name Tensor–Vector–Scalar (TeVeS) (Bekenstein 2004), which also tried to provide consistency with certain cosmological observations, including gravitational lensing.

A different alternative gravity theory with a certain impact is scalar–tensor–vector gravity, known as modified gravity (MOG) (Moffat 2006). Another family of theories is the $f(R)$ gravity group, which modify general relativity by defining a different function of the Ricci scalar (Sotiriou and Faraoni 2010). Other proposals include the dependence of spacetime on curvature in a non-metric theory of gravity (Krasnov and Shtanov 2008), or an interpretation of Mach's principle in which the rotational reference frames for stars in galactic orbits has a relationship to the rotating matter in the local galaxy and/or distant galaxies (Annila 2016). There are many others (Clifton *et al* 2012): Einstein–ether theory, bimetric or general higher-order theories, Hořava–Lifschitz gravity, Galileons, Ghost Condensates, and models of extra dimensions, including Kaluza–Klein, Randall–Sundrum, Dvali–Gabadadze–Porrati model 4D gravity on a brane[2] in 5D Minkowski space, or higher co-dimension braneworlds, Weyl conformal gravity (invariant under Weyl transformations) (Jackson 2014), etc.

The cosmological implications of this change in gravity are important, but in most cases they have not yet been developed. In MOND/TeVeS, there are different views. Felten (1984) does not accept that a MOND-cosmology might be possible, stating that a quasi-Newtonian calculation adapted from Newtonian cosmology suggests that a MOND Universe will recollapse and/or fail to satisfy the cosmological principle of a homogeneous Universe. Other authors claim that a MOND-cosmology can be built (Sanders 1998, 2006) that results in a uniform expansion and homogeneity on the horizon scale consistent with MOND-dominated non-uniform expansion and the development of inhomogeneities on scales out to a substantial fraction of the Hubble radius.

[2] In string theory and related theories such as supergravity theories, a brane is a physical element that generalises the notion of a point particle to higher dimensions. Branes are dynamical objects that can propagate through spacetime according to physical laws of quantum mechanics.

Primordial nucleosynthesis, with its concomitant thermal and dynamical history of the Universe, is identical to that of the standard cosmological model until matter dominates the energy density of the Universe, a moment in which the MOND-cosmology diverges from that of the standard model. Other gravity variations may involve other considerations, but basically they try to preserve the global aspects of the standard cosmology.

- Other Friedmann–Lemaître–Robertson–Walker (FLRW) solutions. Instead of the particular solutions of ΛCDM with fixed parameters $\Omega_m \approx 0.3$, $\Omega_\Lambda \approx 0.7$, the FLRW equations (1.9) and (1.10) may have other solutions. Here we are a couple of examples:

  One remarkable case that has generated a large number of papers in recent years is the zero-active mass condition, also called $R_h = c\ t$. This model was firstly proposed under the name of an 'Ur theory', which relates cosmology to particle physics and quantum theory (Görnitz 1988), or a special (flat) case of an eternal coasting model (John and Joseph 1996), consistent with a scale factor proportional to cosmic time ($a(t) \propto t$) or equivalently an active mass $\rho + 3\frac{p}{c^2}$ equal to zero at all times, which cancels the first term of the right-hand side of equation (1.9) and, together with the ansatz $\Lambda = 0$, makes the acceleration of the expansion equal to zero for all times. The density is fitted to keep the Universe flat.

  The most important researcher defending this model nowadays is Fulvio Melia (1956–), who has produced several tens of papers with theoretical and observational support for the model (e.g., Melia and Shevchuk 2012, Melia and López-Corredoira 2017, Melia 2017, 2019). There is the coincidence that now the deceleration of the Hubble–Lemaître flow is compensated by the acceleration of the dark energy; the average acceleration throughout the history of the Universe is almost nil (Melia and Shevchuk 2012) and the size the Universe is such as if there were constant expansion. This coincidence supports the constant expansion ratio posited as $R_h = c\ t$. According to Melia, this model fits the data pretty well where ΛCDM does; moreover, it offers some further advantages at high z, where the standard model has some difficulties in concealing the existence of objects that usually need a long time to be formed in a very young Universe that does not allow time for such evolution (see chapter 7). $R_h = c\ t$ solves the problem because the age of the Universe at redshift z, $t(z)$, is much greater in this model than with the standard ΛCDM. The Big Bang would have happened $H_0^{-1} = 14.57$ Gyr ago (for $H_0 = 67.4$ km/s/Mpc) (Planck Collaboration 2020), longer than the 13.79 Gyr for ΛCDM with the same Hubble–Lemaître constant. There would have been no inflation. Interestingly, an even more remarkable difference from the standard model is the prediction that the Cosmic Microwave Background Radiation (CMBR) is formed at $z \approx 16$ by dust rethermalization of Population III stellar light (Melia 2020), although with a major difficulty in justifying a perfect black-body shape for that radiation, since we do not know of any kind of dust that produces such a flux shape.

  Melia also points out that the $R_h = ct$ model has a higher probability of being correct than ΛCDM, even with the same quality of fits, based on the fact that $R_h = ct$ has no free parameters in the fits, whereas ΛCDM has. However, we must bear in mind that the election of the $\rho + 3\frac{p}{c^2} = 0\ \forall t$ condition is posited because, *a priori*, we know it gives similar results to the standard models with $\Omega_m = 0.3$ in many cosmological tests at low and intermediate redshift, and non-acceleration of the expansion at present, so there is an implicit choice of parameters. One might have chosen instead $\rho + x\frac{p}{c^2} = 0$ with $x \neq 3$; the fact that $x = 3$ was chosen in this hypothesis is because we know it gives results similar to ΛCDM with $\Omega_m = 0.3$ for low and intermediate redshifts. Melia thinks that this choice of $x = 3$ is not arbitrary and derives from general relativity for FLRW metrics, but this interpretation of general relativity is not supported by other specialists in gravitation (e.g., Kim *et al* 2016, a paper replied to by Melia 2017). Independently of the technical details of the interpretation of general relativity, it looks weird that only one man among the myriads of experts in general relativity should have realised that zero-active mass condition is a 'must' in our Universe rather than only an option within FLRW metrics. I would like to hear the support of this idea by some other expert in general relativity. In any case, whether derived from first theoretical principles or only supported by the observations, it seems that the $R_h = ct$ model works pretty well and may be considered a good model to compete with the standard one.

Another example of an FLRW model is Milne Cosmology, which is obtained by demanding that the energy density, pressure, and cosmological constant are all equal to zero, and that the spatial curvature is negative. Indeed, this can also be considered as a zero active model, but with the particular case of $\rho = 0$. This strictly null density of matter is unrealistic since stars, gas and other components of the Universe have mass and are real, but might possibly consider the density to be low enough to be considered close to zero, in comparison with the critical density in the standard model. Also, a symmetric Milne Universe (Benoit-Lévy and Chardin 2012), composed of matter with positive and negative mass in equal quantities would have net $\rho = 0$. From these assumptions and the FLRW equations, it follows that the scale factor must depend linearly on the time coordinate, $a(t) \propto t$, as in all zero-active mass cases. The model has no expansion of space, but the mathematical equivalence of the zero-energy density version of the FLRW metric to Milne's model implies that a full general relativistic treatment using Milne's assumptions would result in an increasing scale factor and an associated metric expansion of space (Chashchina and Silagadze 2015). The Milne Cosmology does not need dark matter or dark energy; moreover, the model evades horizon, flatness, and cosmological constant problems affecting the standard cosmology, which requires inflation to solve it; it also fits some cosmological data (Benoit-Lévy and Chardin 2012, Vishwakarma 2013).

- Cyclical universes. In the 'Conformal cyclic cosmology' model (Penrose 2010), based in the framework of general relativity, the Universe iterates through infinite cycles, with the future timelike infinity of each previous iteration being identified with the Big Bang singularity of the next. So the Big

Bang (although without inflation) applies to our present Universe, but it is speculated that many other Big Bangs happened previously and will happen after ours. The Big Bang of each is taken to be a smooth conformal continuation of the remote future of the previous one via an infinite conformal rescaling; there is no collapsing phase. The second law of thermodynamics, with the curious nature of its origin, is automatically incorporated, where Hawking evaporation of black holes provides a key ingredient.

Another variation of the cyclical model proposed by Steinhardt and Turok (2008) is an endless Universe without beginning or end, an endless sequence of epochs that starts with a 'Big Bang' (again without inflation) and ends in a 'Big Crunch' (collapse of the Universe due to an excess of mass–energy of the critical density in a closed Universe). Although the model is motivated by M-theory[3], branes, and extra dimensions, the scenario can be described almost entirely in terms of conventional 4D field theory and 4D cosmology, with a continual cycle of expansion and contraction as parallel universes (or 'branes') collide.

In the 'dynamic universe' (Suntola 2020), the Universe is also described as a contraction–expansion cycle from infinity in the past to infinity in the future or in repeated cycles. In the contraction, mass in space derives its energy of motion from its own gravitation and releases it back to gravity in the ongoing expansion phase. This is called the 'zero-energy principle', first applied to space as a spherically closed whole and further to all local interactions in space. Maintaining the zero-energy balance, the build-up of local structures in space converts part of the momentum in the fourth dimension into momentum in space via local bending of space. Such a process occurs in several steps to create a system of nested energy frames linking all local states of motion and gravitation to the state at rest in hypothetical homogeneous space that serves as a universal frame of reference. Phenomena that the theory of general relativity explains in terms of modified spacetime metrics, are here explained as consequences of their different energy states. Local phenomena are linked to the rest of space; motion in space is linked to the motion of space.

All these models are variations based on the Lemaître–Gamow idea, all their supporters are *big-bangists*. But there are however other models that challenge the notions of a state of the Universe with a singularity, unlimited density of matter-energy, or other important tenets of the standard model as proposed in the 1920s–1940s, as I show in the following sections.

2.2 Quasi-steady state cosmology

Among the most developed models, with several professional researchers working on it over several decades, perhaps the alternative hypothesis of highest impact during the last century is 'Quasi-Steady State Cosmology' (QSSC), which was indeed

[3] M-theory unifies all consistent versions of superstring theory.

first called 'Steady State Cosmology' when it was a cosmological model competing at the same level of importance and impact with the Big Bang hypothesis. It was developed and defended for more than sixty years, and even today we cannot declare it dead, although its main supporters have passed away or retired, and the new generations of cosmologists no longer work on it, mainly owing to lack of socio-economic support (see chapter 8).

This model is indeed something beyond a small variation on Lemaître–Gamow ideas, it is a radically different view, in which there is no beginning of the Universe, but an eternal cycle of matter creation. The difference with the cyclical universes of section 2.1 is that there is no singularity of state of infinite or unlimited density, and the moment of 'creation' is not unique, but is continuous and constant (in the Steady State version) or oscillating (in the Quasi-Steady State version).

Fred Hoyle (1915–2001), and independently Hermann Bondi (1919–2005) and Thomas Gold (1920–2004), proposed the hypothesis of the Steady State (Hoyle 1948, Bondi and Gold 1948)[4] in which, contrary to the Big Bang approach, there was no beginning of the Universe. The Universe is expanding, it is eternal, and the homogeneous distribution of matter is being created at a rate of $\sim 10^{-24}$ baryons/cm^3/s owing to the existence of a putative ubiquitous C-field of matter creation, instead of the unique moment of creation in the Big Bang. The perfect cosmological principle of a Universe being observationally the same from anywhere in space and at any time is maintained in this model, whereas the standard model only gives a cosmological principle in space but not in time. There is no evolution. The Universe remains always the same. The matter distribution is homogeneous and the redshift is caused by the expansion of the space with a scale of $a(t) \propto e^{H_0 t}$. Newly created matter forms new galaxies that substitute those that are swept away by the expansion.

During the 1950s, both the Big Bang and Steady State theories held their ground. While there were attempts to explain the abundances of the chemical elements with Gamow *et al*'s theory, the Steady State theory also provided plausible explanations. The abundances of the light elements (helium, lithium, deuterium, and others) were explained in terms of stellar nucleosynthesis and collision with cosmic rays in the remote past of the Universe (Burbidge *et al* 1957). The heaviest elements could also be explained in terms of stellar rather than primordial nucleosynthesis, and the defenders of Big Bang in the end also had to adopt the stellar nucleosynthesis of Burbidge *et al* for the heavy elements.

Nonetheless, the Steady State theory would lose competitiveness by the mid-sixties, because it could not explain certain observational facts. It could not explain why the galaxies were younger at higher redshift. Neither could it explain the excess

[4] Indeed, prior to Hoyle, Bondi, or Gold, Albert Einstein attempted to construct a Steady State model of the Universe, as shown in one of his unpublished manuscripts (O'Raifeartaigh *et al* 2014). The manuscript, which appears to have been written in early 1931, demonstrates that Einstein once explored a cosmic model in which the mean density of matter in an expanding Universe is maintained constant by the continuous formation of matter from empty space. Einstein's Steady State model contained a fundamental flaw and this was possibly the reason why he abandoned this line of research (O'Raifeartaigh *et al* 2014).

of radio sources at large distances (Ryle and Clarke 1961), or the distribution of quasars. Most importantly, it did not explain the CMBR as interpreted in cosmological terms in 1965. This strongly favoured the Big Bang theory.

In 1993–94, Fred Hoyle, Geoffrey Burbidge (1925–2010), and Jayant Narlikar (1938–) published a modification of the model that was called the 'Quasi-Steady State' theory (Hoyle *et al* 1993, 1994, 2000, Narlikar *et al* 2007). The main modification consisted in positing an oscillatory expansion apart from the exponential term:

$$a(t) \propto e^{t/P}[1 + \eta \cos(2\pi\theta(t)/Q)],$$

with a long time scale of expansion of $P \sim 10^{12}$ years, $\theta(t) \sim t$. The exponential factor had already been introduced in the first version of the Steady State model to keep $\frac{\dot{a}}{a}=$ constant and consequently maintain a constant density of matter by invoking the continuous creation of matter. The new term here is the sinusoidal oscillation: smaller time scale oscillations have a period of $Q \sim 4 \times 10^{10}$ years. The exact value of the parameter Q and η would be determined from Hubble–Lemaître's constant, the age of globular clusters, and the maximum observed redshift in the galaxies. The time since the last maximum of $a(t)$ is $0.85Q$. The time since the last minimum of $a(t)$ is 14 Gyr. With the parameters given in the original version of the QSSC theory, the maximum observable redshift should be around 5, although it would be increased later as the parameters changed to adapt to new observations. The creation of matter is confined to epochs with minimum $a(t)$ rather than being continuous. These creation events involve Planck particles and eventually make hydrogen gas plus the lightest elements, deuterium and the two isotopes of helium. Since the overall time scale is very long, many generations of stars (and galaxies) will evolve and die.

With the introduction of the Quasi-Steady variation, some of the problems that affected the original theory of 1948 were solved. This explained why there are younger galaxies at higher redshift, the problem of the radio sources, the distribution of quasars (with lower density for redshifts lower than 2.5), and the formation of large-scale structure (clusters, voids, filaments) (Nayeri *et al* 1999). Wright (1995) complained that Hoyle *et al* had not solved the problem of the radio sources completely, but Hoyle *et al* (1995) later replied that the question might be solved with a change of parameters.

The CMBR and its blackbody spectrum would be explained as the effect of the thermalisation of radiation emitted by stars of the last cycle $P/3$ due to absorption and re-emission that produce iron needle-shaped particles ('whiskers') in the intergalactic medium. Because of the long distances travelled by the CMBR photons in the maxima of the oscillation and the thermalisation that occurs at each minimum, there is no accumulation of anisotropies from one cycle to another. Only the fluctuations of the last minimum survive, which gives fluctuations of temperature comparable to the observed $\Delta T/T \sim 5 \times 10^{-6}$. First, the carbon needles thermalise the visible light from the stars, giving rise to far-infrared photons at $z \sim 5$, thus maintaining the isotropy of the radiation. Afterwards, iron needles dominate, degrading the infrared radiation to produce the observed microwave radiation (Wickramasinghe 2006). Within QSSC, on the other hand, the anisotropies of this radiation would also be explained in terms of interaction of the radiation with clusters of galaxies and other elements (Narlikar *et al* 2003, 2007).

The existence of extragalactic iron whiskers might be formed in a process similar to metallic vapours cooling slowly enough in the laboratory. Whiskers are formed by this type of process during the expansion of the envelopes of supernovae. As a matter of fact, some of the defenders of the QSSC model (Hoyle and Wickramashinghe 1988) have claimed that iron whiskers are observed in the emission spectrum of the Crab pulsar PSR0531+21. Thermalisation was also proposed to be due to the plasma of the intergalactic space, but in this case it takes about 450 Gyr for the starlight to get thermalized (Ibison 2006).

Concerning the origin of the redshift of galaxies, the proposers of this model admit a component due to the expansion $a(t)$, as in the standard model, but further posit the existence of intrinsic redshifts. This allows the solution of problems such as the periodicity of redshift in quasars (see section 4.6.1), and the possible existence of cases with anomalous redshifts (see section 4.6). The total redshift would be the product of both factors, expansion and intrinsic:

$$(1 + z) = (1 + z_{\text{exp.}})(1 + z_{\text{int.}}).$$

The intrinsic redshift is explained by means of the variable mass hypothesis, which is used by the authors of QSSC to explain cases of anomalous redshifts, but it is not part of the main body of the QSSC hypothesis; that is to say, it is optional: QSSC can be conceived without the variable mass hypothesis. Indeed, Bondi and Gold, the other two fathers of the classical steady state theory, publicly disavowed any association with later Hoyle–Narlikar versions of variable mass theory (Ćirković and Perović 2018). Hoyle and Narlikar (1964) derived this hypothesis from a new gravitation theory based on Mach's principle; i.e. it arises from inertial contributions from all other particles in the Universe. The mass of a particle p at a point P on its worldline arises from all other particles b in the Universe: $m_p = \sum_{b \neq p} m_b(P)$, where $m_b(X)$ is the contribution of inertial mass from particle b to any particle situated at a general spacetime point X. $m_b(X)$ satisfies a wave equation. Mass is a scalar function of spacetime that transforms conformally under a conformal transformation, $m \propto t^2$. The intrinsic redshift would be due to variation of the energy of the emitted photon when the masses of protons and electrons vary:

$$(1 + z_{\text{int.}}) = \frac{m_{\text{observer}}}{m_{\text{source}}} = \frac{t_0^2}{(t_0 - r/c)^2}.$$

In the case of quasars, redshift anomalies would be observed because the mass of their constituent particles grows proportionally to $(t - t_{\text{quasar}})^2$ instead of t^2 (Narlikar 1977, Narlikar and Arp 1993).

Summing up, QSSC competes with the standard cosmological model to explain many observations, at least in an approximate way, but with a very different description of the Universe. According to its authors, QSSC can even explain some facts that the standard model cannot, such as possible anomalies in the redshifts of quasars. It also contains predictions that are different from those of the standard model, although these are difficult to test. The predictions include (Narlikar 2006): the existence of faint galaxies ($m > 27$) with small blueshifts ($|\Delta z| < 0.1$), the existence of stars and galaxies older than 14 Gyr, an abundance of baryonic matter in ratios above those predicted by ΛCDM, and gravitational radiation derived from the creation of matter.

2.3 Plasma cosmology

Plasma cosmology is another alternative model that has occupied two or three generations of researchers, some of them still active today. Its proponents include the physics Nobel laureate Hannes Alfvén (1908–1995), Oskar Klein (1894–1977), Anthony L Peratt (1940–), Eric J Lerner (1947–), Ari Brynjolfsson (1926–2013). It has a strong argument against one of the main pillars of the standard model: It proposes an alternative to the belief that gravitation is the fundamental force that controls the dynamics of the Universe. It assumes instead that most of the mass in the Universe is plasma controlled mainly by electromagnetic forces (and also gravity, of course), rather than gravity alone (Alfvén and Klein 1962, Alfvén 1983, 1988, Lerner 1991, Alfvén 1981 [chapter 6]). According to this theory the Universe has always existed, it is always evolving, and it will continue to exist forever.

The plasma in the laboratory, through electric currents and magnetic fields, creates filaments similar to those observed in the large-scale filamentary structure of the Universe. The plasma cosmology model predicts the observer morphological hierarchy: distances among stars, galaxies, cluster of galaxies, and filaments of huge sizes in the large-scale structure. The observed velocities of the streams of galaxies in regions close to the largest superclusters are coincident with those predicted by the model, without the need for dark matter (Lerner 1991). The formation of galaxies and their dynamics would also be governed by forces and interactions of electromagnetic fields (Peratt 1983, 1984, Lerner 1991 [chapters 1, 6]).

Hubble–Lemaître expansion was admitted in the first version of plasma cosmology and was explained by means of the repulsion between matter and antimatter (Alfvén and Klein 1962, Alfvén 1979). A plasma mechanism can separate matter from antimatter and, when an antimatter cloud bumps into an ordinary-matter cloud, they will not totally annihilate each other; instead, only a thin layer will be annihilated (Lehnert 1977), generating a hot, low-density plasma layer which will push the clouds apart. Alfvén proposed his 'fireworks' model, in which a supercluster is repelled by other superclusters; within a supercluster each cluster is repelled by other clusters; and within a given cluster each galaxy is repelled by the other galaxies, and so on, obeying a distribution of matter and antimatter. In each local volume, a small explosion would impose its own local Hubble–Lemaître relationship, and this would explain the variations in the velocities of the Hubble–Lemaître law, i.e. the different values of the Hubble constant measured in the 1970s and 1980s, when Alfvén posited his hypothesis, in different ranges of distances or looking in different directions, all without invoking dark matter. The energy derived from the annihilation of protons and electrons would produce a background radiation of X- and γ-rays. There are some objections against the existence of antimatter based on the absence of γ-rays from annihilation, but they are model-dependent. Many of the objections against antimatter have been analysed (Rogers and Thompson 1980) and it has been shown that none of them is crucial. Nonetheless, some critics remarked that it is not consistent with the isotropy of the x-ray backgrounds (Peebles 1993 [chapter 7]).

Instead of expansion caused by matter–antimatter repulsion, in more recent times, some proponents of plasma cosmology (e.g., Brynjolfsson 2004, Lerner 2006) have stated that there is no expansion, that the Universe is static, and that the redshift of the galaxies would be explained by some kind of tired light effect of the interaction of photons with electrons in the plasma (see section 4.1.1).

With regard to the CMBR, Lerner (1988, 1995) explains it in terms of absorption and re-emission of radiation produced by stars. It is similar to the mechanism proposed by QSSC, but here the thermalisation is due to interaction with electrons. The interaction of photons and electrons produces a loss of direction in the path of the light, giving rise to isotropic radiation.

2.4 Universe as a hypersphere

Another category of models that have appeared in different versions is one that posits that our Universe is a hypersphere of a higher-dimensionality geometrical entity; that is, a set of points at a constant distance from its centre, constituting a manifold with one dimension less than that of the ambient space.

One of the first models maintaining this idea is the chronometric cosmology of Irving E Segal (1918–1998) (Segal 1976, Segal and Zhou 1995). This model assumes that global space structure is a 3D-hypersphere (or 3D-hypersurface, as Segal calls it) in a Universe of four dimensions. Events in the Universe are ordered globally according to a temporal order. This model makes an application of general relativity different from the standard model, getting a relationship of the redshift with distance r: $z = \tan^2 \frac{r}{2R}$. With data about redshifts of galaxies and distances, it is now known this cannot be correct (Sandage and Tammann 1995). His cosmology gives a good fit to the various curves versus redshift: magnitude, counts, angular size, etc, but, as mentioned, with the data of the Hubble–Lemaître law this statement is not sustainable, and there is no explanation for the CMBR. Many other refutations of Segal's claims have also been published (Salpeter and Hoffman 1986, Wright 1987, Sandage 1992, Koranyi and Strauss 1997).

More recent is the hypothesis of the existence of a 5-dimensional spacetime combined. By making some peculiar assignments between coordinates and physical distances and time, a hyperspherical symmetry is made apparent by assigning the hypersphere radius to proper time and distances on the hypersphere to usual 3-dimensional distances in a Euclidean Universe (Almeida 2006), which can explain the Hubble–Lemaître expansion law without appealing to dark matter; an empty Universe will expand naturally at a flat rate in this way. Another variation is the Hypersphere World-Universe Model (Netchitailo 2020, 2021), which claims the existence of a 3-dimensional hypersphere with respect to a 4-dimensional Nucleus of the World. Matter in this Universe is of the ordinary kind with the addition of a multicomponent dark matter (instead of Cold Dark Matter plus Dark Energy). This model has a number of peculiar characteristics: the beginning of the Universe, instead of originating from a singularity, stems from a 4-dimensional Nucleus of the World; the radius of this Nucleus, increasing with speed c, is what produces the expansion; the CMBR stems from the thermodynamic equilibrium of photons with intergalactic plasma; and the nucleosynthesis of light elements occurs inside dark matter cores of Macro-objects.

2.5 Static models and/or non-cosmological redshifts

At the farthest extreme of alternative models with respect to the standard cosmology, we have proposals of Universe that contradict the main interpretation upon which standard cosmology was created from the 1920s onwards: expansion of the Universe and the interpretation of redshifts as cosmological. Some versions of plasma and hypersphere cosmologies figure among these models, but there exist plenty of other models that are characterised by the lack of an origin of time (an eternal Universe), and lack of expansion, in some cases the space even being infinite and Euclidean. The redshift of galaxies given by Hubble–Lemaître's law would be due to some mechanism different from the expansion or Doppler effect, mainly a 'tired light' (loss of energy of photons due to some interaction along their path) effect or others (see section 4.1.1).

Static models are usually rejected by most cosmologists. However, from a purely theoretical point of view, the representation of the Cosmos as Euclidean and static is not excluded. Both expanding and static spaces are possible for the description of the Universe. Before Einstein and the entry of Riemannian and other non-Euclidean geometries into physics, there were attempts to describe the known Universe in terms of Euclidean geometry, but these faced the problem of justifying a stable equilibrium. Within a relativistic context, Einstein (1917) proposed a static model that included a cosmological constant, his 'biggest blunder' according to himself. This model still has problems in guaranteeing stability, but it might be amenable to some kind of solution. Narlikar and Arp (1993) solve it within a variation of the Hoyle–Narlikar conformal theory of gravity, in which small perturbations of the flat Minkowski spacetime would lead to small oscillations about the line element rather than to a collapse. Boehmer *et al* (2007) analyse the stability of the Einstein static Universe by considering homogeneous scalar perturbations in the context of $f(R)$ modified theories of gravity, and it is found that a stable Einstein cosmos with a positive cosmological constant is possible. Other authors solve the stability problem through variation of fundamental constants (Van Flandern 1984, Troitskii 1987). Another idea by Van Flandern (1993) is that hypothetical gravitons responsible for gravitational interaction have a finite cross-sectional area, so that they can only travel a finite distance, however great, before colliding with another graviton. So the range of the force of gravity would necessarily be limited in this way and collapse avoided.

Curved geometry (general relativity and its modifications) does not conserve the energy–momentum of the gravity field. However, Minkowski space does follow the conservation of energy–momentum of the gravitational field. One approach with a material tensor field in Minkowski space is given in Feynman's gravitation (Feynman *et al* 1995), in which space is static but matter and fields can be expanding in that static space. Also worth mentioning is a model related to modern relativistic and quantum field theories of basic fundamental interactions (strong, weak, electromagnetic): the relativistic field gravity theory and fractal matter distribution in static Minkowski space (Baryshev 2008).

Olber's paradox for an infinite Universe is an old problem (e.g., Bondi 1961 [chapter 3]) and also needs subtle solutions, but extinction, absorption, and re-emission of light, fractal distribution of density, and the mechanism which itself produces the redshift of the galaxies might have something to do with its solution.

Of the many cases of static and/or non-cosmological redshift models in the literature, I will mention just a few. Some of them are totally obsolete and with no researcher actively working on them today, but they deserve to be mentioned from a historical point of view. Most of them are proposed by single researchers, they are author's own theories, in some cases produced by non-active professional individuals within astrophysics, or retired researchers or professionals of other fields; nonetheless, they serve to illustrate the range of possible speculations within cosmology. Again, I repeat, there are many other models that are not included here and I apologise to the authors of other models for not having included them here. As said at the beginning of this chapter, it is practically impossible to mention all the different variants of alternative cosmological proposals, given that most of them are published in obscure journals or on private web-pages and blogs, or remain unpublished, merely circulating on the Internet or in e-mail or draft form. In any case, my aim here is not to be encyclopaedic, but rather to illustrate the social phenomenon of this creative way of proposing cosmological model proposals. Just a few examples will suffice:

- Eternal Universe, proposed by Gerald S Hawkins (1928–2003) (Hawkins 1960, 1962a, 1962b): Based on the existence of a negative pressure in a cosmic fluid derived from general relativity (not very different from the role that the cosmological constant has acquired nowadays). The main point which differentiates this model from the standard theory is the proposal that the Universe is static, infinite, without an instant of creation, and without expansion. The redshift of the galaxies is explained as a gravitational effect combined with a slight amount of intergalactic extinction produced by certain particles located in the space surrounding galaxies, which is $\sim 10^{-7}$ times the local interstellar absorption per unit distance. Hawkins (1993) argues that his model is not unstable, having no tendency to collapse or expand, and that the CMBR is due to the emission of galactic and intergalactic dust grains. Olbers' paradox is solved by means of absorption in clouds of dust.
- Curvature Cosmology, proposed by David F Crawford (1937–) (Crawford 2011a, 2011b, 2011c): This cosmology is based on a combination of general relativity and quantum mechanics. The concept of curvature pressure arises

from the idea that the density of particles produces curved spacetime. Then, as a function of the velocity of these particles, there will be a reaction pressure that acts to decrease the local spacetime curvature. In the case of the photons that travel across matter, this produces a 'tired light' effect as a product of the gravitational interaction between wave packets and curved spacetime, which will follow geodesics and be subject to geodesic focusing, giving rise to the observed curvature redshift of galaxies. Since this will alter the transverse properties of the wave, some of its properties such as angular momentum will be altered, which is contrary to quantum mechanics. The result of the interaction of the photon is three new photons: one with almost identical energy and momentum to that of the original photon and two extremely low energy secondary photons. Anomalous redshift cases might be produced by the extra redshift being due to the photons' passage through the cloud around the anomalous object (Crawford 2011a, 2011c). The CMBR comes from the curvature–redshift process acting on the high-energy electrons and ions in the cosmic plasma. The energy loss which gives rise to the spectrum of photons of the CMBR occurs when an electron that has been excited by the passage through curved spacetime interacts with a photon or charged particle and loses its excitation energy.

- Scale Expanding Cosmos, proposed by Carl Johan Masreliez (1939–) (Masreliez 1999, 2012): A Universe is proposed in which not only space expands (therefore, it is not properly a static=non-expanding model) but time also expands: the relationship between space and time could remain constant during the cosmological expansion and all cosmological locations in time and space could be equivalent, if the metrics of both space and time expand. The scale expansion could be eternal, which would eliminate the creation event. Redshift is a tired light effect. The theory is presented based on the proposition that all four metric coefficients of space and time change with cosmological expansion. Such a universal scale expansion would preserve the four-dimensional spacetime geometry and therefore, by general relativity, most physical relationships. In addition, if the scale expansion were exponential with time, all epochs would be equivalent. There are also other proposals on the reinterpretation of space in terms of time expansion or a comoving system of units (e.g., Oliveira and de Abreu 2002, Oliveira 2021).
- Self Creation Cosmology (SCC), proposed by Garth Antony Barber and others (Barber 2002a, 2002b, 2003, 2006), is an adaptation of the Brans Dicke theory in which the conservation requirement is relaxed in order to allow the scalar field to interact with matter. SCC can be thought of as general relativity + Mach's principle + local conservation of energy. In SCC, energy is conserved but energy–momentum is not. Particle masses increase with gravitational potential energy and, as a consequence, cosmological redshift is caused by a secular, exponential increase of particle masses. The Universe is static and eternal in its Jordan frame and linearly expanding in its Einstein frame. Furthermore, as the scalar field adapts the cosmological equations, these require the Universe to have an overall density of only one third of the critical

density while remaining spatially flat. The theory also posits a time-slip between atomic clock time on the one hand, and gravitational ephemeris and cosmological time on the other, which would result in the observed cosmic acceleration.

- Wave System Cosmology, proposed by Thomas B Andrews (1927–) (Andrews 1988, 1998): The Universe is a pure system of waves with mass density and tension parameter proportional to the local intensity of the modes of the waves. The peaks of the constructive interferences are the elementary particles. The redshift is produced by a 'tired light' mechanism. The gravitational force derives from the wave modes originating among protons. The energy input to a single proton from all of the interacting particles in the Universe is found to be equal to the mass energy of the proton.
- Subquantum Kinetics, proposed by Paul A LaViolette (1947–) (LaViolette 2012): This is a unified field theory with the foundations for a new wave theory of matter. Its non-dispersing, periodic structures resolve the wave–particle duality and produce de Broglie wave diffraction effects. The Subquantum kinetics model proposes an open, order-generating Universe, continuously creating matter and energy. It predicts that gravitational potential should have a finite range. It uses 'tired light' redshift in a static Universe, without radiating a secondary photon, angular deflection, or strong wavelength dependence. It works as if intergalactic space were on average endowed with a negative gravitational mass density.

2.6 Caveats/problems in the alternative approaches

Table 2.1 summarises the cosmological models mentioned in this chapter.

We will see in later chapters that the standard model has many problems, many observations not being understood in the light of the predictions of the model. The alternative proposals have their share of difficulties too, and their problems are yet more severe (see, for instance, Edward L Wright's webpage[5]), perhaps because these theories are not as developed and polished as the standard model.
Solution of Olbers' paradox by dust absorption in an infinite Universe, for instance, is not clear. One may wonder, if energy does not disappear, whether the absorbing element (dust) should be heated and re-emit, and, if the energy disappears how that can be consistent with known physical laws. This problem has no easy solution. For expansion in eternal universes, either it is taken as a fact, so the models need speculative elements to argue that there was no beginning of the Universe, or an alternative explanation for the redshift of the galaxies is needed, which raises its own set of difficulties (see section 4.1.1). Also, light element abundances in some cases require very old populations (Population III) that have not been observed yet. The CMBR has alternative explanations to ΛCDM, but with ad hoc elements without direct proof, such as invented particles to thermalise stellar radiation. All proposals to explain a CMBR produced in the intergalactic medium—even assuming that a perfect blackbody shape can be produced—have the problem that the integration

[5] http://www.astro.ucla.edu/~wright/errors.html

Table 2.1. Properties of different cosmological models: the first is the standard model; the rest are those described in this chapter (in order of appearance).

Model	Gravity, forces	Expansion	Age	Redshift	Dark elements	CMBR	Nucleosynth.	Large-scale	Galaxy formation
ΛCDM	Gen. Relat.	Yes	13.8 Gyr	Cosmol.	Inflation, Dark matter, Dark energy	Decoupling matter rad. at $z \approx 1100$	Primordial	Homog.	Evolution of density fluctuations
Inhomog. Univ.	Gen. Relat.	Yes	finite	Cosmol.	Inflation, Dark Matter	Decoupling matter rad. at $z \approx 1100$	Primordial	Inhomog.	Evolution of density fluctuations
Cold Big Bang	Gen. Relat.	Yes	finite	Cosmol.	—	Thermal. rad. Pop. III by IGM	Primor. &/or Pop. III	Homog.	Evolution of density fluctuations
Varying physical constants	several	Yes	finite	Cosmol.	Different solutions	—	—	Homog.	Evolution of density fluctuations
Alternative gravity	non-GR	Yes	finite	Cosmol.	—	Decoupling matter rad.	Primordial	Inhomog.?	Quick evol. of density fluctuations
Zero active mass (ZAM) & Milne	Gen. Relat.	Yes	14.6 Gyr	Cosmol.	Dark matter	Thermal. rad. Pop. III by dust at $z \approx 16$ (ZAM)	Primordial	Homog.	Evolution of density fluctuations
Cyclical universes, no C – field	several	Yes	∞	Cosmol.	Dark matter, Black holes evap. (Conf. C.)	—	Primordial	Homog.	Evolution of density fluctuations

(Continued)

Table 2.1. *(Continued)*

Model	Gravity, forces	Expansion	Age	Redshift	Dark elements	CMBR	Nucleosynth.	Large-scale	Galaxy formation
Quasi Steady State Cosm.	Gen. Relat.	Yes	∞	Cosmol. + Var. mass	C – field, Dark matter, Dark energy	Thermal. rad. Pop. III by whiskers	Pop.III	Homog.	Evolution of density fluctuations
Plasma Cosm.	Electromag.	Yes/No	∞	Doppler repuls., or tired light	Strong magn. fields	Thermal. rad. Pop. III by electrons	Pop.III	Homog.	Plasma
Hyperspheres	Hypersphere	Yes/No	finite	Geometric	Multicomp. dark matter	Thermal. rad. by IGM	In cores Dark Mat. (Netch.20)	Homog.	—
Hawkins' Eternal Cosm.	Gen. Relat.	No	∞	Gravit.	Negative pressure	Dust emission	—	Homog.	—
Crawford's Curvature Cosm.	Gen. Rel. +QM	No	∞	Tired light	—	High energy charged part. redshifted	—	Homog.	—
Masreliez's Scale Expand. Cosm.	Gen. Relat.	Space & Time	finite or ∞	Tired light	—	—	—	Homog.	—
Barber's Self Creation Cosm.	Brans Dicke	No	∞	Var.mass	—	—	—	Homog.	—
Andrews' Wave Syst. Cosmol.	Matter=Wave	No	∞	Tired light	—	—	—	Homog.	—
LaViolette's Subquantum kinetics	Matter=Wave	No	∞	Tired light	Continuous creation of matter	—	—	Homog.	—

along the line of sight gives a superposition of many layers of blackbody radiations, each with a different redshift, giving in total something different from a black body, unless the CMBR originates in the local Universe ($z \approx 0$; in such a case, the problem would be that the space would be too opaque to allow the observations of distant radio sources, Wright 1995) or at a given high redshift $z > 0$ but within a layer with small Δz.

The most elaborate alternative models, such as QSSC, do indeed apply the same methodology as the standard model: they have some basic tenets and a lot of free parameters and ad hoc elements that are introduced every time some observation does not fit their model. The modern version of QSSC is able to explain most of the difficulties of the previous (Steady State) version of the model. They introduce ad hoc elements without observational support in the same way that the standard model introduces ad hoc non-baryonic dark matter, dark energy, inflation, etc. The very idea of continuous creation of matter[6] also necessitates some very exotic physics, and with no empirical support. They also continued ad hoc to skip over the inconsistencies; for instance, the maximum redshift of a galaxy was set to be 5 in the initial version of QSSC; however the authors have free parameters that can be changed conveniently when some new observation does not fit the initial predictions: In the end, then, they can introduce ad hoc corrections that render their theory compatible with any maximum redshift of a galaxy.

My impression is that none of the alternative models has acquired the same level of development as ΛCDM in offering explanations of available cosmological observations. One should not, however, judge any theory according to the number of observations that it can successfully explain; but by the plausibility of its principles and its potential to fit data (provided that we had an army of theoreticians able to correct the theory ad hoc every time new observations need to be accommodated). A pluralist approach to cosmology is a reasonable option when the preferred theory is still under discussion. Therefore, given the number of problems with the standard model (see chapters 3–7), it is quite reasonable to keep a weather eye on alternative ideas that might at least provide better partial explanations of certain observed phenomena. Nonetheless, a global cosmological theory that fully explains everything in the Universe does not yet exist according to either standard or in alternative viewpoints.

References

Aguirre A N 1999 Cold Big Bang nucleogenesis *Astrophys. J.* **521** 17–29

Aguirre A N 2000 The cosmic background radiation in Cold Big Bang *Astrophys. J.* **533** 1–18

Albrecht A and Magueijo J 1999 Time varying speed of light as a solution to cosmological puzzles *Phys. Rev.* D **59** 043516

Alfvén H 1979 Hubble expansion in a Euclidean framework *Astrophys. Space Sci.* **66** 23–7

Alfvén H 1981 *Cosmic Plasma (Astrophysics and Space Science Library, vol. 82)* (Dordrecht: Reidel)

[6] The idea of continuous creation of matter as a result of a modification of Einstein's equation of gravitational field is also independently explored by other alternative cosmological models (e.g., Yang 2016).

Alfvén H 1983 On hierarchical cosmology *Astrophys. Space Sci.* **89** 313–24

Alfvén H 1988 Cosmology in the plasma universe *Laser Part. Beams* **16** 389–98

Alfvén H and Klein O 1962 Matter-antimatter annihilation and cosmology *Ark. Fys.* **23** 187

Almeida J 2006 Geometric drive of the universe's expansion *1st Crisis in Cosmology Conf. (AIP Conf. Ser. 822(1))* ed E J Lerner and J B Almeida (Melville, NY: AIP) pp 110–22

Andrews T B 1988 A wave system theory of quantum mechanics *Physica B+C* **151** 351–4

Andrews T B 1998 Derivation of Newton's law of gravitation and discovery of the unique normal modes of the Universe *Causality and Locality in Modern Physics* ed G Hunter, S Jeffers and J-P Vigier (Dordrecht: Springer) pp 135–42

Annila A 2016 Rotation of galaxies within gravity of the universe *Entropy* **18** 191

Barber G 2002a A new self creation cosmology *Astrophys. Space Sci.* **282** 683–730

Barber G 2002b The principles of self creation cosmology and its comparison with general relativity (arXiv:gr-qc/0212111)

Barber G 2003 The derivation of the coupling constant in the new self creation cosmology (arXiv: gr-qc/0302088)

Barber G 2006 Resolving the degeneracy: experimental tests of the new self creation cosmology and a heterodox prediction for gravity probe b *Astrophys. Space Sci.* **305** 169–76

Baryshev Y V 2008 Field fractal cosmological model as an example of practical cosmology approach *Practical Cosmology 1* (Petersburg: TIN) pp 60–7

Baryshev Y V, Sylos Labini F, Montuori M and Pietronero L 1994 Facts and ideas in modern cosmology *Vistas Astron.* **38** 419–500

Bekenstein J D 2004 Relativistic gravitation theory for the modified Newtonian dynamics paradigm *Phys. Rev.* D **70** 083509

Bekenstein J and Milgrom M 1984 Does the missing mass problem signal the breakdown of Newtonian gravity? *Astrophys. J.* **286** 7–14

Benoit-Lévy A and Chardin G 2012 Introducing the Dirac-Milne universe *Astron. Astrophys.* **537** 12

Boehmer C G, Hollenstein L and Lobo F S N 2007 Stability of the Einstein static universe in $f(r)$ gravity *Phys. Rev.* D **76** 084005

Bondi H 1961 *Cosmology* 2nd edn (Cambridge: Cambridge University Press)

Bondi H and Gold J 1948 The steady-state theory of the expanding universe *Mon. Not. R. Astron. Soc.* **108** 252–70

Brynjolfsson A 2004 Redshift of photons penetrating a hot plasma (arXiv:astro-ph/0401420)

Burbidge E M, Burbidge G R, Fowler W A and Hoyle F 1957 Synthesis of the elements in stars *Rev. Mod. Phys.* **29** 547–650

Chashchina O I and Silagadze Z K 2015 Expanding space, quasars and St+ Augustine's fireworks *Universe* **1** 307–56

Ćirković M M and Perović S 2018 Alternative explanations of the cosmic microwave background: a historical and epistemological perspective *Stud Hist. Phil. Mod. Phys.* **62** 1–18

Clifton T, Ferreira P G, Padilla A and Skordis C 2012 Modified gravity and cosmology *Phys. Rep.* **513** 1–189

Crawford D F 2011a Observational evidence favors a static universe (part I) *J. Cosmol.* **13** 3875–946

Crawford D F 2011b Observational evidence favors a static universe (part II) *J. Cosmol.* **13** 3947–99 (http://thejournalofcosmology.com/crawford2.pdf)

Crawford D F 2011c Observational evidence favors a static universe (part III) *J. Cosmol.* **13** 4000–57 (http://thejournalofcosmology.com/crawford3.pdf)

Einstein A 1917 Kosmologische betrachtungen zur allgemeinen relativitätstheorie *Sitzungsberichte der Königlich Preussischen Akademie der Wissenschaften* (Berlin: Akademie der Wissenschaften) pp 142–52

Ellis G R 2007 Note on varying speed of light cosmologies *Gen. Relativ. Gravit.* **39** 511–20

Famaey B and McGaugh S 2012 Modified Newtonian dynamics (MOND): observational phenomenology and relativistic extensions *Living Rev. Relativ.* **15** 10

Felten J E 1984 Milgrom's revision of Newton's laws–dynamical and cosmological consequences *Astrophys. J.* **286** 3–6

Feynman R P, Morinigo F B and Wagner W G 1995 *Feynman Lectures on Gravitation* (Reading, MA: Addison-Wesley)

Gabrielli A, Sylos Labini F, Joyce M and Pietronero L 2005 *Statistical Physics for Cosmic Structures* (Berlin: Springer)

Gohar H 2017 Cosmology with varying constants from a thermodynamic viewpoint *Universe* **3** 26

Goldman T, Pérez-Mercader J, Cooper F and Nieto M M 1992 The dark matter problem and quantum gravity *Phys. Lett.* B **281** 219–24

Görnitz T 1988 Connections between abstract quantum theory and space-time structure. II. A model of cosmological evolution *Int. J. Theor. Phys* **27** 659–66

Gupta R P 2020 Cosmology with relativistically varying physical constants *Mon. Not. R. Astron. Soc.* **498** 4481–91

Hawkins G S 1960 The redshift *Astron. J.* **65** 52

Hawkins G S 1962a Expansion of the universe *Nature* **194** 563–4

Hawkins G S 1962b The nature of the red-shift *Nuovo Cimento* **23** 1021–7

Hawkins G S 1993 Eternal universe *Encyclopedia of Cosmology* ed N S Hetherington (New York: Garland Publishing) p 196

Hoyle F 1948 A new model for the expanding universe *Mon. Not. R. Astron. Soc.* **108** 372–82

Hoyle F, Burbidge G and Narlikar J V 1993 A quasi-steady state cosmological model with creation of matter *Astrophys. J.* **410** 437–57

Hoyle F, Burbidge G and Narlikar J V 1994 Astrophysical deductions from the quasi-steady state cosmology *Mon. Not. R. Astron. Soc.* **267** 1007–19

Hoyle F, Burbidge G and Narlikar J V 1995 The quasi-steady-state cosmology: a note on criticisms by E. L. Wright *Mon. Not. R. Astron. Soc.* **277** L1–3

Hoyle F, Burbidge G R and Narlikar J V 2000 *A Different Approach to Cosmology* (Cambridge: Cambridge University Press)

Hoyle F and Narlikar J V 1964 A new theory of gravitation *Proc. R. Soc. London* **A282** 191–207

Hoyle F and Wickramashinghe N C 1988 Metallic particles in astronomy *Astrophys. Space Sci.* **147** 245–56

Ibison M 2006 Thermalization of starlight in the steady-state cosmology *1st Crisis in Cosmology Conf. (AIP Conf. Ser. 822)* ed J B Almeida (Melville, NY: AIP) p 171

Jackson L S 2014 *Academic Publishing* (Chisinau: Lambert Academic Publishing)

John M V and Joseph K B 1996 A modified Ozer-Taha type cosmological model *Phys. Lett.* B **387** 466–70

Kim J, Naselsky P and Hansen M 2016 Friedmann–Robertson–Walker models do not require zero active mass *Mon. Not. R. Astron. Soc.* **460** L119–22

Koranyi D M and Strauss M A 1997 Testing the Hubble law with the IRAS 1.2 Jy redshift survey *Astrophys. J.* **477** 36–46

Krasnov K and Shtanov Y 2008 Non-metric gravity: II spherically symmetric solution, missing mass and redshifts of quasars *Class. Quantum Grav.* **25** 025002
LaViolette P A 2012 *ubquantum Kinetics: The Alchemy of Creation* 4th (Niskayana: Starlane Publ)
Layzer D 1990 *Cosmogenesis* (Oxford: Oxford University Press)
Lehnert B 1977 Problems of matter-antimatter boundary layers *Astrophys. Space Sci.* **46** 61–71
Lerner E J 1988 Plasma model of microwave background and primordial elements–an alternative to the big bang *Laser Part. Beams* **6** 457–69
Lerner E J 1991 *The Big Bang Never Happened: A Startling Refutation of the Dominant Theory of the Origin of the Universe* (Toronto: Random House)
Lerner E J 1995 Intergalactic radio absorption and the COBE data *Astrophys. Space Sci.* **227** 61–81
Lerner E J 2006 Evidence for a non-expanding universe: surface brightness data from HUDF *1st Crisis in Cosmology Conf. (AIP Conf. Ser. 822(1))* ed E J Lerner and J B Almeida (Melville, NY: AIP) pp 60–74
Masreliez C J 1999 The scale expanding cosmos *Astrophys. Space Sci.* **266** 399–447
Masreliez C J 2012 *The Progression of Time* (CreateSpace Independent Publishing Platform: North Charleston)
Melia F 2017 The zero active mass condition in Friedmann–Robertson–Walker cosmologies *Frontiers Phys.* **12** 5
Melia F 2019 Cosmological test using the Hubble diagram of high-z quasars *Mon. Not. R. Astron. Soc.* **489** 517–23
Melia F 2020 Reassessing dust's role in forming the CMB Eur *Phys. J. Plus* **135** 511
Melia F and López-Corredoira M 2017 Alcock-Paczyński cosmological test with model-independent BAO data *Int. J. Mod. Phys.* D **26** 1750055
Melia F and Shevchuk A S 2012 The $r_h = ct$ universe *Mon. Not. R. Astron. Soc.* **419** 2579–86
Merritt D 2020 *A Philosophical Approach to MOND. Assessing the Milgromian Research Program in Cosmology* (Cambridge: Cambridge University Press)
Milgrom M 2010 Quasi-linear formulation of MOND *Mon. Not. R. Astron. Soc.* **403** 886–95
Miller R I 202 AltU: The alternative universe (viXra:2102.0009)
Moffat J W 2006 Scalar tensor vector gravity theory *J. Cosmol. Astropart. Phys.* **2006** 004
Narlikar J V 1977 Two astrophysical applications of conformal gravity *Ann. Phys.* **107** 325–36
Narlikar J V 2006 The quasi-steady-state cosmology *Current Issues in Cosmology* ed J C Pecker and J V Narlikar (Cambridge: Cambridge University Press) pp 139–51
Narlikar J V and Arp H C 1993 Flat spacetime cosmology–a unified framework for extragalactic redshifts *Astrophys. J.* **405** 51–6
Narlikar J V, Burbidge G and Vishwakarma R G 2007 Cosmology and cosmogony in a cyclic universe *J. Astrophys. Astron.* **28** 67–99
Narlikar J V, Vishwakarma R G, Hajian A, Souradeep T, Burbidge G and Hoyle F 2003 Inhomogeneities in the microwave background radiation interpreted within the framework of the quasi-steady state cosmology *Astrophys. J.* **585** 1–11
Nayeri A, Engineer S, Narlikar J V and Hoyle F 1999 Structure formation in the quasi-steady state cosmology: a toy model *Astrophys. J.* **525** 10–6
Netchitailo V S 2020 World-universe model-alternative to Big Bang model *J. High Energy Phys.* **6** 133–58
Netchitailo V S 2021 Hypersphere world-universe model *J. High Energy Phys.* **7** 915–41

Oliveira A G 2021 Physics for expanding space–I+ The phenomenon *Manuscript* (private communication)

Oliveira A G and de Abreu R 2002 A relativistic time variation of matter/space fits both local and cosmic data (arXiv:astro-ph/0208365)

O'Raifeartaigh C, McCann B, Nahm W and Mitton S 2014 Einstein's steady-state theory: an abandoned model of the cosmos *Eur. Phys. J. H* **39** 353–67

Peebles P J E 1993 *Principles of Physical Cosmology* (Princeton: Princeton University Press)

Penrose R 2010 *Cycles of Time: An Extraordinary New View of the Universe* (London: Bodley Head)

Peratt A L 1983 Are black holes necessary? *Sky Telesc.* **66** 19–22

Peratt A L 1984 Simulating spiral galaxies *Sky Telesc.* **68** 118–22

Planck Collaboration 2020 Planck 2018 Results. VI. Cosmological parameters *Astron. Astrophys.* **641** A6

Rogers S and Thompson W B 1980 In defense of anti-matter *Astrophys. Space Sci.* **71** 257–60

Romano A E 2007 Lemaître–Tolman–Bondi universes as alternatives to dark energy: does positive averaged acceleration imply positive cosmic acceleration *Phys. Rev.* D **75** 043509

Ryle M and Clarke R W 1961 An examination of the steady-state model in the light of some recent observations of radio sources *Mon. Not. R. Astron. Soc.* **122** 349–62

Salpeter E E and Hoffman G L 1986 The galaxy luminosity function and the redshift-distance controversy (a review) *Proc. Natl Acad. Sci. USA* **83** 3056–63

Sandage A 1992 Evidence that the expansion is real *Phys. Scr.* **T43** 22–30

Sandage A and Tammann G A 1995 Steps toward the Hubble constant. X. The distance of the Virgo cluster core using globular clusters *Astrophys. J.* **446** 1–11

Sanders R H 1998 Cosmology with modified Newtonian dynamics (MOND) *Mon. Not. R. Astron. Soc.* **296** 1009–18

Sanders R H 2006 MOND and cosmology *Mass Profiles and Shapes of Cosmological Structures (EAS Publ. Ser. 20)* ed G A Mamon, F Combes, C Deffayet and B Fort (Les Ulis Cedex: EDP Sciences) pp 231–8

Sanders R H 2015 A historical perspective on modified Newtonian dynamics *Can. J. Phys.* **93** 126–38

Sanders R H and McGaugh S S 2002 Modified Newtonian dynamics as an alternative to dark matter *Annu. Rev. Astron. Astrophys.* **40** 263–317

Segal I E 1976 *Mathematical Cosmology and Extragalactic Astronomy* (New York: Academic)

Segal I E and Zhou Z 1995 Maxwell's equations in the Einstein universe and chronometric cosmology *Astrophys. J. Suppl. Ser.* **100** 307–24

Shojaie H and Farhoudi M 2004 A cosmology with variable *c* *Can. J. Phys.* **84** 933–44

Sotiriou T P and Faraoni V 2010 $f(r)$ theories of gravity *Rev. Mod. Phys.* **82** 451–97

Steinhardt P J and Turok N 2008 *Endless Universe. Beyond the Big Bang* (London: Orion)

Suntola T 2020 Unification of theories requires a postulate basis in common *J. Phys. Conf. Ser.* **1466** 012003

Troitskii V S 1987 Physical constants and evolution of the universe *Astrophys. Space Sci.* **139** 389–411

Unzicker A 2015 *Einstein's Lost Key: How We Overlooked the Best Idea of the 20th Century* (North Charleston: CreateSpace Independent Publishing Platform)

Van Flandern T 1984 Is the gravitational constant changing? *Precision Measurements and Fundamental Constants II* ed B N Taylor and W D Phillips (National Bureau of Standards Special Publication 617) (Gaithersburg: National Bureau of Standards) pp 625–7

Van Flandern T 1993 *Dark Matter, Missing Planets and New Comets* (Berkeley, CA: North Atlantic Books)

Vishwakarma R G 2013 A curious explanation of some cosmological phenomena *Phys. Scr.* **87** 055901

Wickramasinghe N C 2006 Evidence for iron whiskers in the universe *Current Issues in Cosmology* ed J C Pecker and J V Narlikar (Cambridge: Cambridge University Press) pp 152–63

Wiltshire D L 2009 Average observational quantities in the timescape cosmology *Phys. Rev.* D **80** 123512

Wright E L 1987 Source counts in the chronometric cosmology *Astrophys. J.* **313** 551–5

Wright E L 1995 Comments on the quasi-steady-state cosmology *Mon. Not. R. Astron. Soc.* **276** 1421–4

Yang J L 2016 Modification of field equation and return of continuous creation–galaxies form from gradual growth instead of gather of existent matter *Int. J. Adv. Res. Phys. Sci.* **3**(8) 5–32 (https://www.arcjournals.org/pdfs/ijarps/v3-i8/2.pdf)

Chapter 3

CP violation, inflation, dark matter, and dark energy

Four exotic elements have been introduced into the standard cosmological model: CP violation, inflation, non-baryonic dark matter, and dark energy. Success in converting the standard cosmological model into a solid theory depends strongly on whether we are able to confirm the existence of these dark elements and whether the observations that warrant positing them cannot be explained with alternative scenarios. We have not achieved that success yet because of the numerous inconsistencies found and the proposed alternatives to explain the same phenomena. A review of some ideas and facts is given here.

In chapter 1, we introduced the basic characteristics of the standard model in cosmology, but we may expand it here with a brief description of the exotic elements introduced ad hoc in recent decades in order to solve some inconsistencies of the pristine formulation of Big Bang theory.

3.1 Antimatter and CP violation

The topic of antimatter constitutes one of the many consistency problems of the standard cosmological model. There are two possibilities and both are at odds with our present observations and experiments:

1. If there were a symmetry of matter and antimatter at the beginning of the Universe, where is the antimatter corresponding to the matter that we see now? Other galaxies cannot be made of antimatter because that would create a matter–antimatter boundary within the intergalactic medium that would create gamma rays, which are not observed (Taubes 1997).
2. If there were an asymmetry between matter and antimatter, symmetry-breaking would be needed, causing other consequences that so far have not been successfully tested: the violation of charge-conjugation and

doi:10.1088/978-0-7503-3775-5ch3

parity-reversal (CP) symmetry of fundamental particles. This violation of CP symmetry[1] causes most antimatter to annihilate with matter, but leaves much residual matter.

Early models indicated that an asymmetry between matter and antimatter would also imply a finite lifetime for the proton. Indeed, a lifetime of around 10^{30} years (i.e. an average decay rate of one proton per year in 10^{30} protons) was originally predicted by GUT (Grand Unified Theory) (Georgi *et al* 1974) as a necessary condition for resolving the antimatter problem. As experiments ruled out longer and longer lifetimes, these GUT were modified to predict longer lifetimes (López *et al* 1992). To date, however, observations have ruled out a lifetime even larger than 10^{34} yr (Tanaka *et al* 2020). There is no experimental evidence for a finite lifetime of the proton. The lack of such a finite lifetime would rule out the baryon number needed to overcome the 10^{11}-fold gap between Big Bang baryon density predictions and observations. So this massive contradiction of prediction and observation still exists and will continue to exist until we are able to observe a proton decay.

Experimenters are searching for other evidences of this asymmetry. CP violation has been proposed for electroweak baryogenesis from the dynamical Cabibbo–Kobayashi–Maskawa matrix (CKM matrix, quark mixing matrix) (Bruggisser *et al* 2017), CP violation in Higgs boson interactions (Bernlochner *et al* 2019), and in leptons instead of quarks, which could generate the matter–antimatter disparity through a process called leptogenesis (T2K Collaboration 2020), or other mechanisms. So far, none of these ideas has received confirmation, though they are still being actively researched. The sociological environment of this kind of research is the same as in the search for any other dark element: somebody proposes a theory to solve a problem (in this case GUT), this theory contains a certain prediction, this prediction fails, then along comes another theory with more difficult predictions to test, etc. It looks like a competition between theoreticians to produce theories with higher and higher energy limits on their predictions as observers/experimenters boost the capacity of their experiments or observations to reach those limits, with negative results that motivate new theories with even higher limits. Rather like the race between Achilles and the tortoise in Zeno's paradox, with the difference that human patience is not infinite.

3.2 Inflation

The introduction of cosmic inflation into the standard theory is another sign of the weakness of the model. The theory was developed in the late 1970s and early 1980s, with notable contributions by several theoretical physicists, including Alexei Starobinsky (1948–), Alan Guth (1947–), and Andréi D Linde (1948–). It posits

[1] This and other symmetry-breaking phase transitions of the early Universe also determine the nature of the fundamental interactions: the split from a unified force into today's strong, electromagnetic, weak, and gravitational forces. This happened via the Higgs mechanism about 10^{-34} s after the Big Bang at a temperature of 10^{27} K. The transitions in some Grand Unified Theory (GUT) formulations also generate defects in the geometry of space-time (such as cosmic strings).

an exponential expansion of space in the early Universe. The inflationary epoch lasts from 10^{-36} s after the Big Bang singularity to some time between 10^{-33} and 10^{-32} s after it. The inflation paradigm explains the apparent paradox of a superluminal velocity of expansion at these short times. Following the inflationary period, the Universe continued to expand, but the expansion was no longer accelerating (Liddle 1999). It explains why the Universe appears to be the same in all directions (isotropic), why the cosmic microwave background radiation is distributed evenly; two opposite points of the sky at the time of recombination are separated from each other by more than 70 times the distance that light could have travelled till that time. This smooth microwave background would be explained because inflation proceeds far faster than the speed of light; regions at one time in contact with each other, and thus at the same temperature, are blown farther apart from each other than the distance light could have travelled in the duration of the Universe. Moreover, inflation explains why the Universe is flat and the geometry nearly Euclidean ($K = 0$), and why no magnetic monopoles have been observed. The flatness problem is solved because the Universe blew up to such a huge size, far bigger than the part we can observe. It also explains the origin of the large-scale structure of the cosmos: Quantum fluctuations in the microscopic inflationary region, magnified to cosmic size, become the seeds for the growth of structure in the Universe. Different models were also constructed to explain the origin of the initial conditions as seeds for the large-scale structure of galaxies (e.g. the later-refuted baryon isocurvature model by Peebles 1989).

Some authors have argued that the inflation necessary to explain a flat Universe is highly improbable (Steinhardt 2011). If we adopt the idea of a multiverse in an eternal inflation, it has no predictive power because anything may happen with inflation. A theory that can predict anything is a theory that predicts nothing. Many cosmologists see inflation as disconnected from the speculative multiverse hypothesis (Chowdhury *et al* 2019), but even so, whether with or without multiverse, there are ~1000 models of inflation (Martin *et al* 2014), which may lead us to think that almost anything is possible with inflation. Moreover, there are also observations that disprove the most common types of inflation. For instance, the lack of large-angular scale correlations in CMBR anisotropy observations is in considerable conflict with the basic inflationary paradigm (Melia and López-Corredoira 2018).

In 2017, an article very critical of the inflation theory was published. It was written by top cosmologists (Iljas *et al* 2017) and generated fierce debate. The article said that the theory of inflation must be discarded because the only specific forms of inflation of theoretical account no longer fitted recent data. They further argued that the theory is no longer scientific since it had become so all-encompassing and complicated with so many pliable parameters that it could be tweaked and adjusted to fit any kind of observation, and one model of inflation could be replaced by another as new observations appear to contradict it. Guth *et al* (2017) responded to the article by claiming that inflation is untestable because its predictions can be changed. Neither the attack nor the defence look to leave inflation as a solid theory, but merely as a speculative patch without empirical or observational support or that cannot be falsified. Nonetheless, most cosmologists still consider inflation as valid theory with no important difficulties or problems that can be met (Chowdhury *et al* 2019), although important questions still remain open, such as the lack of correlation of large-angular scale correlations in CMBR anisotropies observations, which Chowdhury *et al* (2019) do not even mention.

There is no global consensus among cosmologists concerning the reality of inflation, and we are far from having irrefutable proofs that keep it in line with solid evidence.

3.3 History of the idea of dark matter

The existence of dark or invisible matter detectable only through its gravitational influence has been known by astronomers for a long time now (Bertone and Hooper 2018). Friedrich Wilhelm Bessel (1784–1846) in 1844 argued that the observed proper motions of the stars Sirius and Procyon could be explained only in terms of the presence of faint companion stars (Bessel 1844). In 1846, Urbain Jean Joseph Le Verrier (1811–1877) and John Couch Adams (1819–1892) independently predicted the existence of Neptune based on calculations of the anomalous motions of Uranus. Le Verrier later proposed the existence of the planet Vulcan to explain anomalies in the orbit of Mercury, but he failed this time because the solution was not invisible matter but a change of gravitational laws, as was solved years later by Einstein with General Relativity. The dynamical analysis of dark matter in the form of faint stars in the Milky Way using the motion of stars was carried out by William Thomson (Lord Kelvin; 1824–1907) in 1904, Henri Poincaré (1854–1912) in 1906, Ernst Öpik (1893–1985) in 1915, Jacobus Kapteyn (1851–1922) in 1922, James Hopwood Jeans (1877–1946) in 1922, Bertil Lindblad (1895–1965) in 1926, and Jan Hendrik Oort (1900–1992) in 1932 with different results (Bertone and Hooper 2018).

With regard to extragalactic astronomy, Fritz Zwicky's (1898–1974) paper of 1933 on dark matter in rich clusters applied the virial theorem to these data and found a mass-to-light ratio of ~60 in solar units (rescaled to the present-day value of the Hubble–Lemaître constant) (Zwicky 1933)[2]. In 1959, Franz Daniel Kahn (1926–1998) and Lodewijk Woltjer (1930–2019) determined the mass of the Local Group and obtained a mass-to-light ratio of 43 in solar units (Kahn and Woltjer 1959). In the 1950s, Thornton Leigh Page (1913–1996) also found that pairs of elliptical galaxies had a mass-to-light ratio of 66 in solar units (Page 1952, 1960). This showed that such binaries must either have massive envelopes or be embedded in a massive common envelope. Similar results were obtained in the 1950s for 26 binary galaxies by Erik Holmberg (1908–2000) (Holmberg 1954). In 1939 Horace W Babcock (1912–2003) first showed the need for dark matter for an individual galaxy by measuring the rotation curve of the outer regions of M31 out to 100 arcminutes ($\approx$20 kpc) from its centre (Babcock 1939). However, the majority of astronomers were not

[2] We now know that Zwicky had not taken into account an effect that would reduce the need for dark matter (Křížek *et al* 2014): relativistic effects of high velocities, gravitational redshift, and gravitational lensing in a curved space, which would decrease the Hubble–Lemaître parameter, intergalactic baryonic matter, gravitational aberration, etc.

convinced of the need for dark matter haloes in galaxies until the publication of theoretical papers in the 1970s, such as that on the stability of galactic discs by Jeremiah Paul Ostriker (1937–) and Philip James Edwin Peebles (1935–) (Ostriker and Peebles 1973). Later, rotation curves in the radio by Albert Bosma (Bosma 1978) and in the visible by Vera Rubin (1928–2016), William Kent Ford Jr. (1931–), and Nortbert Thonnard (1943–2014) (Rubin *et al* 1980) more readily persuaded the community. This shows a mindset that is common among astrophysicists: accepting facts only when there is a theory to support them with an explanation, a not-so-empirical approach that has dominated the development of cosmology.

Cosmology has indeed played a very important role in the idea of dark matter on galactic scales. The first predictions based on Cosmic Microwave Background Radiation (CMBR) anisotropies were wrong. It was predicted in the 1960s that $\Delta T/T$ should be one part in a hundred or a thousand (Sachs and Wolfe 1967); however, fluctuations of this amplitude could not be found from observations in the 1970s. In order to solve this problem, non-baryonic dark matter was introduced ad hoc. Shortly afterwards, the connection between particle physics and the missing mass problem in galaxies arose. Many astrophysicists considered dark matter haloes surrounding galaxies and galaxy clusters possibly to consist of a gas of non-baryonic particles rather than faint stars or other astrophysical objects. This was a felicitous notion for which there was no proof; there is no proof that directly connects the problem of the amplitude of CMBR anisotropies with the rotation curves of galaxies or the missing mass in clusters, but the idea was pushed by leading cosmologists, who made the idea fashionable throughout the astrophysical community.

Part of the success of these non-baryonic dark matter scenarios in the haloes of the galaxies was due to the good agreement of simulations of large-scale structure with the observed distributions of galaxies. At first, in the 1980s, with the attempt to fit the data using hot dark matter composed of neutrinos, the simulations showed that very large structures should be formed first and only later go on to form galaxy-sized haloes through fragmentation, which did not match the observations (White *et al* 1983). But later cold dark matter (CDM) models were more successful, at least on large scales (>1 Mpc).

This tendency to sell a failure of a prediction as a success for a model via the ad hoc introduction of some convenient form of unknown dark matter still prevails. A more recent instance of this predilection is the introduction of some peculiar form of dark matter in order to cool the gas at $z \approx 18$ and solving the discrepancies in the measurements of 21 cm line amplitude with respect to *a priori* predictions (Bowman *et al* 2018).

3.4 Dark matter and inconsistencies of the theory on galactic scales

That there is some dark matter, either baryonic or non-baryonic, is clear, but how much is there, and what is its nature? The success of the standard model in converting a hypothesis into a solid theory depends strongly on the answer to these unanswered questions. Stellar and cold gas in galaxies sum to a baryonic matter content of $8^{+4}_{-5}\%$ of the total amount of the baryonic matter predicted with the standard model (Bell *et al* 2003). Where is the rest of the baryonic material? What is the nature of the putative non-baryonic dark matter required to achieve the current value, $\Omega_m \approx 0.3$, of the density of matter?

Current CDM models predict the existence of dark matter haloes for each galaxy whose density profile falls approximately as r^{-2}, although the original idea (White and Rees 1978) concerning hierarchical structures with CDM that gave birth to the present models was that the dark matter was distributed without internal substructure, more like a halo with galaxies than galaxies with a halo, such that, for instance, the Milky Way, Andromeda, and other galaxies of the Local Group are all embedded in the same common halo. The intergalactic matter in this case is not empty but contains the greater part of the mass of groups or clusters of galaxies (Battaner and Florido 2000, López-Corredoira *et al* 1999, 2002).

Some authors have been led to question the very existence of this dark matter on galactic scales since evidence for it is weak (Battaner and Florido 2000, McGaugh 2000, Evans 2001, Tasitsiomi 2003) and the predictions do not fit the observations: CDM has a 'small scale crisis' since there are some features of galaxies that are very different from the predictions of the cosmological model. Problems such as missing satellites, cusped halos, 'too-big-to fail' (the central densities of dwarfs in the simulations are much higher than the observed ones), satellite planes, baryonic/dark matter scaling relations that are too tight within a large diversity in their rotation curves, and other problems challenge the standard picture (Bullock and Boylan-Kolchin 2017). We discuss further some of these problems in this section. Nonetheless, many researchers are eagerly trying to find solutions that make the data and models compatible, assuming *a priori* that the model 'must be' correct.

There is a problem with a density of the halo in the inner galaxy observed to be lower than predicted. ΛCDM (CDM including a Λ term for the cosmological constant; see section 3.7) predicts halo mass profiles with cuspy cores and a low outer density, while lensing and dynamical observations indicate a central core of constant density and a flattish high dark mass density outer profile (Perivolaropoulos 2009). Possible solutions of the core–cusp problem without abandoning the standard model include: bar–halo friction, which reduces the density of the halo in the inner galaxy (Sellwood 2008); haloes around galaxies that may have undergone compression by the stellar disc (Gnedin *et al* 2004) or/and suffered from the effects of baryonic physics (Di Cintio *et al* 2014); destruction of dark matter cusps by supernova feedback (Pontzen and Governato 2012); and galaxies formed through monolithic collapse instead of the usual hierarchical merging model (Sylos Labini and Capuzzo-Dolcetta 2020).

Another problem is that the predicted angular momentum in galaxies is much less than observed. Binney *et al* (2001) claim that the problem of an excess of predicted dark matter within the optical bodies and the fact that the observed discs are much larger than expected can be solved if a considerable mass of low angular momentum baryons is ejected (massive galactic outflows) and discs are formed later from the high angular momentum baryons falling into the Galaxy. The conspiracy problem is also solved if the ejection begins only once $M_{\rm baryons}(r) \sim M_{\rm dark\ matter}(r)$. Another solution within the standard cosmological model for the angular momentum problem is the tidal interaction of objects populating the primordial voids together with the Coriolis force due to void rotation (Casuso and Beckman 2015).

Another fact that could cast doubt upon the existence of very massive haloes of dark matter is that strong bars rotating in dense haloes should generally slow down as they lose angular momentum to the halo through dynamical friction (Debattista and Sellwood 2000), whereas the observed pattern speed of galactic bars indicates that almost all of them rotate quite rapidly (Aguerri *et al* 2015, Roshan *et al* 2021). There should be a net transference of angular momentum from bars to haloes, although friction can be avoided under some special conditions (Sellwood and Debattista 2006), the bars can grow longer than the outward drift of the corotation radius, thus making the situation compatible with dark matter haloes slowing down the rotation of bars on Gyr timescales (Font *et al* 2017). Expected dynamical friction produced by dark matter haloes might also be a problem in groups of galaxies near M81 or Hickson compact groups, because such friction should very rarely give place to the observed compact configurations of major galaxies (Oehm and Kroupa 2018). Also, dynamical friction exerted over long periods of time, as happens with globular clusters, should make them lose their potential energy in dwarf spheroidal galaxies (supposed to have very large ratios of dark matter), and clusters should be concentrated near the centre of the Galaxy; however, this was not observed in Fornax galaxy (Bar *et al* 2021).

The dynamical surface mass density at the solar position with a distance to the Galactic plane between $Z = 1.5$ and 4 kpc was estimated using the kinematics of thick disk stars (Moni Bidin *et al* 2012). The resulting trend matches the expectations of visible mass alone, and no dark component is required to account for the observations, all the current models of a spherical dark matter halo are excluded at a confidence level higher than 4σ.

The enclosed dynamical mass-to-light ratio increases with decreasing galaxy luminosity and surface brightness, a scenario that is not predicted by dark matter scenarios (McGaugh 2014).

Galaxies dominate the halo with little substructure, whereas the standard model predicts that galaxies should be scaled versions of galaxy clusters with abundant substructure (D'Onguia and Lake 2004, Kroupa *et al* 2010). Moreover, ΛCDM simulations predict that the majority of the most massive subhaloes of the Milky Way are too dense to host any of its bright satellites ($L_V > 10^5\ L_\odot$) (Boylan-Kolchin *et al* 2011) ('too-big-to fail'). Also, the satellites are distributed in a plane, which is incompatible with ΛCDM (Kroupa *et al* 2010, 2005, Pawlowski and Kroupa 2013). These are arguments against the standard model to which one cannot make the typical rebuff of incompleteness of knowledge of baryonic physics (Kroupa 2012, Pawlowski *et al* 2015). Furthermore, there is a correlation between bulge mass and the number of satellites (Kroupa *et al* 2010, López-Corredoira and Kroupa 2016, Javanmardi and Kroupa 2020), which is not predicted by the standard model (Javanmardi *et al* 2019), but is predicted by models of modified gravity without dark matter. The disc of satellites and the bulge–satellite correlation suggest that dissipational events forming bulges are related to the processes forming phase-space correlated satellite populations. These events are well known to occur, since in galaxy encounters energy and angular momentum are expelled in the form of tidal tails, which can fragment to form populations of tidal dwarf galaxies and associated star clusters. If Local Group satellite galaxies are to be interpreted as tidal dwarf galaxies, then the substructure predictions of the standard cosmological model are internally in conflict (Kroupa *et al* 2010).

Perhaps, the most severe caveat against retaining the hypothesis of dark matter is that, after the great effort expended in looking for it, it has not yet been found, although non-discovery does not mean that it does not exist. Microlensing surveys (Lasserre *et al* 2000, Tisserand *et al* 2007) constrain the mass of the halo in our Galaxy in the form of dim stars and brown dwarfs to be much less than that necessary for dark matter haloes. In any case, as already mentioned, the primordial nucleosynthesis model constrains baryonic matter to be around 10% of the total mass (Bell *et al* 2003), so these objects could not be compatible with the preferred cosmological model. Some observations are inconsistent with the dominant dark matter component being dissipationless (Moore 1994). Neither massive black hole haloes (Moore 1993) nor intermediate-mass primordial black holes (Mediavilla *et al* 2017, Zumalacárregui and Seljak 2018) provide a consistent scenario. The nature of dark matter has been investigated and there are no suitable astrophysical candidates.

3.5 Dark matter particles

The other possibility is that dark matter is not concentrated in any kind of astrophysical object but in a gas of exotic non-baryonic particles. There are three possible types of candidates (Bertone and Hooper 2018): (1) particles predicted by the supersymmetry hypothesis, which are electrically neutral and not strongly interacting, including superpartners of neutrinos, photons, Z bosons, Higgs bosons, gravitons, and others (neutralinos have been the most recently studied candidates in recent decades); (2) axions, typically with masses between 10^{-6} and 10^{-4} eV, predicted to resolve certain problems in quantum chromodynamics; and (3) Weakly Interacting Massive Particles (WIMPs), which are those particles that interact through the weak force.

Many attempts to search for exotic particles have also ended without success. Technologies used to directly detect a dark matter particle have failed to obtain any positive result (Liu *et al* 2017). Attempts have also been made to detect neutralinos with the MAGIC and HESS Cherenkov telescope systems for very high energy gamma rays through their Cherenkov radiation, but so far without success, and only emission associated with the Galaxy has been found (Aharonian *et al* 2006). Dwarf galaxies are expected to have high ratios of dark matter and low gamma ray emission arising from other astrophysical processes so the search is focussed on these galaxies, but without positive results. The detection of an excess of γ rays in the centre of the Milky Way convinced some physicists that they were seeing evidence of the annihilation of dark matter particles, but this possibility has been ruled out since there is no significant detection of dark matter annihilation (Abazajian *et al* 2020). As usual, the scientists involved in these projects attribute their failure of detection to the inability of the detectors to reach the necessary low cross section of the interaction, or to the possibility that they may be 3–4 orders of magnitude below the possible flux of gamma rays emitted by dark matter (Sánchez-Conde 2009), and they ask for more funding to continue to feed their illusions: a never-ending story, another race between Achilles and the tortoise. As pointed out by Merritt (2017), this will never constitute a falsification of the ΛCDM model because although success of detection would confirm the standard paradigm, non-detection is not applied to discard it.

3.6 Scenarios without non-baryonic cold dark matter

Note also that other dynamical problems in which dark matter has been claimed as necessary can indeed be solved without dark matter: galactic stability (Toomre 1981) or warp creation (López-Corredoira *et al* 2002), for instance. Velocities in galaxy pairs and satellites might also measure the mass of the intergalactic medium filling the space between the members of the pairs rather than the mass of dark haloes supposedly associated with the galaxies (López-Corredoira *et al* 1999, 2002).

Rotation curves in spiral galaxies can be explained with magnetic fields, non-circular orbits in the outer disc (Benhaiem *et al* 2017), alternative gravity theories (see section 2.1), or types of dark matter different from non-baryonic cold dark matter proposed in the standard model (Battaner and Florido 2000).

The most popular alternative to dark matter is the modification of gravity laws proposed in MOND (Sanders and McGaugh 2002, Famaey and McGaugh 2012, Sanders 2015, Merritt 2020), which modifies the Newtonian law for accelerations lower than a constant a_0. This acceleration scale a_0 defines the variation with respect to Newton's law necessary to fit the rotation curves. Its value is very similar in all galaxies (see figure 3.1) and it has been interpreted as a possible sign of confirmation

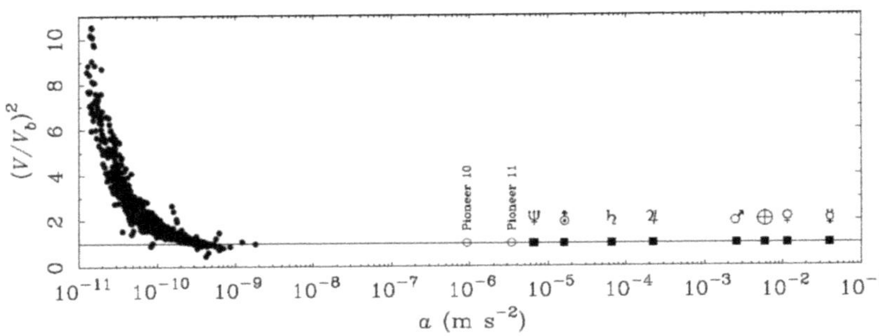

Figure 3.1. MOND theory predictions and data. Newtonian force (factor 1 represented by the solid line) is multiplied by an extra factor that depends of the total acceleration a and a constant a_0, which is the same for all galaxies, represented by points in this plot. It is extended to solar-system scales (each planet is labelled). This illustrates the large gulf in scale between galaxies and the Solar System, where high precision tests are possible. The need for dark matter or modified gravity only appears at very low accelerations. (Reproduced from figure 11 of Famaey and McGaugh (2012), distributed under the terms of the Creative Commons CC BY license.)

of MOND (Lelli *et al* 2017). Rodrigues *et al* (2018) find some small variations of a_0 in different galaxies, thus invalidating the universal application of a MOND-like modification of gravity with constant a_0, but this result was found to rely on galaxies with very uncertain distances and/or nearly edge-on orientations (Kroupa *et al* 2018).

Tentative detections of a departure from Newtonian gravity as predicted by MOND was claimed, for example, for wide binaries (binary stars with distances of several thousands of AU), similar to the expectations of modified gravity theories (Hernández *et al* 2012), although this was done by comparing only the dispersion of radial heliocentric velocities, and the gravitational effects of the surrounding areas are not clear. Moreover, the distribution of separations of wide binaries and their evolution (Poveda and Allen 2004, Jiang and Tremaine 2010) should also have an effect that should be considered carefully in this test. Nevertheless, the successes of MOND and its relativistic version are mostly limited to galactic scales and cannot compete with ΛCDM to explain the large-scale structure and other cosmological predictions.

There are also proposals stating that the dark matter necessary to solve many problems may be baryonic by simply placing baryonic dark matter in the outer disc (Feng and Gallo 2015, Sipols and Pavlovich 2021). Other, more exotic, proposals include positively charged, baryonic particles (protons and helium nuclei) (Drexler 2005), which are massive and weakly interacting, but only when moving at relativistic velocities. Simple composite systems include nucleons but are still bound together by comparable electric and magnetic forces (Mayer and Reitz 2012), making up a three-body system ('tresinos') or four-body system ('quatrinos'), antiparticles, which have negative gravitational charge (Hajdukovic 2014). Then, there is the proposal, among others, of a pressure-free fluid in general relativity (Cooperstock and Tieu 2005). Nonetheless, these proposals have their own set of problems when trying to explain large-scale structure.

In my opinion, the problem of 'dark matter' is not a single problem but a composite of many different problems within astrophysics that might have different solutions. The idea that the same non-baryonic dark matter necessary to explain the low anisotropies in the CMBR is going to solve the large-scale structure distribution, the lack of visible matter in clusters, the dispersion of velocities of their galaxies, the measurements of gravitational lensing, the rotation curves, etc, is a happy fantasy that has dominated astrophysics for the last 40 years. It would be wonderful if we also have a happy ending with the discovery of the particles of dark matter that constitute the dark haloes of galaxies, but, in absence of that outcome, maybe it would be prudent to bet on a combination of different elements to explain the entire set of unexplained phenomena: possibly some baryonic dark matter in some cases, possibly a modification of gravity is part of the explanation for a wide set of events, and maybe with cold dark matter dominating some phenomena and hot dark matter other phenomena. Certainly, a unified picture of a unique non-baryonic type of cold dark matter to explain everything would be a simpler and more elegant hypothesis; the question, however, is not one of simplicity but one of ascertaining how reality is, whether simple or complex.

3.7 Dark energy and the cosmological constant or quintessence

The question of the cosmological constant required to maintain a static Universe was considered by Einstein to be his biggest blunder (Padmanabhan 2003, O'Raifeartaigh *et al* 2018). Indeed, it is equivalent to positing an attractive gravitational acceleration $a(r) = -\frac{GM}{r^2} + Br$, already proposed by Newton for $B < 0$, but with $B > 0$ instead (Komatsu 2011). This is not normal physics but an exotic suggestion, since the usual thermodynamics for fluids with positive heat capacity and positive compressibility is not applicable to dark energy with negative pressure (Barboza *et al* 2015).

After the proposal by Einstein and its later fall from grace in the wake of the discovery of the expanding Universe, the use of the cosmological constant remained marginalised[3], to address specific cosmic puzzles such as the timespan of expansion, the formation of galaxies, and the redshifts of the quasars, until the re-emergence of the term in today's ΛCDM cosmology (O'Raifeartaigh *et al* 2018). Before 1990, most cosmologists did not favour scenarios dominated by the cosmological constant (Fukugita and Lahav 1991). In the eighties, the cosmological constant was either often disregarded as an unnecessary encumbrance or its value was set at zero (Longair 1987), and all observations gave a null or almost null value. However, since other problems in cosmology have risen, many cosmologists at the beginning of the '90s realised that an Ω_Λ ranging from 0.70 to 0.80 could solve many problems in CDM cosmology (Efstathiou *et al* 1990). Years later, evidence for such a value of the cosmological constant began to arrive. A brilliant prediction or a prejudice that conditioned actual measurements?

In the mid-1990s the position of the first peak in the power spectrum of the CMBR was determined to be at $\ell \approx 200$. White *et al* (1996) realised that the preferred standard model at that time (an open Universe with $\Omega = \Omega_m \approx 0.2$ and without dark energy) did not fit the observations—they needed a larger Ω. Between 1997 and 2000 there was a change of mentality in standard cosmology. This was one of the elements, together with Type Ia Supernovae (SNIa) observations and the age problem of the Universe, that would encourage cosmologists to include a new ad hoc element: dark energy. In 1998, it was decided at a scientific congress that the Universe was in an accelerated expansion (Cepa 2007 [p 54]).

All present claims about the existence of dark energy have also measured Ω_Λ through its dependence on the luminosity distance versus redshift dependence (Durrer 2011). Measurements of supernovae in distant galaxies give fainter-than-expected luminosities that can be explained with the introduction of the cosmological constant (see figure 3.2), assuming these supernovae are standard candles with always the same luminosity, but there are other explanations too. It was suggested to be due to intergalactic dust (Aguirre and Haiman 2000, Goobar *et al* 2002, Milne *et al* 2015). The faint luminosities could be due to the presence of grey dust, an idea not necessarily inconsistent with the measure of a supernova at $z = 1.7$

[3] Nonetheless, it was also introduced by Lemaître (1934) in his equations for the evolution of the expanding Universe.

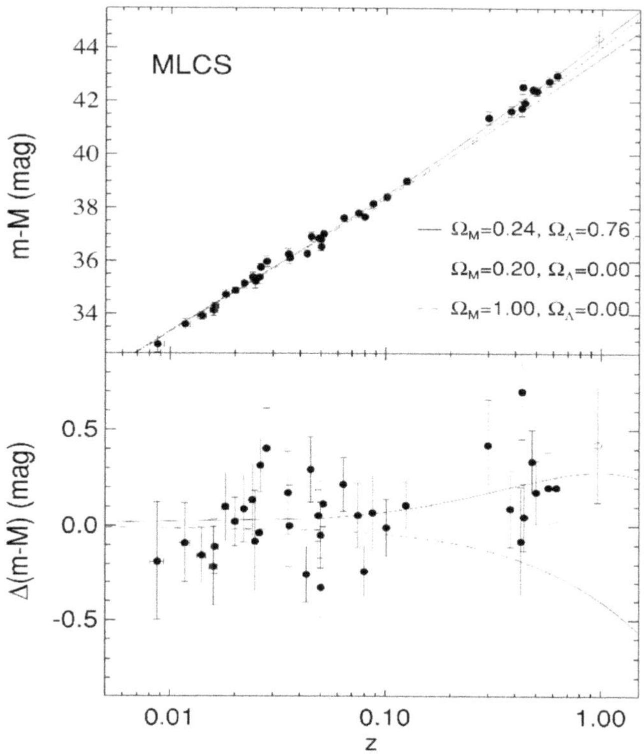

Figure 3.2. Hubble–Lemaître diagram with Type Ia Supernovae. The upper panel shows the Hubble–Lemaître diagram for the low-redshift and high-redshift SNe Ia samples with distances measured. Overplotted are three cosmologies. The bottom panel shows the difference between data and models with the model of $\Omega_m = 0.2$, $\Omega_\Lambda = 0$. (Reproduced by permission of the AAS from figure 4 of Riess *et al* 1998.)

(SN 1997ff) (Goobar *et al* 2002). Dimming by dust along the line of sight, predominantly in the host galaxy of the SN explosion, is one of the main sources of systematic uncertainties (Knop *et al* 2003). The effects of host galaxy extinction was also underestimated: a factor that may contribute to the apparent faintness of high-z supernovae is the evolution of host galaxy extinction with z (Rowan-Robinson 2002); with a consistent treatment of host galaxy extinction and the elimination of supernovae not observed before maximum, the evidence for a positive Λ is not therefore very significant. Fitting the corrected luminosity distances (corrected for internal extinctions) with cosmological models, Balázs *et al* (2006) concluded that the SNe Ia data alone did not exclude the possibility of the $\Lambda = 0$ solution.

SNe Ia also have a possible metallicity dependence and this would imply that the evidence for a non-zero cosmological constant from the SNe Ia Hubble–Lemaître diagram may be subject to corrections for metallicity that are as big as the cosmological effects (Shanks *et al* 2001). This metallicity effect could be substantially larger than had been estimated previously and could quantitatively evaluate the importance of metallicity evolution for determining cosmological parameters (Podsiadlowski *et al* 2006). There is also a significant correlation between SNe Ia luminosity and stellar population age, which causes a serious systematic bias with look-back time able to mimic the dark energy effect (Kang *et al* 2020, Lee *et al* 2020). Therefore, the old supernovae might be intrinsically fainter than the local ones, and the cosmological constant would not be needed (Domínguez *et al* 2000). As a matter of fact, certain cases, such as SNLS-03D3bb, have an exceptionally high luminosity (Howell *et al* 2006). Claims have been made about the possible existence of two classes of normal–bright SNe Ia (Quimby *et al* 2007).

Another caveat in the understanding of dark energy as the reason for supernova dimming is that this putative acceleration should be isotropic, and more recent observations have shown that this is not the case (Colin *et al* 2019).

Furthermore, our limited knowledge of the SNe Ia properties in the *U*-band has been identified as another main source of uncertainty in the determination of cosmological parameters (Knop *et al* 2003). And the standard technique with SNe Ia consists in using spectroscopic templates, built by averaging spectra of well observed (mostly nearby) SNe Ia. Thus, the uncertainty in K-corrections depends primarily on the spectroscopic diversity of SNe Ia.

Even if we accepted that the present-day SNe Ia analyses are correct and without any bias or selection effects, and that these SNe Ia are perfect standard candles, other cosmologies may explain their apparent dimming without introducing a cosmological constant into the standard Einstein field equation, thus negating the necessity for the existence of dark energy (Shu 2015). For instance, an inhomogeneous isotropic Universe described by a Lemaître–Tolman–Bondi solution of Einstein's field equations can provide a positive acceleration of the expansion without dark energy (Romano 2007). Also, in the timescape cosmology (Wiltshire 2009) in an inhomogeneous Universe, time runs slower in overdense regions and more quickly in voids, thus producing in supernovae the same effect as dark energy. Quasi-Steady-State theory predicts a decelerating Universe in the present era; it successfully explains SNe Ia observations (Vishwakarma and Narlikar 2007). Carmeli's cosmology fits data for an accelerating and decelerating universes without dark matter or dark energy (Oliveira and Hartnett 2006). A static Universe can also fit the supernova data without dark energy (Sorrell 2009, Lerner 2009, López-Corredoira 2010, Farley 2010, Marosi 2013).

There are other sources of Ω_Λ measurement such as anisotropies of the CMBR, but they are not free from inaccuracies owing to contamination and anomalies found in it (López-Corredoira 2007, Schwarz *et al* 2016). In the last two decades, many proofs have been presented to the community to convince us that the definitive cosmology has $\Omega_\Lambda \approx 0.7$, which is surprising if we take into account that elsewhere in the history of observational cosmology proofs have been presented for $\Omega_\Lambda \approx 0$. Furthermore, recent tests indicate that other values are available in the literature. For instance, from the angular size versus redshift test for ultracompact radio sources, it is found that Ω_Λ is negative (Jackson and Dodgson 1997). Using the brightest galaxies in clusters, the fit in the Hubble–Lemaître diagram is compatible with a non-accelerated Universe instead of $\Omega_\Lambda = 0.7$ (Vauclair *et al* 2003, Andrews 2006).

The concordance ΛCDM model produces far more high-redshift massive clusters than observed in all existing x-ray surveys (Blanchard 2006).

The current values of Ω_Λ also have a consistency problem in the standard scenario of inflationary ΛCDM. The cosmological constant predicted by quantum field theory has a value much larger than those derived from observational cosmology. This is because the vacuum energy in quantum field theory takes the form of the cosmological constant in Einstein's equations. If inflation took place at the GUT epoch, the present value would be too low by a factor $\sim 10^{-108}$, and if the inflation took place at the quantum gravity epoch, the above factor would be lower still at $\sim 10^{-120}$ (Weinberg 1989). This is derived when the vacuum energy in quantum field theory takes the form of the cosmological constant in Einstein's equations. Furthermore, it is noteworthy that constraints on the variation of the proton to electron mass ratio with redshift indicate that almost all dark energy models using the commonly expected values or parameters are excluded (Thompson 2012). And the standard model has some surprising coincidences. There is the coincidence that now the deceleration of the Hubble–Lemaître flow is compensated by the acceleration of the dark energy; the average acceleration throughout the history of the Universe is almost nil (Melia and Shevchuk 2012) and the size the Universe is as if there were constant expansion. Again, everything is far from being properly understood.

3.8 Grey… neither dark nor luminous

Reality is usually neither black nor white, but something in between, some grey shade with more darkness or more brightness depending on the case. This can applied to hypotheses concerning the existence of dark elements in cosmology, including inflation or CP asymmetry violation as dark in the sense that no direct proofs of its existence can be observed, and it remains a theoretical speculation. To be fair, we must admit the existence of some dark matter at least, some massive objects that do not emit enough light to be detected, but whose gravitational effects are observed in other nearby objects. However, as stated at the beginning and illustrated throughout this chapter, we have not yet achieved the success of the standard cosmological scenario and are left with the four dark knights because of numerous inconsistencies, the non-detection of direct proofs, and the existence of many alternative hypotheses that might substitute the standard picture to explain the cosmological observations. But cosmologists are very attached to these dark elements because they realise that the standard model is a house of cards in which a failure of the weak elements that support the structure would produce a collapse of the theory. This does not mean that the scenario is a fiasco, but it remains in the shadow, within the category of doubtful unconfirmed hypotheses hanging by a few threads.

The future of cosmological research is also grey. All the greatest efforts are being made within the dark sector, and the community cannot prevent the inertia of this never-ending story from reaching an unhappy conclusion. Further sociological considerations will be made in the last chapters of this book.

References

Abazajian K N, Horiuchi S, Kaplinghat M, Keeley R E and Macias O 2020 Strong constraints on thermal relic dark matter from Fermi-LAT observations of the galactic center *Phys. Rev.* D **102** 043012

Aguerri J A L, Méndez-Abreu J and Falcón-Barroso J *et al* 2015 Bar pattern speeds in CALIFA galaxies. I. Fast bars across the Hubble sequence. Hierarchic models for laminated composites *Astron. Astrophys.* **576** 17

Aguirre A N and Haiman Z 2000 Cosmological constant or intergalactic dust? Constraints from the cosmic far-infrared background *Astrophys. J.* **532** 28–36

Aharonian F, Akhperjanian A G and Bazer-Bachi A R *et al* 2006 The HE.S.S. survey of the inner galaxy in very high energy gamma rays *Astrophys. J.* **636** 777–97

Andrews T B 2006 Falsification of the expanding universe model *1st Crisis in Cosmology Conf. (AIP Conf. Ser. 822)* ed E J Lerner and J B Almeida (Melville, NY: AIP) pp 3–22

Babcock H W 1939 The rotation of the Andromeda nebula *Lick Obs. Bull.* **19** 41–51

Balázs L G, Hetesi Z, Regály Z, Csizmadia S, Bagoly Z, Horváth I and Mészáros A 2006 A possible interrelation between the estimated luminosity distances and internal extinctions of type Ia supernovae *Astron. Nachr.* **327** 917–24

Bar N, Blas D, Blum K and Kim H 2021 Assessing the Fornax globular cluster timing problem in different models of dark matter *Phys. Rev.* D **104** 043021

Barboza E M Jr, Nunes R d C, Abrey E M C and Neto J A 2015 Thermodynamic aspects of dark energy fluids *Phys. Rev.* D **92** 083526

Battaner E and Florido E 2000 The rotation curve of spiral galaxies and its cosmological implications *Fundam. Cosm. Phys.* **21** 1–154

Bell F B, McIntosh D H, Katz N and Weinberg M D 2003 A first estimate of the baryonic mass function of galaxies *Astrophys. J. Lett.* **585** L117–20

Benhaiem D, Joyce M and Sylos Labini F 2017 Transient spiral arms from far out-of-equilibrium gravitational evolution *Astrophys. J.* **851** 10

Bernlochner F U, Englert C, Hays C, Lohwasser K, Mildner H, Pilkington A, Price D D and Spannowsky M 2019 Angles on CP-violation in Higgs boson interactions *Phys. Lett.* B **790** 372–9

Bertone G and Hooper D 2018 A history of dark matter *Rev. Mod. Phys.* **90** 045002

Bessel F W 1844 On the variations of the proper motions of Procyon and Sirius *Mon. Not. R. Astron. Soc.* **6** 136–41

Binney J, Gerhard O and Silk J 2001 The dark matter problem in disc galaxies *Mon. Not. R. Astron. Soc.* **321** 471–4

Blanchard A 2006 Evidence for an accelerating universe or lack of? *Current Issues in Cosmology* ed J-C Pecker and J V Narlikar (Cambridge: Cambridge University Press) pp 76–84

Bosma A 1978 The distribution of kinematics of neutral hydrogen in spiral galaxies of various morphological types *PhD thesis* (University of Groningen)

Bowman J D, Rogers A E E, Monsalve R A, Mozdzen T J and Mahesh N 2018 An absorption profile centred at 78 megahertz in the sky-averaged spectrum *Nature* **555** 67–70

Boylan-Kolchin M, Bullock J S and Kaplinghat M 2011 Too big to fail? the puzzling darkness of massive milky way subhaloes *Mon. Not. R Astron. Soc.* **415** L40–4

Bruggisser S, Konstandin T and Servant G 2017 CP-violation for electroweak baryogenesis from dynamical CKM matrix *J. Cosmol. Astropart. Phys.* **2017** 34

Bullock J S and Boylan-Kolchin M 2017 Small-scale challenges to the λCDM paradigm *Annu. Rev. Astron. Astrophys.* **55** 343–87

Casuso E and Beckman J E 2015 On the origin of the angular momentum of galaxies: cosmological tidal torques and Coriolis force *Mon. Not. R. Astron. Soc* **449** 2910–8

Cepa J 2007 *Cosmología física* (Madrid: Akal)

Chowdhury D, Martin J, Ringeval C and Vennin V 2019 Assessing the scientific status of inflation after Planck *Phys. Rev.* D **100** 083537

Colin J, Mohayaee R, Rameez M and Sarkar S 2019 Evidence for anisotropy of cosmic acceleration *Astron. Astrophys.* **631** L13–8

Cooperstock F I and Tieu S 2005 General relativity resolves galactic rotation without exotic dark matter (arXiv:astro-ph/0507619)

Debattista V P and Sellwood J A 2000 Constraints from dynamical friction on the dark matter content of barred galaxies *Astrophys. J.* **543** 704–21

Di Cintio A, Brook C B, Macciò A V, Stinson G S, Knebe A, Dutton A A and Wadsley J 2014 The dependence of dark matter profiles on the stellar-to-halo mass ratio: a prediction for cusps versus cores *Mon. Not. R. Astron. Soc.* **437** 415–23

Domínguez I, Höflich P, Straniero O and Wheeler C 2000 Evolution of type Ia supernovae on cosmological time scales *Mem. Soc. Astron. Ital.* **71** 449–60

D'Onguia E and Lake G 2004 Cold dark matter's small-scale crisis grows up *Astrophys. J.* **612** 628–32

Drexler J 2005 Identifying dark matter through the constraints imposed by fourteen astronomically based 'cosmic constituents' (arXiv:astro-ph/0504512)

Durrer R 2011 What do we really know about dark energy *J. Cosmol* **15** 6065–78

Efstathiou G, Sutherland W J and Maddox S J 1990 The cosmological constant and cold dark matter *Nature* **348** 705–7

Evans N W 2001 No need for dark matter in galaxies? *Proc. of the 3rd Int. Workshop on the Identification of Dark Matter* ed N J C Spooner and V Kudryavtsev (Singapore: World Scientific) pp 85–92

Famaey B and McGaugh S 2012 Modified Newtonian dynamics (MOND): observational phenomenology and relativistic extensions *Living Rev. Relativ.* **15** 10

Farley F J M 2010 Does gravity operate between galaxies? observational evidence re-examined *Proc. R. Soc.* A **466** 3089-96

Feng J Q and Gallo C F 2015 Deficient reasoning for dark matter in galaxies *Phys. Int.* **6** 11–22

Font J, Beckman J E and Martínez-Valpuesta I *et al* 2017 Kinematic clues to bar evolution for galaxies in the local universe: why the fastest rotating bars are rotating most slowly *Astrophys. J.* **835** 279

Fukugita M and Lahav O 1991 Ly-alpha clouds at low redshift and the cosmological constant *Mon. Not. R. Astron. Soc.* **253** 17–20

Georgi H, Quinn H R and Weinberg S 1974 Hierarchy of interactions in unified gauge theories *Phys. Rev. Lett.* **33** 451–4

Gnedin O Y, Kravtsov A V, Klypin A A and Nagai D 2004 Response of dark matter halos to condensation of baryons: cosmological simulations and improved adiabatic contraction model *Astrophys. J.* **616** 16–26

Goobar A, Bergström L and Mörtsell E 2002 Measuring the properties of extragalactic dust and implications for the Hubble diagram *Astron. Astrophys.* **384** 1–10

Guth A H, Kaiser K and Linde A D *et al* 2017 A cosmic controversy *Sci. Am.* **317** 5–7 (https://blogs.scientificamerican.com/observations/a-cosmic-controversy/)

Hajdukovic D S 2014 Virtual gravitational dipoles. The key for the understanding of the universe? *Phys. Dark Universe* **3** 34–40

Hernández X, Jiménez M A and Allen C 2012 Wide binaries as a critical test of classical gravity *Eur. Phys. J.* C **72** 1884

Holmberg E 1954 On the masses of double galaxies *Meddelanden fran Lunds Astronomiska Observatorium Series* I **186** 1–20

Howell D A, Sullivan M and Nugent P E *et al* 2006 The type Ia supernova snls-03d3bb from a super-Chandrasekhar-mass white dwarf star *Nature* **443** 308–11

Iljas A, Steinhardt P J and Loeb A 2017 Cosmic inflation theory faces challenges *Sci. Am.* **316** 32–9

Jackson J C and Dodgson M 1997 Deceleration without dark matter *Mon. Not. R. Astron. Soc* **285** 806–10

Javanmardi B and Kroupa P 2020 A correlation between the number of satellites and the bulge-to-total baryonic mass ratio extending beyond the local group *Mon. Not. R. Astron. Soc.* **493** L44–8

Javanmardi B, Raouf M, Khosroshahi H G, Tavasoli S, Müller O and Molaeinezhad A 2019 The number of dwarf satellites of disk galaxies versus their bulge mass in the standard model of cosmology *Astrophys. J.* **870** 14

Jiang Y-F and Tremaine S 2010 The evolution of wide binary stars *Mon. Not. R. Astron. Soc.* **401** 977–94

Kahn F D and Woltjer L 1959 Intergalactic matter and the galaxy *Astrophys. J.* **130** 705–17

Kang Y, Lee Y-W, Kim Y-L, Chung C and Ree C H 2020 Early-type host galaxies of type Ia supernovae. II. Evidence for luminosity evolution in supernova cosmology *Astrophys. J.* **889** 8

Knop R A, Aldering G and Amanullah R *et al* 2003 New constraints on ω_m, ω_λ, and w from an independent set of 11 high-redshift supernovae observed with the Hubble space telescope *Astrophys. J.* **598** 102–37

Komatsu E 2011 What every dynamicist should know about … cosmology *Bull. Am. Astron. Soc.* **43** 4.01

Kroupa P 2012 The dark matter crisis: falsification of the current standard model of cosmology *Publ. Astron. Soc. Aust.* **29** 395–433

Kroupa P, Banik I and Haghi H *et al* 2018 A common Milgromian acceleration scale in nature *Nat. Astron.* **2** 925–6

Kroupa P, Famaey B and de Boer K S *et al* 2010 Local-group tests of dark-matter concordance cosmology. Towards a new paradigm for structure formation *Astron. Astrophys.* **532** A32

Kroupa P, Theis C and Boily C M 2005 The great disk of Milky-Way satellites and cosmological sub-structures *Astron. Astrophys.* **431** 517–21

Křížek M, Křížek F and Somer L 2014 Which effects of galaxy clusters can reduce the amount of dark matter *Bulg. Astron. J.* **21** 43–65

Lasserre T, Afonso C and Albert J N *et al* 2000 Not enough stellar mass machos in the galactic halo *Astron. Astrophys.* **355** L39–42

Lee Y-W, Chung C, Kang Y and Lee M J 2020 Further evidence for significant luminosity evolution in supernova cosmology *Astrophys. J.* **22** 5

Lelli F, McGaugh S S, Schombert J M and Pawlowski M S 2017 One law to rule them all: the radial acceleration relation of galaxies *Astrophys. J.* **836** 152

Lemaître G 1934 Evolution of the expanding universe *Proc. Natl Acad. Sci. USA* **20** 12–7

Lerner E J 2009 Tolman test from $z = 0.1$ to $z = 5.5$: preliminary results challenge the expanding universe model *Second Crisis in Cosmology Conf. (ASP Conf. Ser. 413)* ed F Potter (San Francisco, CA: Astronomical Society of the Pacific) pp 12–23

Liddle A R 1999 An introduction to cosmological inflation *High Energy Physics and Cosmology* ed A Masiero, G Senjanovic and A Smirnov (Singapore: World Scientific) p 260

Liu J, Chen X and Ji X 2017 Current status of direct dark matter detection experiments *Nat. Phys.* **13** 212–6

Longair M S 1987 Observational cosmology *Observational Cosmology 1986 (IAU Symp. 124)* ed A Hewitt, G Burbidge and L Z Fang (Dordrecht: Reidel) pp 823–40

López-Corredoira M 2007 Some doubts on the validity of the foreground galactic contribution subtraction from microwave anisotropies *J. Astrophys. Astron.* **28** 101–16

López-Corredoira M 2010 Angular-size test on the expansion of the universe *Int. J. Mod. Phys.* **D19** 245–91

López-Corredoira M, Beckman J E and Casuso E 1999 High-velocity clouds as dark matter in the local group *Astron. Astrophys.* **351** 920–4

López-Corredoira M, Betancort-Rijo J and Beckman J E 2002 Generation of galactic disc warps due to intergalactic accretion flows onto the disc *Astron. Astrophys.* **386** 169–86

López-Corredoira M and Kroupa P 2016 The number of tidal dwarf satellite galaxies in dependence of bulge index *Astrophys. J.* **817** 75

López J L, Nanopoulos D V and Pois H 1992 Proton decay and cosmology strongly constrain the minimal SU(5) supergravity model *Proc. of the XXVI Int. Conf. on High Energy Physics, vol II (AIP Conf. Ser. 272)* (Melville, NY: AIP) pp 1395–400

Marosi L A 2013 Hubble diagram test of expanding and static cosmological models: the case for a slowly expanding flat universe *Adv. Astron.* **2013** 917104

Martin J, Ringeval C and Vennin V 2014 Encyclopaedia inflationaris *Phys. Dark Universe* **5** 75–235

Mayer F J and Reitz J R 2012 Electromagnetic composites at the Compton scale *Int. J. Theor. Phys.* **51** 322–30

McGaugh S S 2000 Boomerang data suggest a purely baryonic universe *Astrophys. J. Lett.* **541** L33–6

McGaugh S S 2014 The third law of galactic rotation *Galaxies* **2** 601–22

Mediavilla E, Jiménez-Vicente J, Muñoz J A, Vives-Arias H and Calderón-Infante J 2017 Limits on the mass and abundance of primordial black holes from quasar gravitational microlensing *Astrophys. J. Lett.* **836** L18

Melia F and López-Corredoira M 2018 Evidence of a truncated spectrum in the angular correlation function of the cosmic microwave background *Astron. Astrophys.* **610** 5

Melia F and Shevchuk A S 2012 The $r_h = ct$ universe *Mon. Not. R. Astron. Soc.* **419** 2579–86

Merritt D 2017 Cosmology and convention *Stud. Hist. Phil. Mod. Phys.* **57** 41–52

Merritt D 2020 Philosophical Approach to MOND *Assessing the Milgromian Research Program in Cosmology* (Cambridge: Cambridge University Press)

Milne P A, Foley R J, Brown P J and Narayan G 2015 The changing fractions of type Ia supernova NUV-optical subclasses with redshift *Astrophys. J.* **803** 20

Moni Bidin C, Carraro G, Méndez R A and Smith R 2012 Kinematical and chemical vertical structure of the galactic thick disk. II. A lack of dark matter in the solar neighborhood *Astron. Astrophys.* **751** 30

Moore B 1993 An upper limit to the mass of black holes in the halo of the galaxy *Astrophys. J. Lett.* **413** L93–6

Moore B 1994 Evidence against dissipation-less dark matter from observations of galaxy haloes *Nature* **370** 629–31

Oehm W and Kroupa P 2018 Constraints on the existence of dark matter haloes by the m81 group and the Hickson compact groups of galaxies *Cosmology on Small Scales 2018. Dark Matter Problem and Selected Controversies in Cosmology* ed M Krizek and Y Dumin (Prague: Institute of Mathematics of Czech Academy of Sciences) pp 30–40

Oliveira F J and Hartnett J G 2006 Carmeli's cosmology fits data for an accelerating and decelerating universe without dark matter or dark energy *Fund. Phys. Lett.* **19** 519–35

O'Raifeartaigh C, O'Keeffe M, Nahm W and Mitton S 2018 One hundred years of the cosmological constant: from "superfluous stunt" to dark energy *Eur Phys. J.* H **43** 73–117

Ostriker J P and Peebles P J E 1973 Numerical study of the stability of flattened galaxies: or, can cold galaxies survive? *Astrophys. J.* **186** 467–80

Padmanabhan T 2003 Cosmological constant–the weight of the vacuum *Phys. Rep.* **380** 235–320

Page T 1952 Radial velocities and masses of double galaxies *Astrophys. J.* **116** 63–84

Page T 1960 Average masses and mass-luminosity ratio of the double galaxies *Astrophys. J.* **132** 910–2

Pawlowski M S, Famaey B, Merritt D and Kroupa P 2015 On the persistence of two small-scale problems in ΛCDM *Astrophys. J.* **815** 19

Pawlowski M S and Kroupa P 2013 The rotationally stabilized VPOS and predicted proper motions of the Milky Way satellite galaxies *Mon. Not. R. Astron. Soc.* **435** 2116–31

Peebles P J E 1989 Inflation and the baryon isocurvature model *Large Scale Structure and Motions in the Universe* ed M Mezzetti, G Giuricin, F Mardirossian and M Ramella (Dordrecht: Springer) pp 119–32

Perivolaropoulos L 2009 Six puzzles for LCDM cosmology *The Problems of Modern Cosmology* ed P M Lavrov (Tomsk: Tomsk State Pedagogical University Press) pp 245–54

Podsiadlowski P, Mazzali P A, Lesaffre P, Wolf C and Forster F 2006 Cosmological implications of the second parameter of type Ia supernovae (arXiv:astro-ph/0608324)

Pontzen A and Governato F 2012 How supernova feedback turns dark matter cusps into cores *Mon. Not. R. Astron. Soc.* **421** 3464–71

Poveda A and Allen C 2004 The distribution of separations of wide binaries of different ages *Rev. Mex. Astron. Astrophys. Conf. Ser.* **21** 49–57

Quimby R, Höflich P and Craig Wheeler J 2007 Sn 2005hj: evidence for two classes of normal-bright SNe Ia and implications for cosmology *Astrophys. J.* **666** 1083–92

Riess A G, Filippenko A V and Challis P *et al* 1998 Observational evidence from supernovae for an accelerating universe and a cosmological constant *Astron. J.* **116** 1009–38

Rodrigues D C, Marra V and Popolo A 2018 Absence of a fundamental acceleration scale in galaxies *Nat. Astron.* **2** 668–72

Romano A E 2007 Lemaitre-Tolman-Bondi universes as alternatives to dark energy: does positive averaged acceleration imply positive cosmic acceleration *Phys. Rev.* D **75** 043509

Roshan M, Ghafourian N, Kashfi T, Banik I, Haslbauer M, Cuomo V, Famaey B and Kroupa P 2021 Fast galaxy bars continue to challenge standard cosmology *Mon. Not. R. Astron. Soc* **508** 926–39

Rowan-Robinson M 2002 Do type Ia supernovae prove $\lambda > 0$ *Mon. Not. R. Astron. Soc.* **332** 352–60

Rubin V, Ford W K Jr and Thonnard N 1980 Rotational properties of 21 SC galaxies with a large range of luminosities and radii from NGC 4605 (R=4 kpc) to UGC 2885 (R=122 kpc) *Astrophys. J.* **238** 471–87

Sachs R K and Wolfe A M 1967 Perturbations of a cosmological model and angular variations of the microwave background *Astrophys. J.* **147** 73–90

Sánchez-Conde M A 2009 Gamma-ray dark matter searches in the Milky Way Oral presentation in: Distribution of Mass in the Milky Way, Leiden, Netherlands, 13–17 July 2009

Sanders R H 2015 A historical perspective on modified Newtonian dynamics *Can. J. Phys.* **93** 126–38

Sanders R H and McGaugh S S 2002 Modified Newtonian dynamics as an alternative to dark matter *Annu. Rev. Astron. Astrophys.* **40** 263–317

Schwarz D J, Copi C J, Huterer D and Starkman G D 2016 Huterer D and Starkman G D 2016 CMB anomalies after Planck *Quantum Grav.* **33** 184001

Sellwood J A 2008 Bar-Halo friction in galaxies. III. Halo density changes *Astrophys.* J **679** 379–96

Sellwood J A and Debattista V P 2006 Bar-Halo friction in galaxies. II. Metastability *Astrophys. J.* **639** 868–78

Shanks T, Allen P D, Hoyle F and Tanvir N R 2001 Cepheid Tully-Fisher and SNIa distances (arXiv:astro-ph/0102450)

Shu W-Y 2015 The geometry of the universe *Math. Stat.* **3** 75–88

Sipols A and Pavlovich A 2021 Dark matter Dogma: a study of 214 galaxies *Galaxies* **8** 36

Sorrell W H 2009 Misconceptions about the Hubble recession law *Astrophys. Space Sci.* **323** 205–11

Steinhardt P J 2011 The inflation debate: is the theory at heart of modern cosmology deeply flawed? *Sci. Am.* **304** 18–25

Sylos Labini F and Capuzzo-Dolcetta R 2020 Properties of self-gravitating quasi-stationary states *Astron. Astrophys.* **643** 12

T2K Collaboration 2020 Constraint on the matter-antimatter symmetry-violating phase in neutrino oscillations *Nature* **580** 339–44

Tanaka M, Abe M and Bronner C 2020 Search for proton decay into three charged leptons in 0.37 megaton-years exposure of the Super-Kamiokande *Phys. Rev.* D **101** 052011

Tasitsiomi A 2003 The state of the cold dark matter models on galactic and subgalactic scales *Int. J. Mod. Phys.* **D12** 1157–96

Taubes G 1997 Theorists nix distant antimatter galaxies *Science* **278** 226

Thompson R I 2012 Constraints on quintessence and new physics from fundamental constants *Mon. Not. R. Astron. Soc.* **422** L67–71

Tisserand P, Le Guillou L and Afonso C *et al* 2007 Limits on the macho content of the galactic halo from the EROS-2 survey of the Magellanic clouds *Astron. Astrophys.* **469** 387–404

Toomre A 1981 What amplifies the spirals *The Structure and Evolution of Normal Galaxies* ed D M Fall and D Lynden-Bell (Cambridge: Cambridge University Press) pp 111–36

Vauclair S C, Blanchard A and Sadat R *et al* 2003 The XMM-Omega project. II. Cosmological implications from the high redshift *L–T* relation of x-ray clusters *Astron. Astrophys.* **412** L37–41

Vishwakarma R G and Narlikar J V 2007 Modeling repulsive gravity with creation *J. Astrophys. Astron.* **28** 17–27

Weinberg S 1989 The cosmological constant problem *Rev. Mod. Phys.* **61** 1–23

White M, Viana P T P, Liddle A R and Scott D 1996 Primeval adiabatic perturbation in an expanding universe *Mon. Not. R. Astron. Soc.* **283** 107–18

White S D M, Frenk C S and Davis M 1983 Clustering in a neutrino-dominated universe *Astrophys. J.* **274** L1–5

White S D M and Rees M J 1978 Core condensation in heavy halos–a two-stage theory for galaxy formation and clustering *Mon. Not. R. Astron. Soc.* **183** 341–58

Wiltshire D L 2009 Average observational quantities in the timescape cosmology *Phys. Rev.* D **80** 123512

Zumalacárregui M and Seljak U 2018 Limits on stellar-mass compact objects as dark matter from gravitational lensing of type Ia supernovae *Phys. Rev. Lett.* **121** 141101

Zwicky F 1933 Die rotverschiebung von extragalaktischen Nebeln *Helv. Phys. Acta.* **6** 110–27

IOP Publishing

Fundamental Ideas in Cosmology
Scientific, philosophical and sociological critical perspectives
Martín López-Corredoira

Chapter 4

Redshift and expansion

The standard interpretation of the redshifts of galaxies is that they are due to the expansion of the Universe plus peculiar motions, but there are other explanations, such as the 'tired light' hypothesis, which assumes that the photon loses energy owing to some unknown photon–matter process or photon–photon interaction when it travels some distance. Different observational tests give different results, although none of them so far provides a strong proof in favour of a static Universe. The discussion on anomalous redshifts is also inconclusive.

That redshifts of galaxies are caused by the expansion of the Universe is accepted by almost all present-day cosmologists as an established fact. However, there is no direct proof of the expansion itself; it is not a self-evident fact. Nobody has directly observed a galaxy increasing its distance with respect to us. The most direct argument in favour of expansion is the redshift of galaxies, but the redshift has possible explanations other than expansion.

In this chapter, we examine analyses that lead us to believe that the Universe may be really expanding and is more than merely an established dogma. Theoretically, an alternative cosmology with a static Universe is not impossible, as we shall show in the following section, but various tests based on observations lend further support to expansion.

4.1 Does redshift mean expansion?

Lemaître (1927) and later Hubble (1929) established the redshift (z)-apparent magnitude relation of the galaxies, which provided an observational clue that the Universe is expanding. Hubble was cautious in suggesting this interpretation because, among other reasons, the observational evidence at that time did not clearly favour the expansion scenario and appeared to favour a static Universe (Assis *et al* 2009). Here are some examples of Hubble's quotations that show this scepticism:

> New data to be expected in the future may modify the significance of the present investigation or, if confirmatory, will lead to a solution having many

doi:10.1088/978-0-7503-3775-5ch4

times the weight. For this reason it is thought premature to discuss in detail the obvious consequences of the present results (Hubble 1929).

... if the interpretation as velocity-shifts is abandoned, we find in the redshifts a hitherto unrecognised principle whose implications are unknown (Hubble 1936)

. ... the photons emitted by a nebula lose energy on their journey to the observer by some unknown effect, which is linear with distance and which leads to a decrease in frequency without appreciable transverse deflection and, in particular, without any decrease in rate of arrival at the observer (Hubble 1947).

However, the generations of cosmologists that followed became pretty sure that the redshift of the galaxies following Hubble–Lemaître's law is a definitive proof of expansion in the absence of a good theory to explain possible phenomenological facts proposed by alternative theories. General relativity provided an explanation for cosmological expansion, while alternative proposals were not supported by any well-known orthodox theory. Expansion was preferred and phenomenological approaches that were not supported by present-day theory would be doomed to be forgotten. This position would be permissible if our physics represented all the phenomena in the Universe, but from a deductive–empiricist point of view we should deduce theories from the observations, not vice versa.

4.1.1 Alternative redshift theories

There are other mechanisms that produce redshift apart from space expansion or the Doppler effect (Narlikar 1989, Baryshev *et al* 1994). A bibliographical catalogue collected by Reboul (1981 [section 1]) with hundreds of references before the 1980s reduces these into 17 classes of phenomenologies with non-trivial redshifts, the main idea being the 'tired light' scenario.

A tired light scenario assumes that the photon loses energy owing to some unknown photon–matter process or photon–photon interaction, when it travels some distance: the distance can be either long, if we consider all the intergalactic space between the object and the Earth, or short, for instance taking into consideration only the coronae enveloping the object. Indeed, it is not so much a theory as a possible phenomenological approach to explaining the loss of the energy of the photons in a putative static Universe that could be explained by different theories.

There are several hypotheses that can produce this tired light effect. The idea of loss of energy of the photon in the intergalactic medium was first suggested by Zwicky (1929) and was defended by him for a long time. As late as the mid-twentieth century, Zwicky (1957) maintained that the hypothesis of tired light was viable. Other authors from this epoch also supported the idea (MacMillan 1932, von Nernst 1937, Finlay-Freundlich 1954). But there are two problems (Narlikar 1989): (1) the ϕ-bath smears out the coherence of the radiation from the source, and so all images of distant objects would look blurred if intergalactic space produced scattering, something that is incompatible with present-day observations (Steinbring 2007); (2) the scattering effect and consequent loss of energy would be frequency dependent, which is again incompatible with what we observe in galaxies (Baryshev and Teerikorpi 2012 [section 4.1.1]). Nonetheless, there are proposed solutions to these problems, as we shall see in the following paragraphs.

Vigier (1988) proposed a mechanism in which the vacuum behaves like a stochastic covariant superfluid aether whose excitations can interfere in a dissipative way with the propagation of particles or light waves passing through it, thus avoiding the two former difficulties of blurring and frequency dependence.

The photon interactions with atomic hydrogen (Gallo 1975, Marmet 1988, Laio *et al* 1997, Moret-Bailly 2006) or some processes within the 'plasma redshift' of interaction of photons with electrons (Varshni 1979, Brynjolfsson 2004b, Mamas 2010, Ashmore 2011, Weidner 2014) might produce redshift, some of them even avoiding deviations of photon trajectories and blurring. Potentially, this effect could explain the high redshifts of apparently nearby quasi-stellar objects (QSOs) (see section 4.6), since light travelling through the outer atmosphere of the QSO could be redshifted before leaving it, although the necessary intergalactic matter density should be higher than present observations indicate.

Dynamic multiple scattering theory is also very interesting as a possible tired light mechanism. Results from statistical optics by Wolf (1986) have shown that a shift in the frequency of spectral lines is produced when light passes through a turbulent (or inhomogeneous) medium, owing to multiple scattering effects. The new frequency of the line is proportional to the old one, the redshift not depending on the incident frequency (Roy *et al* 2000). This is analysed by Hansen (2006) in quasar–galaxy associations, requiring a quasar source region whose dielectric susceptibility is a random function of position. This suggests some type of plasma environment or envelope. Redshift could also be explained in terms of non-linear optics, which assumes that the harmonic oscillator model of light has to be extended somewhat by an extremely small anharmonic contribution (Joos and Lutz 2005).

The redshift might be due to a tired light gravitational interaction of wave packets with curved spacetime (Crawford 2006, 2011a, 2011b). The result would be a redshift that is proportional to distance and to the square root of the density. Curvature pressure results from the non-geodesic motion of charged particles in the gravitational interaction.

There is a standard gravitational redshift derived from general relativity (Bondi 1947, Baryshev 1981, 1994, Baryshev *et al* 1994, Su 2006), with its own caveats. It was shown by Bondi (1947) that, in the case of a spherically symmetrical distribution of dust-like matter, the space expansion redshift can be expressed as a sum of two parts: the Doppler effect, and the gravitation effect of the total mass inside the spherical ball due to an increase of gravitational potential (Φ) in highly collapsed matter (Narlikar 1989), with the light source at the centre and the observer at the surface of the ball: $z_{\rm grav} = \frac{\Delta\Phi}{c^2}$. A variation of this idea assuming a fractal structure leads to a cosmological gravitational redshift that is a linear function of distance (Baryshev 1981, Baryshev *et al* 1994, Baryshev 1994, Baryshev and Teerikorpi 2012 [section 9.35.1]). When we consider the mass distribution around the source, but not around the observer, for simple fractal structure with fractal dimension $D = 2$:

$z_{grav} = \frac{2\pi G \rho_0 R_0 d}{c^2}$, where R_0 is the lower cut-off of the fractal structure, ρ_0 is the critical density and d the physical distance, which gives a Hubble–Lemaître's constant of $H_0 = 68.6\left(\frac{\rho_0}{5.2 \times 10^{-24}\ \mathrm{gr\ cm^{-3}}}\right)\left(\frac{R_0}{10\ \mathrm{Kpc}}\right)$ km/s/Mpc.

And there are other theories based on heterodox ideas concerning gravitation. Broberg (2001), for example, pointed out that a comparison between a quantizing scheme inherent in the Schwarzschild metric and observed redshifts from quasars shows that the contracted distances in the gravitational field are quantized in terms of the gravitational radii of the gravitating objects responsible for the field, whereas uncontracted distances are not quantized. Indeed, physicists have observed quantized states of matter under the influence of gravity: cold neutrons moving in a gravitational field do not move smoothly but jump from one height to another (Nesvizhevsky *et al* 2002). In Broberg's view, this might be as important for modern quantum physics as the Michelson–Morley experiment was for the introduction of special relativity. Quantization therefore appears in systems with significant relativistic contraction (gravitational or Lorentz contraction), such as de Broglie waves or, in the most extreme case, photon energy packets moving at the speed of light. In generalised terms, this would mean that quantum physics has its origin in relativistically contracted fields, whereas the gravitational field can be scaled to particle dimensions, with a constant surface energy density as the invariant parameter in the process. This might, for example, lead to the possibility of a topological transformation from the gravitational force to the strong nuclear force.

Ghosh (1997) proposed a phenomenological model of dynamic gravitational interaction between two objects that depends not only on their masses and the distance between them but also on the relative velocity and acceleration between the two. These velocity- and acceleration-dependent interactions are termed 'inertial induction'. The velocity-dependent force leads to a cosmic drag on all objects moving with respect to the mean rest-frame of a Universe treated as infinite and quasi-static. This cosmic drag results in the cosmological redshift of light coming from distant galaxies without any universal expansion. This force model leads to the exact equivalence of gravitational and inertial mass, and explains many not well understood observed celestial and astronomical phenomena.

There is more to be said about photons and gravity or interaction with gravitons. An alternative explanation for cosmological redshift is through the momentum loss of all moving particles due to the finite speed of gravitational interaction (Fischer 2008). In the Quantum Pushing Gravity scenario, instead of Newtonian gravitational pull, it is suggested that quanta generate quantum-pushing forces between particles, and this mechanism provides an explanation for redshifts as well through the collisions of photons with these micro-quanta (Michelini 2012, 2016). In general relativity, dropping the restriction that a global time parameter exists, and assuming instead that the time scale depends on spatial distance, leads to static solutions that exhibit no singularities, need no unobserved dark energy, and can explain the cosmological redshift without expansion (Fischer 2010). Additional redshift can also be explained using Weyl geometry-gravitation (Bouvier and Maeder 1978, Lunsford

2004, Krasnov and Shtanov 2008, Castro Perelman 2009). Interaction between gravitons and photons (Van Flandern 1993, Ivanov 2004, 2006) is a possible cause of the redshift because it provides an energy loss mechanism that is not subject to the usual problem of making distant galaxy images fuzzy and because it varies with $(1 + z)^{-2}$, which is closer to the observed rate of decrease of surface brightness than any Big Bang model.

Instead of new ideas on gravitation, some researchers explored electromagnetism. Maxwell's equations forbid light-by-light scattering, but this can be explained from purely quantum-mechanical process, related to the polarisation of the vacuum, and this scattering has indeed been observed at the LHC (Large Hadron Collider at CERN) (Atlas Collaboration 2017). More problematic is the shift of frequency.

Roscoe (2006) aimed to shed new light on classical electrodynamics. After abstracting from certain well-known symmetries of classical electrodynamics and showing that these symmetries are alone sufficient to recover the classical theory in an exact and unadorned manner, he found that the classical electromagnetic field is then irreducibly associated with a massive vector field. This latter field can be interpreted only as a classical representation of a massive photon (Roscoe 2006, Spallicci *et al* 2021). Massive photons have also been proposed by Cumalat and Bartlett (2011), according to whom gravity has a potential $\phi_g = -(GM/r)[\cos(K_0 r) + \pm i\epsilon \sin(K_0 r)]$, the graviton has imaginary mass, and the photon has real mass, both equal to $\sim 10^{-25}$ eV $= 1.8 \times 10^{-61}$ kg. This massive photon would produce a negative pressure that retards the expansion (Bartlett and Cumalat 2011). Experimentally, there are constraints on the maximum mass of the photon (Ryutov 2007, Retinó and Spallicci 2016, Bonetti *et al* 2017), but these constraints are larger than $\sim 10^{-53}$ kg, so the idea of a massive photon is not totally devoid of real possibilities, although it would conflict with special relativistic principles.

Other new ideas rest not on new physics or new reinterpretations of old theories, but on effects predicted at present with standard physics. Cuesta *et al* (2007) propose a non-cosmological redshift that is non-linear in terms of the electrodynamic description of photon propagation through weak background intergalactic magnetic fields. Indeed, Maxwell may have been the first to consider the possibility of non-conservative photon behaviour. His original electromagnetic wave equation contained the energy damping term $\sigma_0 \mu_0 (\partial\phi/\partial t)$, where σ_0 and μ_0 represented the electrical conductivity and magnetic permeability of background space (Maxwell 1954 [p 431 of reprinted version of 1954], Monti 1988). von Nernst (1921 [p 40]) put forth the idea in an astronomical context by proposing that Olber's paradox might be resolved if photons were assumed to undergo non-conservative energy damping during their journey through intergalactic space.

Furthermore, just to give further examples from the huge literature on the topic, there are explanations of non-cosmological/non-Doppler redshifts in terms of the local shrinkage of the quantum world (Alfonso-Faus 2006); the variation of the speed of light (Alfonso-Faus 2009, Shtyrkov 1993); variation of Planck's constant (Hunter 2010); linear-time variation (Budko 2009); time acceleration or deceleration (Garaimov 2003, Taganov 2008); proposals for a quantum long-time energy redshift

(Urbanowski 2008) in which the energy is smaller owing to a quantum effect when the photon travels over a very long time; chronometric cosmology (Segal 1976, Segal and Zhou 1995) (see section 2.4); the variable mass hypothesis or secular mass variation of particles (e.g., Hoyle and Narlikar 1964, Barber 2002) (see section 2.2 and SCC model of section 2.5); etc.

All these proposed mechanisms show us that it is quite possible to construct a cosmological scenario with non-expansion redshifts. Nonetheless, all these theories are at present just speculation without direct experimental or observational support.

4.2 Conceptual problems of expansion

The very concept of space expansion has its own problems (Francis *et al* 2007, Baryshev 2008). The physics of space expansion contains such puzzling phenomena as the continuous creation of vacuum, violation of energy conservation, violation of the limiting velocity of receding galaxies (Harrison 1981, Baryshev 2015), phenomena that can be assimilated into the context of the standard cosmological picture, but that are nevertheless strange features from the point of view of ordinary physics.

Owing to the creation of space, the total energy of the Universe is not conserved. In cosmology, the equation $dE + pdV = 0$ (E, p and V are respectively energy, pressure, volume) gives us the possibility of calculating by how much the energy increases or decreases inside a finite comoving volume, but it does not tell us where the energy comes from or where it goes. This is because cosmological pressure does not produce work. It has been noted (Harrison 1995) that in a homogeneous unbounded expanding Friedmann–Lemaître–Robertson–Walker model, one may imagine the whole Universe partitioned into macroscopic cells, each of comoving volume V, and all having contents in identical states. The $-pdV$ energy lost (when different from dust: $p \neq 0$) from any single cell cannot reappear in neighbouring cells because all cells experience identical losses. So the usual idea of an expanding cell performing work on its surroundings cannot apply in this case. The energy is not conserved. The same conclusion was reached by Peebles (1993 [p 139]) when he considered the energy loss inside a comoving ball of the photon gas.

4.3 Expansion on small scales

Whether expansion should also affect smaller scales is also a frequent topic of discussion. Willem de Sitter (1872–1934) was the first to raise the question of whether galaxies expand with the Universe and considered the effect to be negligible (de Sitter 1931). In the orthodox interpretation, expansion does not affect galaxies because there is no local effect on particle dynamics from the global expansion of the Universe: the tendency to separate is a kinematic initial condition, and once this is removed all memory of the expansion is lost (Peacock 2008). Indeed, some experiments may show that local space cannot be expanding (Lieu 2006). As a matter of fact, those galaxies at a distance of >1 Mpc from the barycentre of the Local Group all have positive radial velocities, indicating that the expansion is taking place on these scales; however, at distances lower than 1 Mpc there is a random distribution of positive and negative velocities, indicating that on this scale the group is gravitationally bound (Peñarrubia *et al* 2014).

Note, however, that there are other views on whether there is expansion on the galactic scale, or on atomic scales (Bonnor 1999), due to the expansion of the Universe and what the nature is of the expansion itself (Anderson 1995, Domínguez and Gaite 2001, Francis *et al* 2007, Baryshev 2008); indeed, some authors even think there is some kind of evidence for local expansion within the Solar System (Dumin 2002, Krasinsky and Brumberg 2004, Paturel 2008), although it is very small.

Dark energy or the cosmological vacuum might also have some influence on the size of galaxies (Nowakowski and Ashtekar 2001, Sereno and Jetzer 2007), but this effect would be small. Lee (2009) suggests instead that the size of galaxies increases as the scale factor of the Universe, assuming dark matter models based on a Bose–Einstein condensate or scalar field of ultralight scalar particles, which is another heterodox idea to be explored. Nonetheless, apart from these kinds of proposals, within the standard scenarios in the standard interpretation galaxies do not expand with the Universe.

4.4 Hubble–Lemaître constant

The Hubble–Lemaître constant, H_0, relates the distance of galaxies and their redshifts. From the beginning of the discovery of the apparent magnitude relation of the galaxies, first by Lemaître (1927) and later by Hubble (1929)—the so-called Hubble–Lemaître diagram—has decreased the value of that constant by almost an order of magnitude. In the 1980s, two preferred values were defended by different teams: either 50 or 100 km/s/Mpc. Later, in the 1990s and 2000s, a value of around 70 km/s/Mpc became dominant, with preference for the value of 72 km/s/Mpc obtained by the Hubble Space Telescope (HST) Key Project using supernovae (Freedman and Madore 2010).

Nonetheless, discordant values are often still being published. For instance, Paturel (2008) claims that, when one takes into account the period–luminosity bias for the calibration of distances with nearby galaxies, the Hubble–Lemaître constant is reduced to values of around 60 km/s/Mpc. Even supernova data with HST were fitted with these values. Allan R Sandage (1926–2010), and Gustav A Tammann (1932–2019) and his associates have dedicated many decades to the measurement of the Hubble–Lemaître constant, and to analysing the subtleties in its measurement. As to the possible (non-)universality of the Cepheid period–luminosity relation, they have argued that low metallicity Cepheids have flatter slopes, and that the derived distance would depend on what relation is used (e.g., Tammann *et al* 2003). After a correction of the host galaxy distances for the assumed metallicity dependence (as measured by [O/H]), the Type Ia supernovae were calibrated to be somewhat brighter than in the HST Key Project work. Sandage *et al* (2009) say: 'Failure to take the slope differences in the Period--Luminosity relation into account as a function of metallicity using Cepheids as distance indicators results in incorrect Cepheid distances. Part of the 15% difference between our long distance scale [. . .] and that of the HST Key Project short scale is due to the effect of using an inappropriate Period–Luminosity relation'.

We must also bear in mind that the value of H_0 is determined without knowing on which scales the radial motion of galaxies and clusters of galaxies relative to us is completely dominated by the Hubble–Lemaître flow (Matravers *et al* 1995). The homogeneity scale may be much larger than expected (see section 7.3.2), thus giving important net velocity flows on large scales that are incorrectly attributed to cosmological redshifts. Hence, some differences of H_0 measured at different scales might be due to this neglect of net radial motions of galaxies.

Apart from this unsolved dispute over the value of the Hubble–Lemaître constant with Sandage's preferred value, yet another controversy arose in the 2010s. The Hubble–Lemaître constant estimated from the local Cepheid–supernova distance ladder is at odds with the value extrapolated from cosmic microwave background data, assuming the standard cosmological model (74.0 ± 1.4 and 67.4 ± 0.4 km/s/Mpc respectively), which gives an incompatibility at the 4.4σ level (Riess *et al* 2019). In my opinion, given the number of systematic errors that may arise in the measurements, and given that CMB analysis is just an interpretation that might be in error (see chapter 5), I see it as more of a success that a quantity measured by totally different methods should be the same at least to within an order of magnitude and I would not worry about the small differences in the absolute values, possibly arising from an excess of optimism in the evaluation of the error bars. We know indeed that at least 20% of the error bars in the Hubble–Lemaître constant measurements have been underestimated or have not taken into account all the systematic errors (Faerber and López-Corredoira 2020), so it is not surprising that we get this kind of difference when the errors are not correctly estimated.

4.5 Observational tests for the expansion of the Universe

Nobody has ever directly observed a galaxy's distance increase over time. The motion of the putative expansion of the Universe is so slow that it is immeasurable. The variation in the redshift of a galaxy is also very slow and difficult to measure. Sandage (1962) proposed measuring the expected redshift drift $dz/dt = (1 + z)H_0 - H(z)$ (~1 cm/s/yr), but it has not hitherto been possible to measure such a low acceleration. The effect could possibly be observed in the future with a 40 m telescope over a period of ~20 yr using 4000 hours of observing time (Liske *et al* 2008), but the many systematic errors due to internal dynamics of the galaxies will also possibly hinder this enterprise even with the availability of very large telescopes. Apart from the actual redshift of the galaxies, there are various indirect tests to verify whether the Universe is expanding or static:

1. *Microwave background temperature as a function of redshift.* The cosmic microwave background radiation (CMBR) temperature can be detected indirectly at high redshift if suitable absorption lines can be found in high redshift objects. Hot Big Bang cosmology predicts that the temperature of the CMBR required to excite these lines is higher than at $z = 0$ by a factor $(1 + z)$. For a static Universe, there is no specific CMBR model, but one may say that the temperature would be constant and equal to 2.73 K for all redshifts if there were a local CMBR origin, or possibly $T_{\rm CMBR}(z) = 2.73(1 + z)$ K for a

distant CMBR origin produced at a unique distance and with a redshift due to tired light.

The CMBR temperature measured from the rotational excitation of certain molecules as a function of redshift (Molaro *et al* 2012, Noterdaeme *et al* 2011) has been quite successful in proving expansion: the results of Noterdaeme *et al* (2011) with the exact expected dependence on $T = T_0(1 + z)$ are impressive (see figure 4.1). However, there are other results that disagree with this dependence (Krelowski *et al* 2012, Sato *et al* 2013). The discrepancy might be due to a dependence on collisional excitation (Molaro *et al* 2012) or bias due to unresolved structure (Sato *et al* 2013). Nonetheless, there is another piece of evidence of the $T = T_0(1 + z)$ relationship that comes from the Sunyaev–Zel'dovich effect (Luzzi *et al* 2015, López-Corredoira *et al* 2017); namely, the imprint of galaxy clusters on the CMBR. This imprint does not have the problems noted in the previous paragraph. Therefore, if we accept these results, we can at the very least say that a local origin of the CMBR in a static Universe is excluded, but that the possibility of the origin of the CMBR at very high z within a static Universe is not excluded.

2. *Time dilation test.* Clocks observed by us at high redshifts would appear to keep time at a rate $(1 + z)$ times slower when there is expansion. By using sources of known constant intrinsic periodicity, we would expect their light curves to be stretched in the time axis by a factor $(1 + z)$.

   Time dilation tests in Type Ia supernovae (SNe Ia) look to become one of the most successful tests in favour of the expansion of the Universe

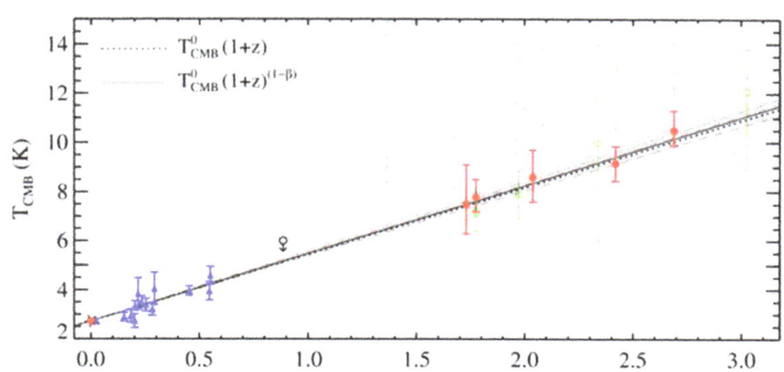

Figure 4.1. CMBR temperature versus redshift. The star represents the measurement at $z = 0$. Measurements based on the rotational excitation of CO molecules are represented by filled circles at $1.7 < z < 2.7$. Other measurements at $z > 0$ are based on: (i) the Sunyaev–Zel'dovich effect (triangles at $z < 0.6$) and (ii) analysis of the fine structure of atomic carbon (open squares). Upper limits come from the analysis of atomic carbon and molecular absorption lines in the lensing galaxy PKS1830-211 (open circle). The dotted line represents the adiabatic evolution of T_{CMB} as expected in standard hot Big Bang models. The solid line with shaded errors is the fit using all the data and the alternative scaling of $T_{CMB}(z)$ yielding $\beta = 0.007 \pm 0.027$. The dashed curve represents the fit and errors using Sunyaev–Zel'dovich + CO measurements. (Credit: Noterdaeme *et al* 2011, reproduced with permission ©ESO.)

(Goldhaber *et al* 2001, Blondin *et al* 2008) (see figure 4.2), but there are still problems with this interpretation. The standard method for calibrating SNe Ia light curves has been criticised as flawed, which might explain why this lack of time dilation has not been observed (Crawford 2018). The fact that SNe Ia light curves are narrower when redder (Nobili and Goobar 2008) is an inconvenience for a clean test free from selection effects. Other selection effects and the possible compatibility of the results with a wider range of cosmological models, including static ones, have also been pointed out (Brynjolfsson 2004a, Leaning 2006, Ashmore 2012, Holushko 2012, LaViolette 2012 [sect. 7.8]).

Moreover, neither gamma-ray bursts (GRBs) (Crawford 2009, Kocevski and Petrosian 2013) nor QSOs (Hawkins 2010) show time dilation, which is puzzling. However, with regard to this last, a variation in the QSO flux has been found (Dai *et al* 2012) in observations of more than 90 days with an average $|dF_V/dt|$ approximately proportional to $(1 + z)^{-1}$, although these authors used a sample of only 13 QSOs without K-corrections. To be able to arrive at any firm conclusion, further research is needed for the time dilation of QSOs.

3. *Cosmic chronometers.* Measurements of the Hubble–Lemaître parameter $\dot{z}$ that use the variation of the age of elliptical galaxies (from the Balmer break in their spectra) with redshift as cosmic chronometers (Moresco *et al* 2012, 2016) might

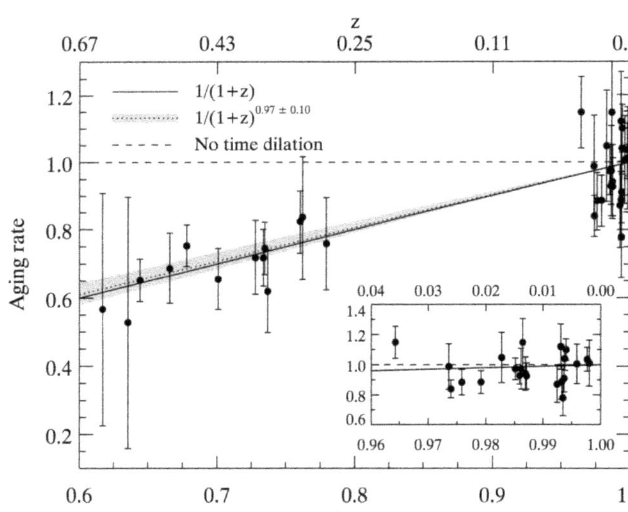

Figure 4.2. Apparent aging rate versus $1/(1 + z)$ for the 13 high redshift ($0.28 \leqslant z \leqslant 0.62$) and 22 low redshift ($z < 0.04$) SNe Ia in the sample of Blondin *et al* (2008). Overplotted are the expected $1/(1 + z)$ time dilation in an expanding Universe (solid line) and the best-fit $1/(1 + z)^b$ model (with $b = 0.97 \pm 0.10$; dotted line and grey area). The dashed line corresponds to no time dilation, as expected in the tired light model, and is clearly inconsistent with the data. Inset: close-up view of the low redshift sample. (Reproduced by permission of the AAS from figure 8 of Blondin *et al* 2008.)

be able to differentiate among the various cosmological models. For the standard cosmological model $-\dot{z}/(1+z) = H(z) = H_0\sqrt{\Omega_m(1+z)^3 + \Omega_\Lambda}$, whereas for a static model with tired light $-\dot{z}/(1+z) = H_0$, and for a static model with a linear extrapolation of Hubble–Lemaître's law with redshift $-\dot{z}/(1+z) = H_0/(1+z)$, assuming no creation or destruction of elliptical galaxies. The differences in the predictions are huge, and the actual data (Moresco *et al* 2012, 2016) favour the standard model and exclude a static Universe. Regrettably, however, the assumptions made for cosmic chronometers are subject to uncertainties. Elliptical galaxies at redshift ~2 have not a single stellar population, but two distinct stellar populations, a young and an old one, and the use of the Balmer break to measure the age of the galaxy reflects an average age of both populations, not the maximum stellar age corresponding to the Galaxy (López-Corredoira and Vazdekis 2018). Moresco *et al* (2018) replied to this criticism (although without explicitly citing López-Corredoira and Vazdekis 2018) by claiming that their sample of elliptical galaxies is free from contamination by the young population[1], but they analyse a stacked spectrum instead of individual objects, so the conclusions apply to the average but not necessarily to each galaxy; moreover, the average redshift of their sample is ~0.5, while they cannot say anything for the highest redshift galaxies ($z \sim 2$) that are crucial for discriminating among different cosmological models. This problem poses a serious challenge to the use of cosmic chronometers as currently applied. Therefore, so long we have no accurate means of determining the age of galaxies, this method is inapplicable.

4. *The Hubble–Lemaître diagram.* The apparent magnitude (taking K-corrections into account) versus redshift diagram has been used for different sources that were at first thought to be free from evolutionary effects. The Hubble–Lemaître diagram for the brightest galaxies in clusters better fits a static model according to some authors (LaViolette 1986) or the standard cosmological model according to others (Collins and Mann 1998). In any case, nowadays it is thought that brightest cluster galaxies also undergo some evolution in luminosity (Chu *et al* 2021), so any possible disagreement with the standard model could be solved with it.

   For SNe Ia (Scolnic *et al* 2018) and GRBs (Wei 2010, Marosi 2019), it is supposed that there is no evolution. The standard model works to fit the Hubble–Lemaître diagram for $z < 2$ (not for GRBs with greater redshifts, Marosi 2019), provided that an ad hoc dark energy constant is included, and that we assume zero evolution and negligible extinction or selection effects, neither of which assumption is universally accepted (Balázs *et al* 2006, Podsiadlowski *et al* 2008, Bogomazov and Tutukov 2011). Nonetheless, a

[1] Figure 1 of Moresco *et al* (2018) shows a small difference in the ratio of H and K lines of CaII when the young population contamination of the Balmer break is 17% of the total, but they claim to discard this scenario based on the measurement of this ratio in the stacked spectrum (within 15σ). I wonder whether such a high precision might be due to an underestimate of uncertainties and systematic errors of the stellar population synthesis model.

static Universe without evolution may also fit those data at $z < 2$ (Sorrell 2009, Lerner 2009, López-Corredoira 2010a, Farley 2010, Marosi 2013, 2019) and it is a better prospect for GRBs at higher redshift (Marosi 2019). Another unexpected problem for the standard model is that there is anisotropy in the Hubble–Lemaître diagram (values of H_0 and Ω_Λ) when comparing SNe Ia data within $z < 0.2$ from both Galactic hemispheres (Schwarz and Weinhorst 2007).

5. *The Tolman surface brightness test.* Hubble and Tolman (1935) proposed the so-called Tolman test, based on the measurement of surface brightness. A galaxy at redshift z varies in surface brightness proportionally to $(1 + z)^{-n}$, with $n = 4$ for expansion and $n = 1$ for the static case.

   Lubin and Sandage (2001), using the Tolman test up to $z = 0.9$, claimed to have a definitive proof of the expansion of the Universe. However, their claim, rather than being a Tolman test, was that the evolution of galaxies can explain the difference between the results of the Tolman test and their preferred model, which included expansion. Lerner (2006) observed that Lubin and Sandage used a very involved evolutionary K-correction scheme with many adjustable assumptions and parameters to correct observed high-z surface brightness. These evolutionary and K-corrections are subject to uncertainty and cannot be used convincingly. Crawford (2011a) also pointed out that Lubin and Sandage performed an erroneous analysis to exclude the static solution, mixing Big Bang and tired light models.

   Furthermore, more recent Tolman tests (Lerner 2006, Andrews 2006b, Brynjolfsson 2006, Crawford 2011a, Lerner *et al* 2014), some of them up to redshifts of ~5 and with different wavelength filters so that no K-corrections are necessary, favour a static Universe with no need for galaxy evolution, but they cannot exclude the standard scenario if evolution is allowed.

6. *Angular size versus redshift test.* The angular size (θ) of a galaxy of a given linear size (Hoyle 1959) differs greatly according to whether we assume the standard model with expansion or a static Universe. Another variety of this test is Etherington's distance-duality equation: the relationship between the luminosity distance and the angular-diameter distance, which is a factor $(1 + z)^2$ in the standard model.

   Tests have been made by several authors (LaViolette 1986, Kapahi 1987, Andrews 2006a, López-Corredoira 2010a), and all of them, either in the radio, near infrared or visible, show, over a wide range of redshifts, a dependence $\theta \sim z^{-1}$, a static Euclidean effect over all scales. This result cannot be reconciled with the standard cosmological model unless we assume a strong evolution of galactic radii that just coincidentally happens to compensate for the difference (Nabokov and Baryshev 2008): galaxies with the same luminosity should be six times smaller at $z = 3.2$ than at $z = 0$ (López-Corredoira 2010a); 10 times smaller at $z = 6$ than at $z = 0$ (Shibuya *et al* 2015); 20 times smaller at $z = 10$ than at $z = 0$ (Shibuya *et al* 2015). It is usually further argued that elliptical and disc galaxies have different evolutionary paths; that is right, but this difference has been analysed and found to be compatible with a null one if we take into account the error bars of the selection effects (López-Corredoira 2010a).

The angular size test for massive disc-like galaxies also produces a good fit for static universes without evolution according to some authors (LaViolette 2012 [section 7.4], Lerner 2018). However, with these alternative explanations, there still remains to be explained why the evolution depends on the environment of the galaxies too (Andreon 2018) because, if the angular galaxy size change is a purely cosmological effect without evolution, it should not depend on its environment or type. It might be a combination of both using an erroneous cosmological model and some slight linear size evolution of the Galaxy.

With the standard cosmological model, the hypothesis that galaxies which formed earlier have much higher densities, their luminosity evolution, merger ratio, massive outflows due to a QSO feedback mechanism are, taken together, insufficient to justify such a strong size evolution (López-Corredoira 2010a). The velocity dispersion at high z would also be much higher than observed (López-Corredoira 2010a). A static Universe is fitted without any ad hoc elements. However, we need to be cautious with this interpretation, because of the uncertainty in galaxy size evolution.

There is yet another idea to be found in the literature about such a strong apparent evolution of galaxies: at high redshift, owing to the Tolman factor being proportional to $(1 + z)^{-4}$, the surface brightness of the galaxies is much lower, and only very compact galaxies are detected (with high surface brightness) for a given stellar mass/luminosity (Disney and Lang 2012). It is a selection effect. This will give rise to a certain 'apparent' evolution in the average size of galaxies (smaller galaxies being observed at high redshifts), thereby producing an illusion of downsizing (dwarf galaxies apparently 'lift themselves above the sky' in recent epochs).

So far, there are no standard rods serving as physical objects without evolution. Double radio sources have been proposed as standard rods (Valtonen 1991), but there is also a strong inverse correlation between their absolute brightness and linear size, and some degree of evolution cannot be totally excluded (Valtonen 1991, Nilsson *et al* 1993). Ultracompact radio sources have been claimed in the past to be free from evolutionary effects, but they are now thought to show some signs of evolution as well (López-Corredoira 2010a, Pashchenko and Vitrishchak 2011). The huge size evolution necessary to fit an angular size test with an expanding Universe is not well understood (López-Corredoira 2010a), but still depends on our understanding of the galaxies rather than on pure cosmological approaches. The ratio between gas mass fraction in clusters measured through x-ray and gas mass fraction (in the same clusters) measured through the Sunyaev–Zel'dovich effect is proportional to the ratio of luminosity distance and angular distance, thus providing a method of indirectly deriving the angular size, although this method assumes certain mass distributions in the cluster (Holanda *et al* 2012).

7. *The galaxy number count magnitude test.* Differential galaxy number counts versus galaxy K magnitude fit both the standard cosmological model (Totani *et al* 2001) and the tired light Universe (LaViolette 2012 [section 7.7]). The same problem of degeneracy between evolution and cosmological geometry, however, once again raises its head.
8. *UV surface brightness test.* Lerner (2006) proposed a test of the evolution hypothesis that is also useful in the present case. There is a limit on the ultraviolet surface brightness (UV SB) of a galaxy because, when the surface density of hot bright stars and thus supernovae increases, large amounts of dust are produced to absorb all the UV except that from a thin layer. Further increase in the surface density of hot bright stars beyond a given point just produces more dust and a thinner surface layer, not an increase in UV SB. Based on this principle, there should be a maximum UV (at-rest) SB that is independent of redshift. This has been analysed at high redshift (Lerner 2006, López-Corredoira 2010a) with the result that the intrinsic UV SB magnitude of some galaxies would be prohibitively low (i.e. much brighter) than 18.5 mag $_{\rm AB}$/arcsec^2 with the standard model. For a static model, however, it would be within the normally expected range. Lerner (2006) also argues why alternative explanations (lower production of dust at high redshift, winds, or other scenarios) are inconsistent. Nonetheless, Lerner's hypothesis of a maximum UV SB might be incorrect, so this should be further explored before reaching definite conclusions about this test.
9. *The Alcock–Paczyński test.* Given a distribution of spherically symmetric objects with a radius along the line of sight $s_{\parallel} = \Delta z(d\ d_{\rm com}(z)/dz)$ and a radius perpendicular to the line of sight $s_{\perp} = \Delta\theta(1 + z)^m d_{\rm ang}(z)$ ($m = 1$ with expansion, $m = 0$ for static), the ratio $y \equiv \Delta z/(z\Delta\theta)(s_{\perp}/s_{\parallel})$ depends on the cosmological comoving distance ($d_{\rm com}(z)$) and the angular distance ($d_{\rm ang}(z)$), and is independent on the evolution of galaxies, but it depends also on the redshift distortions produced by the peculiar velocities of gravitational infall (López-Corredoira 2014).

   The variable $y(z)$ was measured by means of an analysis of the anisotropic correlation function of sources in several surveys (López-Corredoira 2014), using a technique to disentangle the dynamic and geometric distortions that also took other values available from the literature. Of seven different cosmological models, only two fitted the data of Alcock and Paczyński's test: concordance ΛCDM and a static Universe with tired light redshift. The rest were excluded at a >95% confidence level (CL). Analyses with further data using Baryonic Acoustic Oscillations (BAO) (Melia and López-Corredoira 2017) improve the test and give us a more accurate constraint: ΛCDM with standard values of parameters is excluded at 98% CL, whereas static with tired light fits the data, but an expanding Universe with zero active mass or wCDM with $\omega_{\rm dark\ energy} = -0.60^{+0.30}_{-0.27}$ and $\Omega_m = 0.45^{+0.21}_{-0.19}$ also fits the data.

Table 4.1 summarises the analyses of this subsection. The first two tests favour expansion, whereas the 4–8th tests get a less ad hoc fit with the static solution,

Table 4.1. Cosmological tests of the expansion.

Test	Expansion	Static
$T_{CMBR}(z)$	Good fit	Tired light redshift of a CMBR coming from very high z.
Time dilation	Good fit for SNeIa. Unexplained absence of time dilation for QSOs and GRBs.	Selection effects, ad hoc modification of the theory the zero point calibration, or evolution of SNIa periods.
Cosmic chronometer	Good fit but by chance, measurements of differential age are incorrect.	Bad fit but measurements of differential age are incorrect.
Hubble–Lemaître diagram	Good fit but requires the introduction of dark energy and/or evolution of galaxies.	Good fit for galaxies. Good fit with some models.
Tolman (SB)	Requires strong an SB evolution.	Good fit
Angular size	Requires too strong evolution of angular sizes.	Good fit
Galaxy counts	Good fit for galaxies with evolution.	Good fit
UV SB limit	Anomalously high UV SB at high z	Within the constraints
Alcock-Paczyński	Bad fit for the standard model but good fit with other wCDM models.	Good fit for tired light

although this is insufficient cause for rejecting expansion. Most of the cosmological tests are entangled with the evolution of galaxies and/or other effects. Tolman and angular size tests need to assume a very strong evolution of galaxy sizes to fit the data with the standard cosmology, whereas the Alcock–Paczyński test is independent of the evolution of galaxies but does not yet show any preference for an expanding or static Universe.

4.6 Anomalous redshifts

Although, as discussed above, doubt might be cast upon the reality of expansion, even in the case that it were definitively proven, this does not mean that all galaxies have a cosmological redshift, i.e. that all galaxies have a redshift due to expansion. There might be some exceptions, known as 'anomalous redshift' cases (Narlikar 1989, Arp 1987, 2003b, Burbidge 2001, Bell 2002b, 2002c, 2006, 2007, López-Corredoira and Gutiérrez 2006, López-Corredoira 2010b); that is, a redshift produced by a mechanism different from the expansion of the Universe or the Doppler effect. The most important supporters of this idea of anomalous redshift were Halton Christian Arp (1927–2013) and Geoffrey Ronald Burbidge (1925–2010).

There are plenty of individual cases of galaxies with an excess of QSOs with high redshifts near the centre of nearby galaxies, mostly AGN (active galactic nuclei).

In some cases, the QSOs are only a few arcseconds away from the centre of galaxies with different redshifts. Examples are NGC 613, 2237+0305, and 3C 343.1. In some cases there are even filaments/bridges/arms apparently connecting objects with different redshift: in NGC 4319+Mrk 205 (see figure 4.3), QSO 1327-206, NGC 3067+3C232 (in the radio). The probability of chance projections of background/ foreground objects within a short distance of a galaxy or onto the filament is very low. For instance, even if we think that the filament connecting both objects of figure 4.3 is an artefact of the image, the chance probability of having Mrk 205 within 40 arcsec from the centre of any galaxy as bright as NGC 4319 in the V-filter (0.028 galaxies per square degree on average) is only $\sim 10^{-5}$ (we might add a factor 4 to take into account the *a posteriori* selection of the angular separation and the brightness of the Galaxy; (López-Corredoira 2010b)) and there are only 84 QSOs in the whole sky brighter than, or as luminous as, Mrk 205 ($V = 15.2$). Furthermore,

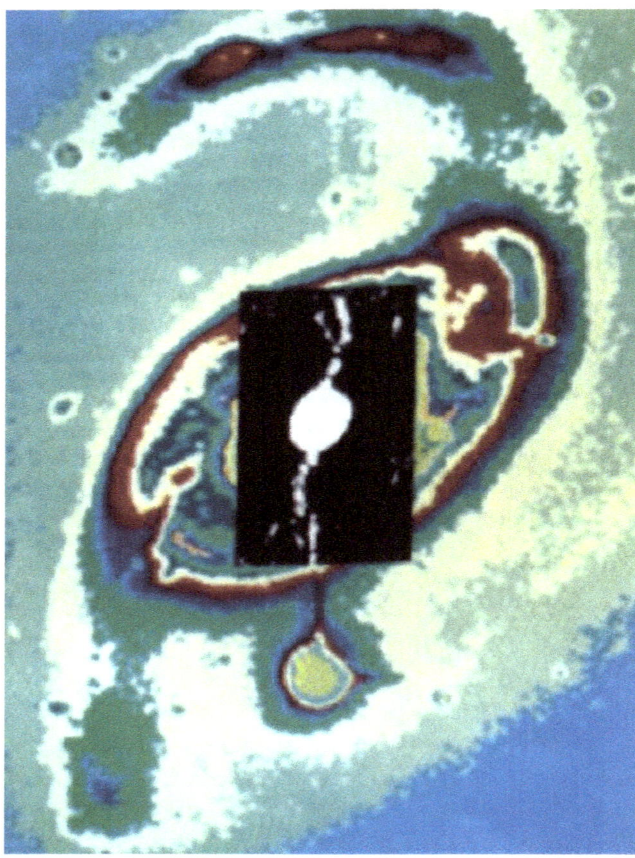

Figure 4.3. A remarkable example of an apparent connection between a galaxy (NGC 4319; with a superposed a high-pass filtered image of its inner disc) and a QSO (Mrk 205; the object connected by a filament below the Galaxy) with different redshifts: $z = 0.006$ and $z = 0.070$ respectively. (Reproduced by permission of the AAS from figure 5 of Sulentic and Arp 1987.)

we have to add the other casual coincidences that were observed in this system (Sulentic and Arp 1987).

The alignment of sources with different redshifts also suggests that they may have a common origin, and that the direction of alignment is the direction of ejection (Arp and Hazard 1980, Arp 1999, 2003a, Arp and Russell 2001). This happens with some configurations of QSOs around NGC 4235, NGC 5985, GC 0248+430, etc. In my opinion, many of these statistical analyses by Arp are wrong. For instance, I have carried out some rough calculations with Arp (2003a)'s data and I could see that these alignments are not so surprising: six out of eleven QSOs within 15′ have a difference of less than 32 degrees from the average position angle (i.e. each one has a probability of 128/360, which is not a low probability). By chance, one would expect $11 \times 128/360 = 3.9$ QSOs in this position angle range, and we have six QSOs, which is quite probable.

There are also plenty of statistical analyses (Chu *et al* 1984, Burbidge *et al* 1985, Zhu and Chu 1995, Burbidge 1996, 2001, Harutyunian and Nikogossian 2000, Benítez *et al* 2001, Gaztañaga 2003, D'Elia *et al* 2004, Nollenberg and Williams 2005, Russell *et al* 2006, Bukhmastova 2007, Burbidge and Napier 2009, Sandrinelli *et al* 2018) showing an excess of high redshift sources near low redshift galaxies, positive and very significant cross-correlations between surveys of galaxies and QSOs, an excess of pairs of QSOs with very different redshifts, etc. Other proofs presented in favour of the QSO/galaxy association with different redshifts is that no absorption lines have been found in QSOs corresponding to foreground galaxies, or distortions in the morphology of isolated galaxies. The non-cosmological redshift hypothesis also affects galaxies differently from QSOs. Cases such as NGC 7603, AM 2004-295, AM 2052-221, etc, present statistical anomalies that also suggest that the redshift of some galaxies different from that of QSOs might have non-cosmological causes. Not all supporters of the non-cosmological redshift agree with this idea; for instance, Halton Arp claimed that galaxies might have non-cosmological redshift because they derive from an evolution of ejected QSOs, while Geoffrey Burbidge only defended the non-cosmological redshifts in QSOs. There are also proposals for two different types of non-cosmological redshifts, for non-active galaxies and QSOs respectively (Marmet 2009). Fred Hoyle, Geoffrey Burbidge, and Jayant Narlikar have agreed that the Ly-α forest might be intervening gas clouds, but that many of them were ejected by the QSO itself.

In addition to the statistical anomalies, QSOs are still objects that are not completely understood in other aspects. There are many pending problems, inconsistencies, and caveats in QSO research (López-Corredoira 2011). The standard paradigm model based on the existence of very massive black holes that are responsible for QSOs' huge luminosities, resulting from their cosmological redshifts, leaves many facts without explanation. There are several observations that lack a clear explanation; for instance, the absence of bright QSOs at low redshifts, a mysterious evolution not properly understood (López-Corredoira 2011). Nonetheless, the cosmological nature of QSO redshifts is supported at least in those cases where galaxies with the same redshift are in the same position or at a small angular distance from the QSO (Stockton 1978).

The solution to the problem is not clear. Calculations of the probabilities of the anomalous redshift cases being normal background sources with cosmological redshifts indicate that some of the examples of apparent associations of QSOs and galaxies with different redshifts may be just fortuitous cases in which background objects are close to the main galaxy although the statistical mean correlations remain to be explained, and some lone objects have a very low probability of being a projection of background objects. Nevertheless, these very low probabilities (down to 10^{-8} or even lower, assuming that the calculations are correct) are not extremely low and, if the anomaly is real, one wonders why we do not find very clearly anomalous cases with probabilities as low as 10^{-20}.

Amplifications due to gravitational lensing might help explain statistical anomalies (Scranton *et al* 2005, López-Corredoira 2010b), but gravitational lensing does not seem to be a general solution for all the anomalous cases (Arp 1990). The microlensing-induced QSO number enhancement is independent of the mass function of microlenses and spatial distribution of microlensed objects, and it relies only on the total mass and redshift of the lenses; microlenses are more effective than smoothly distributed singular isothermal lenses for nearby galaxies (Tang and Zhang 2005b). Komberg and Pilipenko (2008) has suggested the existence of large number of globular or proto-globular clusters in the intergalactic medium of clusters of galaxies and their gravitational lensing effect as an explanation of the correlations of galaxies and QSOs, a hypothesis that would not explain why there are many field galaxies also with an excess of QSOs.

The possibilities are few: either all cases of associations are lucky coincidences or with a higher probability than expected for some still unknown reason, or there are at least some few cases of non-cosmological redshifts. The debate has lasted a very long time, around 50 years, and it is high time to consider making a last-ditch effort to finish with the problem. However, every time the problem is mentioned, supporters of the standard view just smile or talk about '*a posteriori*' calculations, manipulations of data, fringe ideas, without even bothering to read any paper on the theme. The Arp–Burbidge hypothesis has become a topic concerning which everybody has an opinion without having read the papers or knowing the details of the problem, because some leading cosmologists have declared it to be bogus. On the other hand, the main supporters of the hypothesis of non-cosmological redshifts produce analyses of cases in favour of their ideas with insufficient care, pictures without rigorous statistical calculations in many cases, or with wrong identifications, underestimated probabilities, biases, use of incomplete surveys for statistics, etc, in many other cases (López-Corredoira 2010b). There are, however, many papers in which no objections are found in the arguments and they present quite controversial objects, but, owing to the bad reputation of the topic, the community simply ignores them.

4.6.1 Periodicity of redshifts

The redshift distribution of extragalactic objects is expected to approximate to a continuous and aperiodic distribution. However, Tifft (1976, 1977, 1980) affirmed that there was a periodicity of 70–75 km s^{-1} in the redshift of galaxies. In an improved correction for the optimum solar vector, the periodicity is found to be 37 km s^{-1} (which includes Tifft's result for twice this velocity) with a probability of 2.7×10^{-5} of finding this period by chance (Guthrie and Napier 1996). A galactocentric periodicity of 37 km s^{-1} was confirmed to exist in the redshift distribution of nearby spiral galaxies (Napier 2006). The periodicity is seen in all the datasets examined. Using the distance information, there is some evidence that the periodicity is not strictly with respect to a fixed galactocentric velocity vector, but is referred to a vector that varies with increasing distance.

A periodicity with $\Delta z = 0.031$ or 0.062 was also found for QSOs (Burbidge and O'Dell 1972, Bell 2002a) and, on even larger scales, a 0.089 periodicity in $\log_{10}(1 + z)$ (Napier 2006). A $\Delta z = 0.031$ is equivalent to $\Delta r \sim 100\ h^{-1}$ Mpc, and coincidently a periodicity for clusters of galaxies at scales is found on scales $\sim 100\ h^{-1}$ Mpc (Einasto *et al* 1997, 1997), so one may wonder whether there may be some connection between both periodicities; if so, we would be tracing the net of the large-scale structure, Other works have also found similar periodicities (Bell and Comeau 2003a, 2003b, Bell and McDiarmid 2006, Hartnett and Hirano 2008, Hartnett 2009, Fulton and Arp 2012, Fulton *et al* 2018, Mal *et al* 2020, Ryabinkov and Kaminker 2021). Some authors have related this redshift periodicity with possible non-cosmological redshifts, but they might instead be due to processes that we do not know concerning the formation of galaxies. Nonetheless, other authors (Hawkins *et al* 2002, Tang and Zhang 2005a, 2010, Bajan *et al* 2007, Repin *et al* 2012) have found no significant periodicities in the distribution of redshifts, which is another solution to the problem.

Some exotic proposals have emerged within the scenario in which the periodicities are real and for which there are no explanations in the standard picture of the formation of galaxies in the framework of ΛCDM. An oscillating scalar field model (Hirano and Komiya 2010) with mass for particles of mass 3.2×10^{-31} eV has been proposed to explain the periodic structure in galaxy counts versus redshift. The periodicity is also explained by a photon interactions with atomic hydrogen, linking it with the quantization of energy levels in hydrogen (Moret-Bailly 2006). Also, in 1990 the Finnish physicist Ari Lehto had come up with a general formula for the quantization of physical quantities, without knowing of Tifft's astronomical studies (Baryshev and Teerikorpi 2012 [section 9.5.2]). He found that his formula describes well the redshift periods 36 and 72 km s^{-1} that Tifft had derived. If the energy of any photon is quantized, as Lehto suggests, then so are the frequencies, hence the redshifts of light would occur at preferred values. Lehto had searched for a common rule for the properties of the micro- and the macroworld and he found that the ratios of the quantities involving lengths or energies may be expressed as $2^{n/3}$ ($n = 1, 2, 3, \ldots$). What he did was to take the ratios of observed values of various stationary discrete systems (such as the Planck energy/electron rest mass etc) and noted that the exponents of two thus obtained systems seemed to group near 0, 1/3 and 2/3 (Lehto 1990). In a more extensive study (Lehto 2009), involving over 40 quantities, it was shown that it is quite unlikely that the groupings could be due to chance. The pattern is best revealed by quantities that have been most accurately measured (laboratory physics), but it seems to be visible also for cosmic quantities.

Lehto's system has the Planck units as natural starting points for making up the physical world. The Planck scale is absolute, based on constants of nature. For example, the observed temporary periods take values $t = t_{\mathrm{Pl}}2^{n/3}$, and the electron mass is obtained as $m_e = m_{\mathrm{Pl}}2^{227/3}$. The redshift steps of about 72 and 36 km s^{-1} would come from $V = 2^{n/3}c$ where n/3 is 12 and 13.

4.6.2 Proposed tests

Tests have been proposed to continue research into the possibility that some redshifts are not cosmological or Doppler redshifts, and that some objects (usually QSOs) are at closer distances than indicated by their redshifts.

Popowski and Weinzierl (2004) have proposed that, if QSOs are nearby, then they should have detectable proper motions. Andrei *et al* (2009) have indeed measured variations of astrometric positions in QSOs (up to 10 mas, in 415 days, which is equivalent to a variation of 0.044 kpc yr^{-1} for a QSO 1512-0905 [$z = 0.361$]), but they found tentatively significant correlations with photometric variations of the QSOs too, so they attributed the variation of the position not to a true motion but to the optical emission from the core region usually exceeding the other components by a factor of a hundred, so that the variability of those components must surely imply in some measure variability of the astrometric barycentre. This is the crux of the problem: it is not easy to disentangle the variability of the QSO from its proper motions.

Zackrisson (2005) also proposed studying the host galaxy of those QSOs suspected of having non-cosmological redshifts. Recently ejected QSOs should not have a host, or only a host with ionised gas and/or young stars, whereas in most detected host galaxies of other QSOs the energy distribution is consistent with populations of stars 1 Gyr old. Zackrisson (2005) claims that his test can be carried out for QSOs with $z < 0.7$ by doing photometry in the optical and near infrared of the host. As far as I know, nobody has carried out this test. In any case, QSOs without hosts (Magain *et al* 2005) are already known and it does not seem to be a problem for the standard interpretation of QSOs at large distances.

According to Visbal and Croft (2008), the ultraviolet radiation from a QSO leaves a characteristic pattern in the distribution of ionised hydrogen throughout the surrounding space. This pattern, or light echo, propagates through the intergalactic medium at the speed of light and can be observed by its imprint on the Lyman-α forest spectra of background sources. As the echo persists after the QSO has switched off, it offers the possibility of searching for dead QSOs, constraining their luminosities and lifetimes, and testing whether the luminosity of the QSO is as high as expected from its cosmological redshift distance. Indeed, a lack of change in the HI absorption due to ultraviolet photons has been observed (Kirkman and Tytler 2008), but explanations are given in terms of the shorter time of the ultraviolet luminosities of QSOs produced. Therefore, this is not the definitive test of the non-cosmological redshifts of QSOs.

A method of determining with high precision the distance of a high redshift objects might be to use a low frequency gravitational wave detector (such as LISA), applying it to black hole binaries in other galaxies (MacLeod and Hogan 2008, Kyutoku and Seto 2017). This looks like a feasible project for the coming years.

Another test could be measuring QSO parallax using the angular size of QSO broad emission line regions using time-lag reverberation techniques, which gives linear size, combined with an interferometer, which gives angular size (Karovska and Elvis 2003, Petrov *et al* 2012), but these angular sizes are of the order of few microarcsec or milliarcsec at most, which is too small to be spatially resolved with today's optical long-baseline interferometers; they can be resolved only by differential visibility and phase. The dust continuum emission, however, is larger by a factor of ~4; as a matter of fact, infrared interferometers have now managed to resolve some AGN with $z < 0.2$ (Hönig 2014). Moreover, using dust emission requires only photometric reverberation mapping instead of the spectral resolution of emission lines.

Dai *et al* (2012) do find a variation of the QSO flux in observations of more than 90 days with an average $|dF_V/dt|$ approximately proportional to $(1 + z)^{-1}$, and they claim that this flux variation can be used as a way to use QSOs as standard candles. In my opinion, the work by Dai *et al* is insufficient: they only use 13 QSOs, they do not carry out K-corrections, and there are other selection effects that may explain this variation of fluctuations. A more appropriate method of using QSOs as standard candles would be using a non-linear relation between the ultraviolet and x-ray luminosities of QSOs (Risaliti and Lusso 2015, 2019). The error bars in this method, however, are quite large, which makes it more useful for a statistical analysis rather than for individual objects. So far, no application of this method has shown that a QSO may have a distance very different from that derived from its cosmological redshift. Interesting enough, Risaliti and Lusso (2019) find that the Hubble–Lemaître diagram for QSOs is not fitted by the ΛCDM model, with a statistical significance of the deviation of 4σ. However, they used as external calibrators Type Ia SNe to fix the data, which may produce small biases, whereas calibration with quasars alone can remove the conflict (Melia 2019).

Measuring the corresponding redshifts of neutrinos from the same source for which we know a photon redshift could give an answer to our question (Gallo 2003). If the photons and neutrinos yield the same redshift, then we may conclude that the redshift mechanism is of the Doppler/expansion variety. However, if the photon redshifts are substantially greater than the corresponding neutrino redshifts, then the photon redshift must be due to an interaction between the photons and intervening matter. This would follow since neutrinos do not interact as strongly with intervening matter as photons. This effect would not reveal the exact nature of the photon–matter energy loss mechanism (that would require further tests). However, such measurements would definitively distinguish between Doppler/ expansion redshifts versus photon energy-loss redshifts with intervening matter. This experimental test requires neutrino spectroscopy and neutrino telescopes, which are not presently available.

Summing up, none of these tests could be carried out yet, or if they could, no signs of support for anomalous redshifts have been discovered.

4.6.3 Anomalous redshift in the laboratory and in the solar system

There are not many experiments that try to measure anomalous (neither Doppler nor gravitational) redshifts in the laboratory. One of the few that has attracted my attention: Chen *et al* (2009) reported the observation of line shift increase with plasma electron densities: roughly $\Delta\lambda(\mathrm{nm}) \sim N_e(10^{18}\mathrm{cm}^{-3})$ for the HgI line with $\lambda = 435.83$ nm. The dependence is also influenced by temperature. This is justified by the authors in terms of the difference in atomic energy levels for higher electron density, leading to a longer wavelength. The Hg atomic levels embedded in a density environment are influenced by the free electron density. The electronic fields generated from free electrons compressed inside an atom screen the Coulomb potential of the atomic nucleus. The nuclear forces on the bound electrons are then diminished, while the mutual repulsion of free and bound electrons is enhanced, so that the energy levels outside the nucleus are raised and the intervals of excited energy level $7s^3S$ to $6p^3P_1^0$ are diminished. Previous studies had also pointed out this kind of redshift (Nguyen *et al* 1986). My impression is that this emission redshift affects only some lines and not the spectrum as a whole; in principle one would anticipate a dependence on wavelength. Moreover, it is not shown here that this can explain, as Ashmore (2011) thinks, the absorption of a photon and its emission with redshift (such as cosmological redshift).

The centre-to-limb variation of the wavelengths of the solar spectrum was discovered by Jakob Karl Ernst Halm (1866–1944) in 1907. Redshifts in the radio of background stars when they are near the Sun were measured (Mérat 1974a, 1974b), crossing the solar coronal layers, and is strong near the solar limb. General relativity predicts a deviation in the light path of $1.75''$. However, very near the solar limb, the deviation is $2.03''$, whose excess over $1.75''$ might be linked with an excess of redshift. Further research was carried out later by several defenders of anomalous redshifts (Marmet 1989, Brynjolfsson 2004b). Marmet (1989) proposes inelastic scattering in the photon–matter interaction for which, each time a photon meets an individual hydrogen atom or molecule on its path, it loses 10^{-13} of its energy due to bremsstrahlung arising from momentum transfer to hydrogen in the solar chromosphere. However, solar physicists consider the effect to be well explained in terms of standard physics and our knowledge of convection in the Sun (Dravins 1982). The properties of solar line asymmetries and wavelength shifts can be traced back to small scale inhomogeneities in the photosphere (Dravins 1982): granulation (irregular polygonal shape of 1000–2000 km separated by narrow dark lanes).

4.7 So … is the Universe expanding or not?

In this chapter, I have shown that there are possible theoretical scenarios in which a static Universe with galaxy redshifts produced by a mechanism different from expansion or the Doppler effect is plausible. Nonetheless, some observational tests favour an expanding Universe rather than a static one, and the few arguments against expansion are not very strong yet. Does this mean that we can be 100% sure that the expansion of the Universe is truly established beyond speculative hypotheses? I would not go so far. In cosmology, very few ideas are worth staking one's life on them. I would bet on expansion, but I think definitive proof of it has yet to be produced.

References

Alfonso-Faus A 2006 Mass-boom versus big-bang: an alternative model *1st Crisis in Cosmology Conf. (AIP Conf. Ser. 822(1))* ed E J Lerner and J B Almeida (Melville, NY: AIP) pp 107–219

Alfonso-Faus A 2009 The case for a non-expanding universe (arXiv:0908.1539)

Anderson J L 1995 Multiparticle dynamics in an expanding universe *Phys. Rev. Lett.* **75** 3602–4

Andrei A H, Bouquillon S and de Camargo J I B *et al* 2009 Astrometric and photometric variability in quasars *Proc. of the Journées "Systèmes de référence spatio-temporels" & X. Lohrmann-Kolloquium: Astrometry, Geodynamics and Astronomical Reference Systems* ed M Soffel and N Capitaine (Paris: Lohrmann-Observatorium and Observatoire de Paris) pp 199–202

Andreon S 2018 The cosmic epoch dependence of environmental effects on size evolution of red-sequence early-type galaxies *Astron. Astrophys.* **617** A53

Andrews T B 2006a Derivation of the Hubble redshift and the metric in a static universe *1st Crisis in Cosmology Conf. (AIP Conf. Ser. 822)* ed E J Lerner and J B Almeida (Melville, NY: AIP) pp 123–43

Andrews T B 2006b Falsification of the expanding universe model *1st Crisis in Cosmology Conf. (AIP Conf. Ser. 822)* ed E J Lerner and J B Almeida (Melville, NY: AIP) pp 3–22

Arp H C 1987 *QSOs, Redshifts and Controversies* (Berkeley, CA: Interstellar Media)

Arp H C 1990 Quasar numbers as a function of apparent magnitude–gravitational lensing or quasars associated with nearby galaxies? *Astron. Astrophys.* **229** 93–8

Arp H C 1999 A QSO 2.4 arcsec from a dwarf galaxy–the rest of the story *Astron. Astrophys.* **341** L5–8

Arp H C 2003a Alignment of radio emission and quasars across NGC 613 and NGC 936 and radio ejection from NGC 941 (arXiv:astro-ph/0312198)

Arp H C 2003b *Catalogue of Discordant Redshift Associations* (Montreal: Apeiron)

Arp H C and Hazard C 1980 Peculiar configurations of quasars in two adjacent areas of the sky *Astrophys. J.* **240** 726–36

Arp H C and Russell D 2001 A possible relationship between quasars and clusters of galaxies *Astrophys. J.* **549** 802–19

Ashmore L 2011 Intrinsic plasma redshifts now reproduced in the laboratory–a discussion in terms of new tired light (arXiv:1105.0010)

Ashmore L 2012 Supernovae Ia light curves show a static universe (arXiv:1207.0015)

Assis A K T, Neves M C D and Soares D S L 2009 Hubble's cosmology: from a finite expanding universe to static endless universe *econd Crisis in Cosmology Conf. (ASP Conf. Ser. 413)* ed F Potter (San Francisco, CA: Astronomical Society of the Pacific) pp 255–67

Atlas Collaboration 2017 Evidence for light-by-light scattering in heavy-ion collisions with the atlas detector at the LHC *Nat. Phys.* **13** 852–8

Bajan K, Flin P, Godlowski W and Pervushin V P 2007 On the investigations of galaxy redshift periodicity *Phys. Part. Nuclei Lett.* **4** 5–10

Balázs L G, Hetesi Z, Regály Z, Csizmadia S, Bagoly Z, Horváth I and Mészáros A 2006 A possible interrelation between the estimated luminosity distances and internal extinctions of type Ia supernovae *Astron. Nachr.* **327** 917–24

Barber G 2002 The principles of self creation cosmology and its comparison with general relativity (arXiv:gr-qc/0212111)

Bartlett D F and Cumalat J P 2011 How a massive photon retards the universal expansion until galaxies form *Bull. Am. Astron. Soc.* **43** 4.03

Baryshev Y V 1981 Hierarchical structure of metagalaxy–problem review *Astrofiz. Issled.-Izv. Spets. Astrofiz. Obs* **14** 24

Baryshev Y V 1994 On the fractal nature of the large-scale structure of the universe *Astron. Astrophys. Trans* **5** 15–23

Baryshev Y V 2008 Expanding space: the root of conceptual problems of the cosmological physics *Practical Cosmology, 1* ed Y V Baryshev, I N Taganov and P Teerikorpi (St. Petersburg: TIN) pp 20–30

Baryshev Y V 2015 Paradoxes of the cosmological physics in the beginning of the 21st century *Particle and Astroparticle Physics, Gravitation and Cosmology: Predictions, Observations and New Projects* ed R R Anatolievich and P V Alexeevich (Singapore: World Scientific) pp 297–307

Baryshev Y V, Sylos Labini F, Montuori M and Pietronero L 1994 Facts and ideas in modern cosmology *Vistas Astron.* **38** 419–500

Baryshev Y V and Teerikorpi P 2012 *Fundamental Questions of Practical Cosmology* (Dordrecht: Springer)

Bell M B 2002a Discrete intrinsic redshifts from quasars to normal galaxies (arXiv:astro-ph/0211091)

Bell M B 2002b Further evidence for large intrinsic redshifts *Astrophys. J.* **566** 705–11

Bell M B 2002c On quasar distances and lifetimes in a local model *Astrophys. J.* **567** 801–10

Bell M B 2006 Evidence that quasars and related active galaxies are good radio standard candles and that they are likely to be a lot closer than their redshifts imply (arXiv:astro-ph/0602242)

Bell M B 2007 Further evidence that the redshifts of AGN galaxies may contain intrinsic components *Astrophys. J. Lett.* **667** L129–32

Bell M B and Comeau S P 2003a Further evidence for quantized intrinsic redshifts in galaxies: is the great attractor a myth? (arXiv:astro-ph/0305112)

Bell M B and Comeau S P 2003b Intrinsic redshifts in QSOS near NGC 6212 (arXiv:astro-ph/0306042)

Bell M B and McDiarmid D 2006 Six peaks visible in the redshift distribution of 46,400 SDSS quasars agree with the preferred redshifts predicted by the decreasing intrinsic redshift model *Astrophys. J.* **648** 140–7

Benítez N, Sanz J and Martínez-González E 2001 Quasar-galaxy associations revisited *Mon. Not. R. Astron. Soc.* **320** 241–8

Blondin S, Davis T M and Krisciunas K *et al* 2008 Time dilation in type Ia supernova spectra at high redshift *Astrophys. J.* **682** 724–36

Bogomazov A I and Tutukov A V 2011 Type Ia supernovae: Non-standard candles of the universe *Astron. Rep.* **55** 497–504

Bondi H 1947 Spherically symmetrical models in general relativity *Mon. Not. R. Astron. Soc.* **107** 410–25

Bonetti L, Ellis J, Mavromatos N E, Sakharov A S, Sarkisyan-Grinbaum E K and Spallicci D A M 2017 FRB 121102 casts new light on the photon mass *Phys. Lett.* B **768** 326–9

Bonnor W B 1999 Size of a hydrogen atom in the expanding universe *Class Quantum Grav.* **16** 1313–21

Bouvier P and Maeder A 1978 Consistency of Weyl's geometry as a framework for gravitation *Astrophys. Space Sci.* **54** 497–508

Broberg H 2001 The geometry of acceleration in space-time: application to the gravitational field and particles *Gravitation, Electromagnetism and Cosmology: Toward a New Synthesis* ed K Rudnicki (Montreal: Apeiron) p 9

Brynjolfsson A 2004a Plasma redshift, time dilation, and supernovas Ia (arXiv:astro-ph/0406437)

Brynjolfsson A 2004b Redshift of photons penetrating a hot plasma (arXiv:astro-ph/0401420)

Brynjolfsson A 2006 Surface brightness in plasma-redshift cosmology (arXiv:astro-ph/0605599)

Budko N V 2009 Exponential cosmological redshift in a linearly expanding universe (arXiv:0904.3280)

Bukhmastova Y L 2007 Quasars lensed by globular clusters of spiral and elliptical galaxies *Astron. Lett.* **33** 355–67

Burbidge G R 1996 The reality of anomalous redshifts in the spectra of some QSOS and its implications *Astron. Astrophys.* **309** 9–22

Burbidge G R 2001 Noncosmological redshifts *Publ. Astron. Soc. Pac.* **113** 899–902

Burbidge G R and Napier W M 2009 Associations of high-redshift quasi-stellar objects with active, low-redshift spiral galaxies *Astrophys. J.* **706** 657–64

Burbidge G R, Narlikar J V and Hewitt A 1985 The statistical significance of close pairs of QSOS *Nature* **317** 413–5

Burbidge G R and O'Dell S L 1972 The distribution of redshifts of quasi-stellar objects and related emission-line objects *Astrophys. J.* **178** 583–606

Castro Perelman 2009 On dark energy, Weyl geometry and Brans-Dicke-Jordan scalar field (viXra:0901.0001)

Chen C S, Zhou X L, Man B Y, Zhang Y Q and Guo J 2009 Investigation of the mechanism of spectral emission and redshifts of atomic line in laser-induced plasmas *Optik* **120** 473–8

Chu A, Durret F and Márquez I 2021 Physical properties of brightest cluster galaxies up to redshift 1.80 based on HST data *Astron. Astrophys.* **649** 19

Chu Y, Zhu X, Burbidge G and Hewitt A 1984 Statistical evidence for possible association between QSOS and bright galaxies *Astron. Astrophys.* **138** 408–14

Collins C A and Mann R G 1998 The k-band Hubble diagram for brightest cluster galaxies in x-ray clusters *Mon. Not. R. Astron. Soc.* **297** 128–42

Crawford D F 2006 *Curvature Cosmology* (Boca Raton, FL: BrownWalker Press)

Crawford D F 2009 No evidence of time dilation in gamma-ray burst data (arXiv:0901.4169)

Crawford D F 2011a Observational evidence favors a static universe (part I) *J. Cosmol.* **13** 3875–946

Crawford D F 2011b Observational evidence favors a static universe (part III) *J. Cosmol.* **13** 4000–57 (http://thejournalofcosmology.com/crawford3.pdf)

Crawford D F 2018 Is the universe static? (arXiv:1804.10274)

Cuesta M J H, Salim J M and Novello M 2007 Cosmological redshift and nonlinear electrodynamics propagation of photons from distant sources (arXiv:0710.5188)

Cumalat J P and Bartlett D F 2011 Einstein, Schwinger, and the Sinusoidal potential *Bull. Am. Astron. Soc.* **43** 4.02

Dai D-C, Starkman G D, Stojkovic B, Stojkovic D and Weltman A 2012 Using quasars as standard clocks for measuring cosmological redshift *Phys. Rev. Lett.* **108** 231302

de Sitter W 1931 Do the galaxies expand with the universe? *Bull. Astron. Inst. Neth.* **6** 146

D'Elia V, Fiore F, Elvis M, Cappi M, Mathur M, Mazzotta P, Falco E and Cocchia F 2004 The faint x-ray source population near 3c 295 *Astron. Astrophys.* **422** 11–21

Disney M J and Lang R H 2012 The galaxy ancestor problem *Mon. Not. R. Astron. Soc.* **426** 1731–49

Domínguez A and Gaite J 2001 Influence of the cosmological expansion on small systems *Europhys. Lett.* **55** 458–64

Dravins D 1982 Photospheric spectrum line asymmetries and wavelength shifts *Annu. Rev. Astron. Astrophys.* **20** 61–89

Dumin Y V 2002 On a probable manifestation of Hubble expansion at the local scales, as inferred from LLR data (arXiv:astro-ph/0203151)

Einasto J, Einasto M and Gottlöber S *et al* 1997 A 120-Mpc periodicity in the three-dimensional distribution of galaxy superclusters *Nature* **385** 139–41

Einasto M, Tago E, Jaaniste J, Einasto J and Andernach H 1997 The supercluster-void network. I. The supercluster catalogue and large-scale distribution *Astron. Astrophys. Suppl. Ser* **123** 119–33

Faerber T and López-Corredoira M 2020 A chi-squared analysis of the measurements of two cosmological parameters over time *Universe* **6** 114

Farley F J M 2010 Does gravity operate between galaxies? observational evidence re-examined *Proc. R. Soc.* A **466** 3089–96

Finlay-Freundlich E 1954 Red shifts in the spectra of celestial bodies *Proc. Phys. Soc.* A **67** 192–3

Fischer E 2008 Energy loss by gravitational viscosity (arXiv:0805.1638)

Fischer E 2010 Homogeneous cosmological solutions of the Einstein equation *Astrophys. Space Sci.* **325** 69–74

Francis M J, Barnes L A, James J B and Lewis G F 2007 Expanding space: the root of all evil? *Publ. Astron. Soc. Aust.* **24** 95–102

Freedman W L and Madore B F 2010 The Hubble constant *Annu. Rev. Astron. Astrophys.* **48** 673–710

Fulton C C and Arp H C 2012 The 2df redshift survey. i. physical association and periodicity in quasar families *Astrophys. J.* **754** 134

Fulton C C, Arp H C and Hartnett J G 2018 Physical association and periodicity in quasar families with SDSS and 2MRS *Astrophys. Space Sci.* **363** 134

Gallo C F 1975 A new red shift mechanism with possible applications to astrophysical problems such as quasars *Int. J. Theor. Phys.* **13** 417–8

Gallo C F 2003 Redshifts of cosmological neutrinos as definitive experimental test of Doppler versus non-Doppler redshifts *IEEE Trans. Plasma Sci.* **31** 1230–1

Garaimov V I 2003 Time and entropy *The Emergence of Cosmic Structure (AIP Conf. Proc. 666)* ed S S Holt and C S Reynolds (Melville, NY: AIP) pp 361–4

Gaztañaga E 2003 Correlation between galaxies and quasi-stellar objects in the Sloan Digital Sky Survey: a signal from gravitational lensing magnification? *Astrophys. J.* **589** 82–99

Ghosh A 1997 Velocity dependent inertial induction: A possible mechanism for cosmological red shift in a quasi static infinite universe *J. Astrophys. Astron.* **18** 449–54

Goldhaber G, Groom D E and Kim A *et al* 2001 Timescale stretch parameterization of type Ia supernova b-band light curves *Astrophys. J.* **558** 359–68

Guthrie B and Napier W M 1996 Redshift periodicity in the local supercluster *Astron. Astrophys.* **310** 353–70

Hansen P M 2006 Redshift components of apparent quasar-galaxy associations: a parametric model *Apeiron* **13** 17–33

Harrison E 1981 *Cosmology: The Science of the Universe* (Cambridge: Cambridge University Press)

Harrison E R 1995 Mining energy in an expanding universe *Astrophys. J.* **446** 63–6

Hartnett J G 2009 Unknown selection effect simulates redshift periodicity in quasar number counts from Sloan Digital Sky Survey *Astrophys. Space Sci.* **324** 13–6

Hartnett J G and Hirano K 2008 Galaxy redshift abundance periodicity from Fourier analysis of number counts n(z) using SDSS and 2DF GRS galaxy surveys *Astrophys. Space Sci.* **328** 13–24

Harutyunian H A and Nikogossian E H 2000 Quasars in regions of rich clusters of galaxies *Astrophysics* **43** 391–402

Hawkins E, Maddox S and Merrifield M 2002 No periodicities in 2df redshift survey data *Mon. Not. R. Astron. Soc.* **336** L13–6

Hawkins M R S 2010 On time dilation in quasar light curves *Mon. Not. R. Astron. Soc.* **405** 1940–6

Hirano K and Komiya Z 2010 Observational tests for oscillating expansion rate of the universe *Phys. Rev.* D **82** 103513

Holanda R F L, Goncalves R S and Alcaniz J S 2012 A test for cosmic distance duality *J. Cosmol. Astropart. Phys.* **2012** 022

Holushko H 2012 Tired light and type Ia supernovae observations (viXra:1203.0062)

Hönig S F 2014 Dust reverberation mapping in the era of big optical surveys and its cosmological application *Astrophys. J.* **784** L4

Hoyle F 1959 The relation of radio astronomy to cosmology *Radio Astronomy (IAU Symp. 9)* ed R N Bracewell (Stanford: Stanford University Press) pp 529–33

Hoyle F and Narlikar J V 1964 A new theory of gravitation *Proc. R. Soc. London* A **282** 191–207

Hubble E P 1929 A relation between distance and radial velocity among extra-galactic nebulae *Proc. Natl Acad. Sci. USA* **15** 168–73

Hubble E P 1936 Effects of red shifts on the distribution of nebulae *Proc. Natl Acad. Sci. USA* **22** 621–7

Hubble E P 1947 The 200-inch telescope and some problems it may solve *Publ. Astron. Soc. Pac.* **59** 153–67

Hubble E P and Tolman R C 1935 Two methods of investigating the nature of the nebular redshift *Astrophys. J.* **82** 302–37

Hunter J 2010 Redshifts and scale factors: a new cosmological model *J. Cosmol* **6** 1485–97

Ivanov M A 2004 Another origin of cosmological redshifts (arXiv:astro-ph/405083)

Ivanov M A 2006 Low-energy quantum gravity leads to another picture of the universe *1st Crisis in Cosmology Conf. (AIP Conf. Ser. 822(1))* ed E J Lerner and J B Almeida (Melville, NY: AIP) pp 187–99

Joos C and Lutz J 2005 Quantum redshift *Paper presented at the Crisis in Cosmology Conf.-I, Monaco, Portugal, 23–25 June 2005*

Kapahi V K 1987 The angular size-redshift relation as a cosmological tool *Observational Cosmology (IAU Symp. 124)* ed A Hewitt, G Burbidge and L-Z Fang (Dordrecht: Reidel) pp 251–65

Karovska M and Elvis M 2003 Distances on cosmological scales with VLTI *Astrophys. Space Sci.* **286** 261–6

Kirkman D and Tytler T 2008 The transverse proximity effect in the $z \sim 2$ lyman α forest suggests quasi-stellar object episodic lifetimes of ~1 MYR *Mon. Not. R. Astron. Soc.* **391** 1457–71

Kocevski D and Petrosian V 2013 On the lack of time dilation signatures in gamma-ray burst light curves *Astrophys. J.* **765** 116

Komberg B V and Pilipenko S V 2008 A possible explanation for the Arp-Burbidge paradox *Astron. Rep.* **52** 517–25

Krasinsky G A and Brumberg V A 2004 Secular increase of astronomical unit from analysis of the major planet motions, and its interpretation *Celest. Mech. Dyn. Astron.* **90** 267–88

Krasnov K and Shtanov Y 2008 Non-metric gravity. II. Spherically symmetric solution, missing mass and redshifts of quasars *Class. Quantum Grav.* **25** 025002

Krelowski J, Galazutdinov G and Gnacinski P 2012 CN rotational excitation *Astron. Nachr.* **333** 627–33

Kyutoku K and Seto N 2017 Gravitational-wave cosmography with LISA and the Hubble tension *Phys. Rev.* D **95** 083525

Laio A, Rizzi G and Tartaglia A 1997 Quantum theory of frequency shifts of an electromagnetic wave interacting with a plasma *Phys. Rev.* E **55** 7457–61

LaViolette P A 1986 Is the universe really expanding? *Astrophys. J.* **301** 544–53

LaViolette P A 2012 *Subquantum Kinetics: The Alchemy of Creation* 4th edn (Niskayuna, NY: Starlane Publ)

Leaning S P 2006 New analysis of observed high redshift supernovae data show that a majority of SN1a decay lightcurves can be shown to favourably compare with a non dilated restframe template *1st Crisis in Cosmology Conf. (AIP Conf. Ser. 822)* ed E J Lerner and J B Almeida (Melville, NY: AIP) pp 48–59

Lee J-W 2009 Are galaxies extending? *Phys. Lett.* B **681** 118–21

Lehto A 1990 Periodic time and the stationary properties of matter *Chin. J. Phys.* **28** 215–25

Lehto A 2009 On the Planck scale and properties of matter *Nonlinear Dyn.* **55** 279–98

Lemaître G 1927 Un univers homogène de masse constante et de rayon croissant rendant compte de la vitesse radiale des nébuleuses extra-galactiques *Ann. Soc. Sci. Brux.* **A47** 49–59

Lerner E J 2006 Evidence for a non-expanding universe: surface brightness data from HUDF *1st Crisis in Cosmology Conf. (AIP Conf. Ser. 822(1))* ed E J Lerner and J B Almeida (Melville, NY: AIP) pp 60–74

Lerner E J 2009 Tolman test from $z = 0.1$ to $z = 5.5$: preliminary results challenge the expanding universe model *Second Crisis in Cosmology Conf. (ASP Conf. Ser. 413)* ed F Potter (San Francisco, CA: Astronomical Society of the Pacific) pp 12–23

Lerner E J 2018 Observations contradict galaxy size and surface brightness predictions that are based on the expanding universe hypothesis *Mon. Not. R. Astron. Soc.* **477** 3185–96

Lerner E J, Falomo R and Scarpa R 2014 UV surface brightness of galaxies from the local universe to $z \sim 5$ *Int. J. Mod. Phys.* D **1450058**

Lieu R 2006 The outermost gravitationally bound orbit around a mass clump in an expanding universe: implication on rotation curves and dark matter halo sizes (arXiv:astro-ph/0605611)

Liske J, Grazian A and Vanzella E *et al* 2008 Cosmic dynamics in the era of extremely large telescopes *Mon. Not. R. Astron. Soc.* **386** 1192–218

López-Corredoira M 2010a Angular-size test on the expansion of the universe *Int. J. Mod. Phys.* D **19** 245–91

López-Corredoira M 2010b Apparent discordant redshift QSO-galaxy associations *Evolution of Cosmic Objects through their Physical Activity* ed H A Harutyunian, A M Mickaelian and Y Terzian (Yerevan: Gitutyun Publ. House of NAS RA) pp 196–205

López-Corredoira M 2011 Pending problems in QSOS *Int. J. Astron. Astrophys* **1** 73–82

López-Corredoira M 2014 Alcock-paczyński cosmological test *Astrophys. J.* **96** 15 id 96

López-Corredoira M and Gutiérrez C M 2006 Research on candidates for non-cosmological redshifts *1st Crisis in Cosmology Conf. (AIP Conf. Ser. 822)* ed E J Lerner and J B Almeida (Melville, NY: AIP) pp 75–92

López-Corredoira M, Gutiérrez C M and Génova R 2017 Analysis of the amplitude of the Sunyaev-Zel'dovich effect out to redshift $z = 0.8$ *Astrophys. J.* **840** 62

López-Corredoira M and Vazdekis A 2018 Impact of young stellar components on quiescent galaxies: deconstructing cosmic chronometers *Astron. Astrophys.* **614** 11

Lubin L M and Sandage A 2001 The Tolman surface brightness test for the reality of the expansion. IV. A measurement of the Tolman signal and the luminosity evolution of early-type galaxies *Astron. J.* **122** 1084–103

Lunsford D R 2004 Gravitation and electrodynamics over SO(3,3) *Int. J. Theor. Phys.* **43** 161–77

Luzzi G, Génova-Santos R T, Martins C J A P, De Petris M and Lamagna L 2015 Constraining the evolution of the CMB temperature with SZ measurements from Planck data *J. Cosmol. Astropart. Phys.* **9** 011

MacLeod C L and Hogan C J 2008 Precision of Hubble constant derived using black hole binary absolute distances and statistical redshift information *Phys. Rev.* D **77** 043512

MacMillan W D 1932 Velocities of the spiral nebulae *Nature* **129** 93

Magain P, Letawe G, Courbin F, Jablonka P, Jahnke K, Wisotzki L and Meylan G 2005 Discovery of a bright quasar without a massive host galaxy *Nature* **473** 381–4

Mal A, Palit S, Bhattacharya U and Roy S 2020 Periodicity of quasar and galaxy redshift *Astron. Astrophys.* **643** A160

Mamas D L 2010 An explanation for the cosmological redshift *Phys. Essays* **23** 326–9

Marmet L 2009 Optical forces as a redshift mechanism: the "spectral transfer redshift" *2nd Crisis in Cosmology Conf., CCC-2 (ASP Conf. Ser. 413)* ed F Potter (San Francisco, CA: Astronomical Society of the Pacific) pp 268–76

Marmet P 1988 A new non-Doppler redshift *Phys. Essays* **1** 24–32

Marmet P 1989 Red shift of spectral lines in the Sun's chromosphere *IEEE Trans. Plasma Sci.* **17** 238–44

Marosi L A 2013 Hubble diagram test of expanding and static cosmological models: the case for a slowly expanding flat universe *Adv. Astron.* **917104**

Marosi L A 2019 Extended Hubble diagram on the basis of gamma ray bursts including the high redshift range of $z = 0.033\ 1 - 8.1$ *Int. J. Astron. Astrophys.* **90158** 11

Matravers D R, Ellis G F R and Stoeger W R 1995 Complementary approaches to cosmology–relating theory and observations *Q. J. R. Astron. Soc.* **36** 29–45

Maxwell J C 1954 *A Treatise on Electricity and Magnetism* vol II (New York: Dover)

Melia F 2019 Cosmological test using the Hubble diagram of high-z quasars *Mon. Not. R. Astron. Soc.* **489** 517–23

Melia F and López-Corredoira M 2017 Alcock-paczyński cosmological test with model-independent BAO data *Int. J. Mod. Phys.* D **26** 1750055

Mérat P 1974a Possible interpretation of an anomalous redshift observed on the 2292 MHz line emitted by pioneer-6 in the close vicinity of the solar limb *Astron. Astrophys.* **30** 167–74

Mérat P 1974b Observed deflation of light by the sun as a function of solar distance *Astron. Astrophys.* **32** 471–5

Michelini M 2012 Beyond the special and general relativity theories. the micro-quanta paradigm *Appl. Phys. Res.* **4** 210–25

Michelini M 2016 Discussion on fundamental problems of physics hidden in cosmology *Appl. Phys. Res.* **8** 19–43

Molaro P, Levshakov S A, Dessauges-Zavadsky M and D'Odorico S 2012 The cosmic microwave background radiation temperature at $z_{abs} = 3.025$ toward QSO 0347-3819 *Astron. Astrophys.* **381** L64–7

Monti R 1988 The electric conductivity of background space *Problems in Quantum Physics, Gdansky 87—Recent and Future Experiments and Interpretations* ed L Kostro, A Posiewnik, J Pykacz and M Zukowski (Singapore: World Scientific) p 640

Moresco M, Cimatti A and Jiménez R *et al* 2012 Improved constraints on the expansion rate of the universe up to $z \sim 1.1$ from the spectroscopic evolution of cosmic chronometers *J. Cosmol. Astropart. Phys.* **2012** 006

Moresco M, Jiménez R, Verde L, Pozzetti L, Cimatti A and Citro A 2018 Setting the stage for cosmic chronometers. I. Assessing the impact of young stellar populations on Hubble parameter measurements *Astrophys. J.* **868** 84

Moresco M, Pozzetti L and Cimatti A *et al* 2016 A 6% measurement of the Hubble parameter at $z \sim 0.45$: direct evidence of the epoch of cosmic re-acceleration *J. Cosmol. Astropart. Phys.* **2016** 14

Moret-Bailly J 2006 The parametric light-matter interactions in astrophysics *1st Crisis in Cosmology Conf. (AIP Conf. Ser. 822)* ed E J Lerner and J B Almeida (Melville, NY: AIP) pp 226–38

Nabokov N V and Baryshev Y 2008 Classical cosmological tests for galaxies of the Hubble ultra deep field *Astrophys. Bull.* **63** 244–58

Napier W M 2006 Statistics of redshift periodicities *Current Issues in Cosmology* ed J-C Pecker and J V Narlikar (Cambridge: Cambridge University Press) pp 207–16

Narlikar J V 1989 Noncosmological redshifts *Space Sci. Rev.* **50** 523–614

Nesvizhevsky V V, Börner H G and Petukhov A K *et al* 2002 Quantum states of neutrons in the earth's gravitational field *Nature* **415** 297–9

Nguyen H, Koenig M, Benredjem D, Caby M and Coulaud G 1986 Atomic structure and polarization line shift in dense and hot plasmas *Phys. Rev.* A **33** 1279–90

Nilsson K, Valtonen M J, Kotilainen J and Jaakkola T 1993 On the redshift-apparent size diagram of double radio sources *Astrophys.* J **413** 453–76

Nobili S and Goobar A 2008 The colour-lightcurve shape relation of type Ia supernovae and the reddening law *Astron. Astrophys.* **487** 19–31

Nollenberg J G and Williams L R 2005 Galaxy-quasar correlations between APM galaxies and Hamburg-ESO QSOS *Astrophys. J.* **634** 793–805

Noterdaeme P, Petitjean P, Srianand R, Ledoux C and López S 2011 The evolution of the cosmic microwave background temperature. Measurements of t_{CMB} at high redshift from carbon monoxide citation *Astron. Astrophys.* **526** L7

Nowakowski M and Ashtekar A 2001 The consistent Newtonian limit of Einstein's gravity with a cosmological constant *Int. J. Mod. Phys.* **D10** 649–61

Pashchenko I N and Vitrishchak V M 2011 The use of ultra-compact radio sources for the angular size-redshift cosmological test *Astron. Rep.* **55** 293–301

Paturel G 2008 Some difficulties for measuring and interpreting the expansion of the universe *Practical Cosmology* 2 edn (Petersburg: TIN) pp 178–84

Peñarrubia J, Ma Y-Z, Walker M G and McConnachie A 2014 A dynamical model of the local cosmic expansion *Mon. Not. R. Astron. Soc.* **443** 2204–22

Peacock J A 2008 A diatribe on expanding space (arXiv:0 809.457 3)

Peebles P J E 1993 *Principles of Physical Cosmology* (Princeton, NJ: Princeton University Press)

Petrov R G, Millour F, Lagarde S, Vannier M, Rakshit S, Marconi A and Weigelt G 2012 Optical and infrared interferometry iii *SPIE* **8445** 84450W

Podsiadlowski P, Mazzali P, Lesaffre P, Han Z and Förster F 2008 The nuclear diversity of type Ia supernova explosions *New Astron. Rev.* **52** 381–5

Popowski P and Weinzierl W 2004 A test for the origin of quasar redshifts *Mon. Not. R. Astron. Soc.* **348** 235–40

Reboul H J 1981 Untrivial redshifts–a bibliographical catalogue *Astron. Astrophys. Supp. Ser.* **45** 129–44

Repin S V, Komberg B V and Lukash V N 2012 Absence of a periodic component in the quasar z distribution *Astron. Rep.* **56** 702–9

Retinó A and Spallicci A D A 2016 Solar wind test of the de Broglie-Proca massive photon with cluster multi-spacecraft data *Astropart. Phys.* **82** 49–55

Riess A G, Casertano S, Yuan W, Macri L M and Scolnic D 2019 Large Magellanic Cloud cepheid standards provide a 1% foundation for the determination of the Hubble constant and stronger evidence for physics beyond ΛCDM *Astrophys. J.* **876** 85

Risaliti G and Lusso E 2015 A Hubble diagram for quasars *Astrophys. J.* **815** 33

Risaliti G and Lusso E 2019 Cosmological constraints from the Hubble diagram of quasars at high redshift *Nature Astron.* **3** 277

Roscoe D 2006 Maxwell's equations: new light on old problems *Apeiron* **13** 206–39

Roy S, Kafatos M and Datta S 2000 Shift of spectral lines due to dynamic multiple scattering and screening effect: implications for discordant redshifts *Astron. Astrophys.* **353** 1134–8

Russell D M, Ellison S L and Benn C R 2006 An excess of damped lyman α galaxies near quasi-stellar objects *Mon. Not. R. Astron. Soc.* **367** 412–22

Ryabinkov A I and Kaminker A D 2021 Traces of anisotropic quasi-regular structure in the SDSS data *Universe* **7** 289

Ryutov D D 2007 Using plasma physics to weigh the photon *Plasma Phys. Control. Fusion* **49** B429–38

Sandage A 1962 The change of redshift and apparent luminosity of galaxies due to the deceleration of selected expanding universes *Astrophys. J.* **136** 319–33

Sandage A, Tammann G A and Reindl B 2009 New period-luminosity and period-color relations of classical cepheids. III. Cepheids in SMC *Astron. Astrophys.* **493** 471–9

Sandrinelli A, Falomo R, Treves A, Scarpa R and Uslenghi M 2018 Overdensity of galaxies in the environment of quasar pairs *Mon. Not. R. Astron. Soc.* **474** 4925–36

Sato M, Reid M J, Menten K M and Carilli C L 2013 On measuring the cosmic microwave background temperature at redshift 0.89 *Astrophys. J.* **132** 8

Schwarz D J and Weinhorst B 2007 An)isotropy of the Hubble diagram: comparing hemispheres *Astron. Astrophys.* **474** 717–29

Scolnic D M, Jones D O and Rest A *et al* 2018 The complete light-curve sample of spectroscopically confirmed SNe Ia from pan-starrs1 and cosmological constraints from the combined pantheon sample *Astrophys. J.* **859** 101

Scranton R, Ménard B and Richards G T *et al* 2005 Detection of cosmic magnification with the Sloan Digital Sky Survey *Astrophys. J.* **633** 589–602

Segal I E 1976 *Mathematical Cosmology and Extragalactic Astronomy* (New York: Academic)

Segal I E and Zhou Z 1995 Maxwell's equations in the Einstein universe and chronometric cosmology *Astrophys. J. Supp. Ser.* **100** 307–24

Sereno M and Jetzer P 2007 Evolution of gravitational orbits in the expanding universe *Phys. Rev.* D **75** 064031

Shibuya R, Ouchi M and Harikane Y 2015 Morphologies of ∼190 000 galaxies at z = 0-10 revealed with HST legacy data. 1. Size evolution *Astrophys. J.* **219** 15

Shtyrkov E I 1993 A new interpretation of cosmological redshifts: variable light velocity *Progress in New Cosmologies* ed H C Arp, C R Keys and K Rudnicki (Boston, MA: Springer) pp 327–32

Sorrell W H 2009 Misconceptions about the Hubble recession law *Astrophys. Space Sci.* **323** 205–11

Spallicci A D A M, Helayël-Neto J A, López-Corredoira M and Capozziello S 2021 Cosmology and the massive photon frequency shift in the standard-model extension *Eur. Phys.* C **81** 4

Steinbring E 2007 Are high-redshift quasars blurry? *Astrophys. J.* **655** 714–7

Stockton A 1978 The nature of QSO redshifts *Astrophys. J.* **223** 747–57

Su C-C 2006 A proposed mechanism for the intrinsic redshift and its preferred values purportedly found in quasars based on the local-ether theory (arXiv:physics/0608164)

Sulentic J-W and Arp H-C 1987 The galaxy-quasar connection–NGC 4319 and Markarian 205. I–Direct imagery. II–Spectroscopy *Astrophys. J.* **319** 687–708

Taganov I N 2008 *Quantum Cosmology: Deceleration of Time* (Petersburg: TIN)

Tammann G A, Sandage A and Reindl B 2003 New period-luminosity and period-color relations for classical cepheids: I. Cepheids in the galaxy *Astron. Astrophys.* **404** 423–48

Tang S M and Zhang S N 2005a Critical examinations of QSO redshift periodicities and associations with galaxies in Sloan Digital Sky Survey data *Astrophys. J.* **633** 41–51

Tang S M and Zhang S N 2005b QSO-galaxy association and gravitational lensing Chin *J. Astron. Astrophys* **5** 147–52

Tang S and Zhang S N 2010 Evidence against non-cosmological redshifts of QSOS in SDSS data *Redshifts in Spectral Lines of Quasi Stellar Objects* ed Quasi (Kerala: Research Signpost) pp 125–36

Tifft W G 1976 Discrete states of redshift and galaxy dynamics. I–Internal motions in single galaxies *Astrophys. J.* **206** 38–56

Tifft W G 1977 Discrete states of redshift and galaxy dynamics. II–Systems of galaxies *Astrophys. J.* **211** 31–46

Tifft W G 1980 Periodicity in the redshift intervals for double galaxies *Astrophys. J.* **236** 70–4

Totani T, Yoshii Y, Maihara T, Iwamuro F and Motohara K 2001 Near-infrared faint galaxies in the subaru deep field: comparing the theory with observations for galaxy counts, colors, and size distributions to $k \sim 24.5$ *Astrophys. J.* **559** 592–605

Urbanowski K 2008 On a possible quantum contribution to the red shift *Practical Cosmology* vol 1 ed Y Baryshev, I N Taganov and P Teerikorpi (St. Petersburg: TIN) pp 117–22

Valtonen M, Nilsson K, Kotilainen J and Jaakkola T 1991 Double radio sources as standard rods of testing cosmological models *Proc. of the 25th Annual Conf. of the Finnish Phys. Soc. Oulu* Univ. Oulu

Van Flandern T 1993 *Dark Matter, Missing Planets and New Comets* (Berkeley, CA: North Atlantic Books)

Varshni Y P 1979 The physics of quasars *Phys. Can.* **35** 11–7

Vigier J P 1988 Alternative interpretation of the cosmological redshift in terms of vacuum gravitational drag *New Ideas in Astronomy* ed F Bertola, B Madore and J Sulentic (Cambridge: Cambridge University Press) pp 257–74

Visbal E and Croft R A C 2008 On the search for quasar light echoes *Astrophys. J.* **674** 660–7

von Nernst W 1921 *The Structure of the Universe in Light of our Research* (Berlin: Springer) von Nernst W 1937 Weitere prüfung der annahme lines stationären zustandes im weltall *Z. Phys.* **106** 633–61

Wei H 2010 Observational constraints on cosmological models with the updated long gamma-ray bursts *J. Cosmol. Astropart. Phys.* **8** 20

Weidner H 2014 The size and energy loss of a wave packet (arXiv:1408.0139)

Wolf E 1986 Invariance of the spectrum of light on propagation *Phys. Rev. Lett.* **56** 1370–2

Zackrisson E 2005 On quasar host galaxies as tests of non-cosmological redshifts *Mon. Not. R. Astron. Soc.* **359** 1193–200

Zhu X F and Chu Y Q 1995 The association between quasars and the galaxies of the Virgo cluster *Astron. Astrophys.* **297** 300–4

Zwicky F 1929 On the red shift of spectral lines through interstellar space *Proc. Natl Acad. Sci. USA* **15** 773–9

Zwicky F 1957 *Morphological Astronomy* (Berlin: Springer)

IOP Publishing

Fundamental Ideas in Cosmology
Scientific, philosophical and sociological critical perspectives
Martín López-Corredoira

Chapter 5

The cosmic microwave background radiation

Cosmic Microwave Background Radiation (CMBR) has been interpreted as relict light released at an early stage of the Universe at $z \approx 1100$. Its blackbody spectrum of 2.7 K reveals a very small dependence on sky position, of the order of $\Delta T/T \sim 10^{-5}$ apart from the dipole due to the motion of the Earth. Measurements of the anisotropies carried out by several teams of researchers over the last three decades have been claimed to provide accurate information on cosmology. But doubt is cast upon that precision cosmology owing to the difficulties in making maps totally free from foreground contamination. Moreover, many alternative explanations of the origin of the CMBR are found in the technical literature, and certain observed anomalies in CMBR maps, such as the lack of low multipole signal, alignment of quadrupole and octopole and others, are at odds with the standard model, which opens the door on possible fundamental errors in the standard cosmological description of the CMBR.

The Cosmic Microwave Background Radiation (CMBR) has been interpreted as relict light emitted at $z \approx 1100$ from the decoupling of matter and radiation. Its blackbody spectrum of 2.7 K reveals a very small dependence on sky position, of the order of $\Delta T/T \sim 10^{-5}$, excluding the dipole due to the motion of the Earth. According to the standard interpretation, measurements of the anisotropies carried out by several teams of researchers over the last three decades provide information on the structural formation of the Universe, inflation in its early stages, quantum gravity, topological defects if any (strings, etc), dark matter type and abundance, dark energy, the geometry and dynamics of the Universe, the thermal history of the Universe at the recombination epoch, etc. The CMBR has become a source of accurate values for the parameters of this standard model of cosmology. Quantitative science is pursued for which the fitting of the power spectrum of the distribution of CMBR anisotropies and a some other data give us numerical values for the parameters in the equations governing the entire past, present, and future of our Universe.

We may say that CMBR research is today one of the most important pillars sustaining the edifice of the standard model of cosmology, provided that the

doi:10.1088/978-0-7503-3775-5ch5

foundations of cosmology based on the present-day interpretation of CMBR are well established. But how sure can we be of the sturdiness of this pillar and that our interpretation of its nature is the right one?

5.1 Early predictions and observations

George Gamow (1904–1968), Ralph Alpher (1921–2007), and Robert Herman (1914–1997) predicted an early stage of the Universe in the Big Bang model that would produce a relic radiation from a fireball that would produce a uniform background in all directions (Alpher and Herman 1949, Gamow 1953). However, Gamow and his coworkers were of the opinion that its detection was completely unfeasible (Novikov 2001). The first published recognition of the relic radiation as a detectable microwave phenomenon was given by Andrei Doroshkevich (1937–) and Igor Dmitryevich Novikov (1935–) (Doroshkevich and Novikov 1964). None of the predictions of the background temperature, which ranged between 5 K and 50 K based on the Big Bang, matched observations (Van Flandern 1993), the worst being Gamow's upward-revised estimate of 50 K made in 1961. In 1965, the year of the official discovery, Robert H. Dicke (1916–1997) and collaborators[1] calculated a temperature of 30 K for the observed amount of helium production (Dicke *et al* 1965). As the energy is proportional to T^4, the observed energy of CMBR associated to $T = 2.7$ K is several thousand times less than the predicted energy, but it was correctly predicted that it has a perfect blackbody spectrum, as observed later with great accuracy (Mather *et al* 1994).

Another discrepancy between observations and first expectations is that the number of CMBR photons is much higher (10^9 times) than the number of cosmic baryons, thus indicating that cosmic evolution violates baryon number conservation; a heavy baryon–antibaryon annihilation? Curiously, the CMBR photon density implies that the mean distance of photons is 0.2 cm, which, to the surprise of some, is just about identical to the maximum wavelength of the CMBR blackbody emission (Fahr and Zönnchen 2009).

Although the official discovery is attributed to Arnos Allias Penzias (1933–) and Robert Woodrow Wilson (1936–), who found a temperature of 3.5 K in the radiation (Penzias and Wilson 1965) and won the Nobel Prize in Physics in 1978 for their discovery, the radiation had indeed been previously discovered, although not interpreted in terms of cosmological radiation (see section 1.3). In any case, the radiation is there and the question is the origin of this radiation not the discoverer. Is a high-energy primordial Universe the only possible scenario?

5.2 Alternative explanations for the temperature of 2.7 K

There were predictions of CMBR temperature with origins different from the Big Bang (Assis and Neves 1995, Meyers 2003). Charles-Édouard Guillaume (1861–1938; Nobel laureate in Physics 1920) predicted that the radiation of stars alone

[1] Among these collaborators, was a young James Peebles (1935–), who would receive a Nobel laureate in Physics in 2019 for several contributions in cosmology.

would maintain a background temperature of 6.1 K (Guillaume 1896). The expression 'the temperature of space' is the title of chapter 13 of *Internal Constitution of the Stars* by Arthur S Eddington (1882–1944) (Eddington 1926), who calculated the minimum temperature any body in space would cool to, given that it is immersed in the radiation of distant starlight. With no adjustable parameters, he derived a temperature of 3 K, essentially the same as that observed for the CMBR. Other early predictions, given by Erich Rudolf Alexander Regener (1881–1955) and Walther Nernst (1864–1941), gave a temperature of 2.8 K for a blackbody that absorbed the energy of the cosmic rays arriving at Earth (Regener 1933, von Nernst 1937).

It was countered that Eddington's argument for the 'temperature of space' applied at most to our Galaxy. But Eddington's reasoning applies also to the temperature of intergalactic space, for which a minimum is set by the radiation of galaxy and quasi stellar object (QSO) light. The original calculations by Erwin Finlay-Freundlich (1885–1964) showed that this limit probably fell in the range 1–6 K (Finlay-Freundlich 1954). That was before QSOs were discovered and before we knew the modern space density of galaxies. In the same way, the existence of a microwave background in a tired light scenario was also deduced by the same author and Max Born (1882–1970) (Finlay-Freundlich 1954, Born 1954). But in a tired light model there are other caveats, as we shall see in the next section.

In the fifties, Hermann Bondi (1919–2005), Thomas Gold (1920–2004), Fred Hoyle (1915–2001), and Geoffrey R Burbidge (1925–2010) pointed out that, if the observed abundance of He comes from hydrogen fusion in stars (from a hypothetical Population III), there must have been a phase in the history of the Universe when the radiation density was much higher than the energy density of starlight today (Bondi *et al* 1955, Burbidge 1958). If the average density of the visible matter in the Universe is $\rho \sim 3 \times 10^{-31}$ g cm^{-3} and the observed He/H ratio by mass is 0.244 (see chapter 6), then the energy that must have been released in producing He is 4.39×10^{-13} erg cm^{-3}. If this energy is thermalized, the blackbody temperature would turn out to be $T = 2.76$ K, very close to the temperature for the CMBR observed several years after the prediction. Hence, there is a likely explanation of the energy of the microwave radiation in terms of straightforward astrophysics involving hydrogen fusion in stars (Burbidge 1999). Hoyle *et al* (1968) also pointed out the suspect coincidence between the microwave background temperature and that of hydrogen condensation on grains. They postulated that galaxy and star formation proceed very readily when hydrogen is condensed onto grains but not when it is gaseous, and that the universal microwave background temperature is that associated with galaxy formation. Within the alternative cosmology of a Cold Big Bang, Aguirre (2000) also maintains the non-primordial CMBR formed by Population III stars at redshifts greater than 100.

5.3 Alternative origin of the CMBR

The production of radiation at a temperature of 2.7 K is not difficult to explain, as we have seen. The hardest problem, however, is explaining its blackbody energy distribution, associated with a mechanism of thermalisation. Indeed, the thermal spectrum of the CMBR was considered to be the most direct evidence that redshift is a result of expansion (Fixsen *et al* 1996). For instance, in a tired light model in a static Universe the photons suffer a redshift that is proportional to the distance travelled, but in the absence of absorption or emission the photon number density remains constant. Furthermore, the Universe cannot have an optical depth large enough to preserve a thermal background spectrum in a tired light model (Peebles 1998) because we could not observe radio galaxies at $z \sim 3$ with the necessary optical depth. There are, however, a number of ideas that attempt to surmount the difficulties of the tired light model (see review in Ćirković 2018).

Dust emission differs substantially from that of a pure blackbody. Moreover, dust grains cannot be the source of the blackbody microwave radiation because there are too few of them to produce the opacity needed to produce a blackbody spectrum. A solution for the blackbody emission shape might be the unusual properties of the dust particles: carbon needles or whiskers, multiple explosions or big bangs, the energy release of massive stars during the formation of galaxies, and so forth. Quasi-Steady State Cosmology (Hoyle *et al* 1993, 1994) argues that there is a distribution of whiskers of approximate length 1 mm and diameter 10^{-6} cm with average density $\sim 10^{-35}$ g cm^{-3} providing optical depth $\tau \sim 7$ up to redshift ~ 4. However, the presence of a huge dust density to make the Universe opaque is forbidden by the observed transparency up to $z \sim 4$ or 5. A solution might be an infinite Universe: an opaque Universe is required only in a finite space, an infinite Universe can achieve thermodynamic equilibrium even if it is transparent out to very large distances.

A previous similar proposal by Rana (1981) suggested a thermalisation with whiskers of normal starlight in the range $5 \leqslant z \leqslant 10$, and yet more recent proposals by Vavryčuk (2018), Melia (2020) suggest that the CMBR is produced by dust rethermalization at $z < 20$–30, presumably supplied to the interstellar medium by Population III stars existing in that epoch. Somewhat similar to the proposal of whiskers is the proposal of the thermalisation of 'cosmoids', cosmic meteoroids that are also observed in the Solar System (Soberman and Dubin 2001), and the proposal of emission by millimetre black holes (Alfonso-Faus 2010).

Another possible explanation is the existence of an aether (Clube 1980, Sorrell 2008), i.e. a material vacuum, whose emission gives the microwave background emission itself. This aether would be an incompressible fluid according to Lorentz's theory and its existence is in opposition to Einstein's special relativity. Lorentz's invariant laws of mechanics (Lorentz 1904) in a flat space-time reproduce the standard observational tests of general relativity provided that rest mass and light speed are not constant, so the final choice between Einstein's and Lorentz's theories cannot yet be regarded as settled according to Clube (1980).

In first versions of Plasma Cosmology (Lerner 1991), the CMBR was produced from the slow annihilation of matter and antimatter, which are contained symmetrically in the Universe. Later versions of the model speculated about other possible origins. The CMBR may be produced by the scattering of photons by electrons in a plasma of temperature 2 million K (Brynjolfsson 2004). Lerner (1988, 1995) proposes that electrons in intergalactic magnetic fields emit and absorb microwave radiation. There is no relation between the direction in which the radiation was moving when it was absorbed and its direction on re-emission, so the microwaves would be scattered. After a few scatterings, the radiation would be smoothed out. Magnetic fields much stronger than the average field between galaxies would be needed; perhaps the jets emitted from galactic nuclei would provide them. The background radiation would be distorted by this intergalactic absorption against isotropy observations, so the radiation must instead come from the intergalactic medium itself in equilibrium. This prediction agrees with the fact that the number of radio sources increases much more slowly than the number of optical sources with distance, presumably owing to this absorption of radio waves in the intergalactic medium. Lerner claims that the intergalactic absorption occurs in narrow filaments, with tiny holes scattered about randomly so that distant compact radio sources such as QSOs and radio galaxies can be seen through the holes. Observational evidence has also been presented that the intergalactic medium is absorbing radio and microwaves because farther radio sources with a given constant infrared emission are fainter in the radio (Lerner 1990, 1993). However, there are sources with strong radio emission and faint infrared flux at high redshift (Orenstein *et al* 2019), and on average there is a constant FIR/radio emission up to $z = 1.5$ (Garrett 2002, Mao *et al* 2011), which casts doubt upon Lerner's argument.

Related to electrons, there are other discussions of alternative explanations of the CMBR in the literature too. For instance, the radiation of orbital electrons in atoms (Shpenkov and Kreidik 2002). It is proposed that the hydrogen atoms emit energy while in equilibrium owing to the orbiting of the electrons around protons (which goes against the principles of quantum mechanics), thereby generating the background spectrum of 2.73 K. The atoms emit and absorb this radiation creating a background field with a blackbody shape.

Krishan (2009) points out that, in addition to Thomson scattering, absorption due to electron–electron, electron–ion, and electron–atom collisions in a partially ionised cosmic plasma would also contribute to the optical depth of the CMBR. The absorption depth depends on the plasma temperature and the frequency of the CMBR. The absorption effects are prominent at the low frequency part of the CMBR spectrum. These effects, when included in the interpretation of the CMBR spectrum, may require a revised view of the ionisation of the Universe.

In 'Curvature Cosmology' (Crawford 2011) the CMBR comes from the curvature–redshift process acting on the high-energy electrons and ions in the cosmic plasma. The energy loss that gives way to the spectrum of photons of the CMBR occurs when an electron that has been excited by its passage through curved spacetime interacts with a photon or charged particle and loses its excitation energy.

In order to preserve its blackbody shape, the origin of CMBR must be non-redshifted, which seems problematic, or come from a very narrow redshift range, or even be totally local, which is even more problematic. Some proposals have placed the origin of the CMBR within our own Galaxy, in the local bubble within 100 pc (Pecker *et al* 2015), positing that the microwave background radiation stems in very large part from re-radiation of thermalized Galactic starlight. But the high level of isotropy, or the Sunyaev–Zel'dovich imprint of clusters of galaxies in the CMBR

(Planck Collaboration 2011a, López-Corredoira *et al* 2017), are strong arguments against this hypothesis.

Attempts to defend a local origin of the CMBR within 50 Mpc have also been made. Navia *et al* (2007) find that there is an isotropic distribution of cosmic rays with energy $>6 \times 10^{19}$ eV, so they should come from distances greater than 50 Mpc. Given that the cosmological interpretation of the CMBR does not allow these cosmic rays to travel distances $\gtrsim 50$ Mpc (Greisen 1966, Zapsepin and Kuzmin 1966), this would indicate that the CMBR originates at distances closer than 50 Mpc. However, Abraham *et al* (2007) claim that an isotropic distribution of these cosmic rays is rejected within >99% confidence level (CL) and that the most probable sources are nearby AGN. Kashti and Waxman (2008) also claim that the distribution of these high-energy cosmic rays is inconsistent with isotropy at ~98% CL, and consistent with a source distribution that traces large-scale structure, with preference to a source distribution that is biased with respect to galaxy distribution.

In my opinion, at present there is no satisfactory alternative scenario exempt from problems in explaining the microwave background radiation, and the standard scenario seems to be the best solution. In any case, one's mind should not be definitively closed against other options. The hot primordial Big Bang is only a hypothesis that should not be definitively considered as a solid theory just because of the microwave background radiation.

5.4 Microwave background radiation anisotropies

Another fact to consider is the existence of certain anisotropies in the CMBR. There are still some authors who doubt the existence of such anisotropies, pointing out that there are spurious temperature anisotropies that are comparable with the entire signal (Li *et al* 2009, Roukema 2010, Liu *et al* 2011), or testing the null hypothesis that the Time-Ordered-Data (TOD) were consistent with no anisotropies when hourly calibration parameters were allowed to vary, i.e. that sky maps with no anisotropies outside the Galactic band other than the dipole were a better fit to the uncalibrated TOD than those from the official analysis (Cover 2009). In my opinion, these analyses cannot be correct. The fact that the same anisotropies have been found by many different experiments and by different teams leaves no doubt in my mind that the anisotropies are real.

The first predictions of CMBR anisotropies were wrong. It was predicted $\Delta T/T$ to be one part in hundred or thousand (Sachs and Wolfe 1967); however, this value did not fit the observations, which gave values hundreds of times smaller, so non-baryonic dark matter was introduced ad hoc to solve the question (see section 3.3).

Another element important to take note of is the ad hoc introduction of inflation (see section 3.2). This idea was meant to solve the inconsistency of having photons come from regions that had not been causally connected before reaching the same temperature with very small anisotropies. Without inflation that would be a paradox.

CMBR analyses in the last two decades have concentrated on anisotropies on small angular scales, smaller than the angular resolution of several degrees achieved by COBE-DMR in the 1990s. The power spectrum C_ℓ of the anisotropies of the

BOOMERANG and MAXIMA-1 experiments showed a first peak at the Legendre multipole $\ell \approx 200$ (Bernardis *et al* 2000, Hanany *et al* 2000) corresponding to angular scales of $\approx 1.2°$, which was interpreted as a discovery of the previously predicted acoustic peaks in a flat Universe with $\Omega = \Omega_m + \Omega_\Lambda = 1$, the greatest contribution coming from a dark energy component of $\Omega_\Lambda = 0.7$ (Hu and Dodelson 2002 [and references therein])[2].

Indeed, there are two outcomes in the comparison of theoretical predictions and observations: a successful one, and a half-successful/half-failed one. The totally successful prediction was the qualitative shape of the power spectrum containing several peaks: For instance, Peebles and Yu (1970) had predicted a CMBR power spectrum with fluctuations, the acoustic peaks, from the hypothesis of primaeval adiabatic perturbations in an expanding Universe. However, the first peak was not at the expected position when first measured: already in the mid-1990s, with small sky coverage surveys, the position of the first peak at $\ell \approx 200$ did not fit the standard model of the epoch (with $\Omega = \Omega_m \approx 0.2$, without dark energy), so a larger Ω was needed. This was one of the elements, together with SNIa observations and the age problem of the Universe, that would encourage cosmologists to include the new ad hoc element of dark energy. This change of mentality in standard cosmology was produced between 1997 and 2000; then, in 2000, with the results of the new BOOMERANG and MAXIMA-1 experiments, cosmologists were proud to announce that the new observations were giving exactly the results they had expected. However, by just paying attention to the first analyses of the first peak in the mid-1990s we should attribute a half-success to the prediction: even though they failed to fit the observations with the preferred standard model of a curved open Universe at that time, the idea of a flat Universe was also presented as a possibility in the 1990s (this was indeed the possibility preferred by inflationary paradigms).

The position of the other peaks would also serve to constrain the cosmological models. The acoustic peaks on angular scales of 1 deg and 0.3 deg were predicted with the second peak nearly as high as the first one in $\ell(\ell + 1)C_\ell/(2\pi)$ (Bond and Efstathiou 1987, Jorgensen *et al* 1995). Later, as more data became available, its amplitude was reduced and the positions of the other peaks and their relative heights were also constrained from the model with a higher agreement with the data (Hu *et al* 2001).

More recently, the analysis of CMBR polarization anisotropies (Larson *et al* 2011) has also provided strong support for the standard cosmology. This indicates the way in which the CMBR is polarized owing to the scattering of free electrons. Photon diffusion into regions of different temperatures is possible only when the plasma becomes sufficiently optically thin; these diffused photons could then scatter only while there are still free electrons left. Since photons could not diffuse too far, polarization cannot vary much over very large angular scales. Nonetheless, the success of the predictions is not so extraordinary. Roughly speaking, the relationship between temperature–temperature and the polarization–temperature or the polarization–polarization expected since they are different ways of seeing the same light with different filters.

[2] More recent data (di Valentino *et al* 2021) have shown that adding a simple dark energy term and preserving the flatness of the Universe is excluded at >99% CL.

Even so, the agreement between different sets of data is not perfect: there is a moderate inconsistency (around 2.3 σ) between CMBR temperature data and polarization data sets (Lin and Ishak 2017).

Given this history of CMBR analyses, it is difficult to accept that our observations reproduce 'predictions'. Nonetheless, whichever comes first, theory or observation, the fact remains that we now have a cosmological model that is able to fit the CMBR power spectrum C_ℓ quite accurately. It is usually thought that this may not happen by chance, given the apparent complexity of the power spectrum shape, whereas the six-parameter cosmological model is relatively simple (Peiris 2005, Larson *et al* 2011) but it is not totally correct, as I explain in the following paragraphs.

First, there are more than six parameters, there are other parameters that introduce second-order changes. And there is also a degeneracy in the solutions with different values of H_0 and Ω_Λ: CMBR data, and the large-scale structure of galaxies could be reproduced without explicitly requesting the existence of dark energy (Blanchard *et al* 2003), i.e. with $\Lambda = 0$, this degeneracy being broken by adding cosmological information from other sources, for instance, from SNIa data. Also, in order to fit the temperature–polarization cross power spectrum and the polarization–polarization power spectrum (Larson *et al* 2011), one would need an extra parameter (optical depth), so at least seven free parameters are necessary.

But the most important question is raised by certain critics of the standard model; Disney (2007), for example, claim that there is little value in the fact that the standard cosmological model might fit some observational cosmological data if the number of free parameters in the model were comparable to the number of independent parameters characterising the observations. For instance, Ptolemaic geocentric astronomy may roughly fit the observations of the orbits of the planets but with too high a number of free parameters in the theory. The principle of Occam's razor tells us that a theory is better when the number of free parameters is low. Occam's razor can indeed be understood through a Bayesian analysis, in which it is tested how probable a model is with a given number of free parameters with respect to other models with a higher number of free parameters (Berger and Jefferys 1992, Jefferys and Berger 1992). Otherwise, if a theory has a high number of free parameters, it loses credibility because it is always possible to create a 'false' model to fit some data when the number of free parameters is comparable to the number of degrees of freedom in the data. Philosophers of science, when talking about cosmology, associate this approach with 'instrumentalism' (e.g., Soler Gil 2012).

Saying that the power spectrum contains hundreds of independent parameters for a given resolution is incorrect, because the different values of C_ℓ for each ℓ are not independent in the same sense that hundreds of observations of the position and velocity of a planet do not indicate hundreds of independent parameters. Indeed, the information on the orbit of planet is reduced to only six Keplerian parameters.

For the reasons just given, understanding how much information is in the power spectrum is important for key questions when discussing the fundamentals of cosmology. There are indeed two main points which should be addressed:

1. Are the oscillations atypical in a power spectrum? That is, should we consider the fact the power spectrum contains oscillations a successful prediction of the standard cosmological model that cannot be produced by any other means?

2. How much information is contained in the power spectrum? That is, how many free parameters in a function are necessary to fit the CMBR power spectrum? Should we consider fitting the power spectrum with a model of six free parameters as a validation of the standard cosmological model?

For the answer to the first question, we must bear in mind that the presence of peaks in the power spectrum is a rather normal characteristic expected from any fluid with clouds of overdensities that emit/absorb radiation or interact gravitationally with the photons, and with a finite range of sizes and distances for those clouds. Apart from the standard cosmological model, other scenarios may also follow these conditions. The interpretation of 'acoustic' peaks is just a particular case; peaks in the power spectrum may be generated in scenarios that have nothing to do with oscillations due to gravitational compression (López-Corredoira and Gabrielli 2013, López-Corredoira 2013).

The fit of the theoretical predictions to the observed CMBR power spectrum peaks is at present very good on small scale (<5 deg.), where the bumps and wiggles in the radiation power spectrum are interpreted in terms of sound waves propagating in the baryon–photon fluid before recombination. According to the standard model, the different peaks and troughs correspond to photon density and velocity perturbations that had an integer number of half-oscillations before the Jeans scale reached their sizes (Scott *et al* 1995). It is surprising and fascinating to see a coincidence in the position and amplitude of the peaks of the power spectrum as predicted by the standard cosmological model, but this is mostly a success in our understanding of the propagation of acoustic waves in a fluid, a question of local physics, rather than a direct proof of this cosmological model.

For the second question, we must note that a simple polynomial function with no physical interpretation with six free parameters is able more or less to reproduce the two-point correlation function (the Fourier transform of the power spectrum), giving very good fit for at least the first two peaks of the power spectrum (López-Corredoira 2013). The fact the standard model is able to fit the CMBR anisotropies with a model with six free parameters is astonishing but it would be much more surprising if it could fit it without free parameters or with only one or two free parameters fitted to the power spectrum. The same parameters which are used to fit the CMBR power spectrum are certainly able to fit other cosmological data, but more than six parameters are necessary, as well as additional information, such as initial conditions, conditions of stellar formation, galaxy formation, how dark matter is distributed in galaxies, etc. A global analysis of cosmological models would require an examination of all of the available independent sources of cosmological data (nucleosynthesis, supernovae, gravitational lensing, etc) and it would need to be checked whether they are comparable to the number of free parameters in the model.

In cosmologies different from the standard model, there have also been attempts to fit the CMBR power spectrum with models with a similar number of free parameters. Narlikar *et al* (2003, 2007) fitted C_ℓ with five–six parameters apart from the amplitude within a model that has nothing to do with the origin of the CMBR in the standard cosmology: see figure 5.1. Power spectrum peaks are shown to be

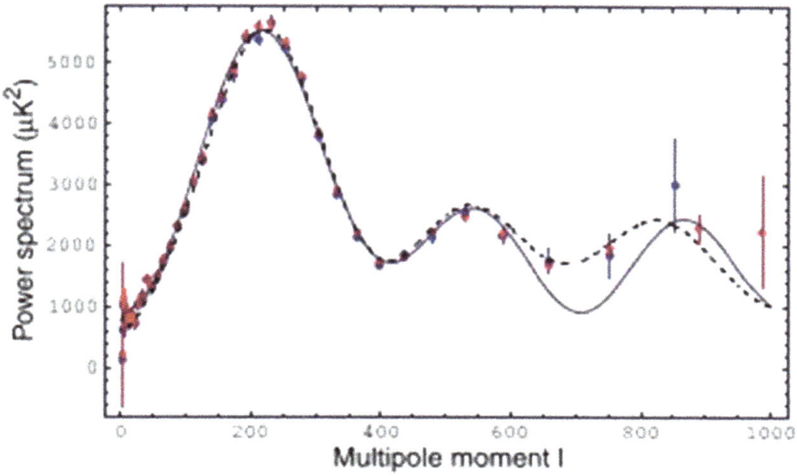

Figure 5.1. Plot of the best-fitting angular power spectrum curves to the WMAP-3 year data averaged into 39 bins. The continuous curve corresponds to the QSSC with six parameters and the dashed one to the big bang model with $\Omega_m = 0.2$, $\Omega_\Lambda = 0.8$. (Reproduced with permission from figure 1 of Narlikar *et al* 2007.)

related in their Quasi-Steady State cosmology to curvature effects at the last minimum of the scale factor, clusters, and groups of galaxies. Starlight originating from the last generation of supercluster formation, which cannot be entirely thermalized, stands out as inhomogeneities in the overall uniform background. It had been previously calculated that clusters of cold clouds in the halo produce anisotropies, which peak at $\ell \approx 50$ and can represent some anisotropies (Walker *et al* 2002). For a MOND (Modified Newtonian Dynamics) cosmology without cold dark matter, it is also possible to fit the power spectrum with few parameters (McGaugh 2004, Angus and Diaferio 2011, Skordis and Zlosnik 2021): see figure 5.2. It is not clear, however, that these alternative scenarios can explain the phase coherence needed to account for the clear peak/trough structure observed in CMBR anisotropies, as predicted by an inflationary scenario in the standard model (Dodelson 2003).

5.5 Some doubts on the validity of the foreground Galactic contribution subtraction from microwave anisotropies

Many authors have studied the different components of the Galactic microwave foreground radiation, but in my view these efforts are still insufficient to separate the components appropriately. The errors in the foreground emission subtraction are small but not negligible. At least, some doubts on the validity of the foreground Galactic subtraction from microwave anisotropies may be voiced (López-Corredoira 2007).

The question of the Galactic foregrounds has been present for a long time now. However, an analysis of the papers analysing the problem reveals that it has been usually underestimated. For instance, in the 1990s, it was thought that the Tenerife

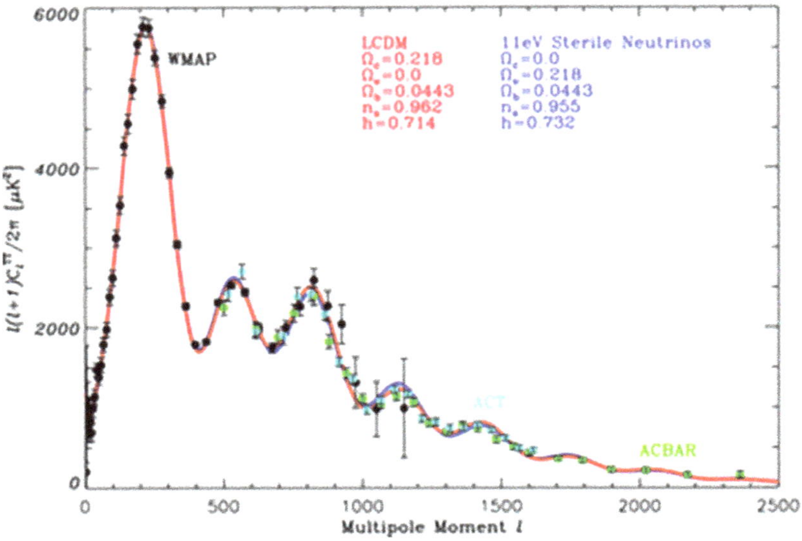

Figure 5.2. The CMB angular power spectrum for the MOND with added sterile neutrinos cosmological model, compared with the ΛCDM model. (Reproduced from figure 1 of Angus & Diaferio 2011.)

data at 15 GHz were likely to be dominated by cosmological fluctuations, and that the 33 GHz data were scarcely affected by the Galaxy (~4 μK on scales of 5–15 degrees) (Gutiérrez de la Cruz *et al* 1995, Davies *et al* 1996). Later analyses with the Wilkinson Microwave Anisotropy Probe (WMAP) (Bennett *et al* 2003 [figure 10]), however, showed that for comparable angular scales the Galaxy is dominant in the anisotropies at 22.8 GHz (and the Galactic contamination at 15 GHz is not much smaller than at 22.8 GHz), and at 33 GHz the Galactic anisotropies are of the same order as those from the CMBR. The WMAP microwave emission at frequencies lower than the W band is much higher than expected from thermal dust emission (Paladini *et al* 2007 [figure 8]). Also, at small scales, less than a degree, the foreground Galactic emission dominates (Leitch *et al* 2000).

An explanation for this high Galactic contamination in microwaves is that it is mainly due to the existence of a kind of emission unknown in the early 1990s that is correlated with dust that was not discovered previously; the synchrotron and or free–free emission might also have been underestimated. Positive correlations between the microwave anisotropies, including the region around 15 GHz and far-infrared maps, which trace Galactic dust, were found (Leitch *et al* 2000, Finkbeiner *et al* 2004, Fernández-Cerezo *et al* 2006). In the Helix region the emission at 31 GHz and 100 μm are well correlated (Casassus *et al* 2004); the 100 μm-correlated radio emission, presumably due to dust, accounts for at least 20% of the 31 GHz emission in the Helix (so the total dust emission is higher than 20% because there is an uncorrelated component too). The anomalous emission at around 23 GHz of the Perseus molecular cloud with a temperature of ~1 mK (Watson *et al* 2005) is an order of magnitude higher than the emission expected from synchrotron + free–free + thermal dust.

There is also some correlation of 10, 15 GHz maps with H_α maps, but it is very low, so free–free emission is detected at levels far lower than the dust correlation (de Oliveira-Costa *et al* 1999, 2002).

The most likely current explanation for this emission correlated with dust around 15–50 GHz is spinning dust grain emission (Draine and Lazarian 1998) and/or magnetic dipole emission from ferromagnetic grains (Draine and Lazarian 1999). However, Mukherjee *et al* (2001) expressed some doubts about the spinning dust hypothesis, since the evidence for those correlations is low and compromised by a systematic effect in the Galactic plane signal. More recently, a new foreground was discovered at low frequencies: 'microwave haze' emission around the Galactic centre (Pietrobon *et al* 2012). Whatever it is, it is now clear that the 15–40 GHz range is dominated by the Galaxy. A question might arise as to whether the contamination in the remaining frequencies (40–200 GHz) is being correctly accounted for. Milky Way emission, at least on large scales, is far from negligible; indeed, the Galactic quadrupole is comparable in amplitude to the quadrupole in the CMBR (Kogut *et al* 1996). Typical calculations for dust contamination claim that it should not be predominant, and that its contribution can be accurately subtracted, but how sure can we be of this statement? How accurate is the subtraction of the dust foreground signal?

A first difficulty in the subtraction of the dust component is knowing exactly how much dust emission there is in each line of sight. First approximations came with extrapolations from the IRAS far-infrared and DIRBE data (with the zodiacal light and cosmic infrared background subtracted) used to model the dust thermal emission in microwaves. Templates were taken from these infrared maps and extrapolated in amplitude by a common factor for all pixels. The problem is that 'a template approach is often carelessly used to compare observations with expected contaminants, with the correlation amplitude indicating the level of contamination. (…) These templates ignore well-measured variation in dust temperature and variations in dust/gas ratio.' (Finkbeiner *et al* 1999). The growing contrast of colder clouds in the background of the diffuse interstellar medium will produce much higher microwave anisotropies than the product of the template extrapolation (López-Corredoira 1999). Neither is it a good strategy to subtract a scaled IRAS template to remove the spinning dust in multifrequency data since it produces large residual differences (Leitch *et al* 2000). There is a presence of residual foreground emission not traced by the templates (Bottino *et al* 2008). A better approximation to this dust emission in the microwave region came with the adoption of an extrapolation with colour corrections in each pixel. This method assigns a different temperature to each pixel. This colour correction, together with a ν^2 emission emissivity (Schlegel *et al* 1998), gave a much tighter agreement with FIRAS data. However, it was still inconsistent with the FIRAS data below 800 GHz in amplitude (Finkbeiner *et al* 1999): a 14% error in the mean amplitude at 500 GHz. Indeed, no power-law emissivity function fits the FIRAS data in the 200–2100 GHz region (Finkbeiner *et al* 1999). Furthermore, laboratory measurements suggest that the universality of ν^2 emissivity is an oversimplification, with different species of grains having different emissivity laws (Finkbeiner *et al* 1999). A better approximation is an extrapolation with two components ($\nu^{1.7}$, $\langle T \rangle = 9.5$ K and $\nu^{2.7}$, $\langle T \rangle = 16$ K); each pixel has an assignation of two temperatures (Finkbeiner *et al* 1999) with correction factors that are a function of these temperatures. This still fails in the predictions of FIRAS from the IRAS–DIRBE extrapolation by 15% in zones dominated by atomic gas (Finkbeiner *et al* 1999).

The approach of Finkbeiner *et al* (1999), although much better than a direct extrapolation of the template, was insufficient. The problem is difficult to solve because it is an extrapolation by a factor ~15–60 in frequency (from 1250 GHz [240 μ m] or 3000 GHz [100 μ m] to 50–90 GHz). This is equivalent, for instance, to the attempt to derive a map of stellar emission at 12 μm as an extrapolation of the emission in the optical B filter. Frankly, when I see the IRAS 12 μm map of point sources and the Palomar plates in blue filters, I observe huge differences, and I do not know how we can extrapolate the second map to obtain the first one. Each star has a different colour, and stars which are very bright in blue may be very faint at mid-infrared and vice versa. The same thing happens with the diffuse + cloud emission: there are hot regions, cold regions, different kinds of emitters (molecular gas, atomic gas), and we have to integrate all this into each line of sight. The assumption that with only two temperatures we can extrapolate the average flux, and that a colour term can correct the pixel-to-pixel differences is comparable to the assumption that a model with only two kinds of stars and a knowledge of $(B - V)$ for each star we can extrapolate the star counts from optical to mid-infrared for the whole Galaxy. It is also very common (e.g., Masi *et al* 2001, Fernández-Cerezo *et al* 2006) to calculate the Galactic dust contribution in some microwave data by just making a cross-correlation between these data and some far-infrared map of the sky. This is simply wrong and not even valid for ascertaining the order of magnitude of such contamination.

Therefore, any method of foreground subtraction that uses templates will have serious credibility problems with regard to the goodness of the subtraction. This applies not only to those methods that use templates directly with coupling coefficients derived from cross-correlation but also to the MEM (maximum entropy method) (Bennett *et al* 2003), which uses in the initial stage templates for the dominant foreground components and also establishes some *a priori* conditions of their spectral behaviour. Moreover, any calculation of the limits of such contamination based on cross-correlations will not be totally accurate.

The presence of different dust-emitting components causes the spectral indices of the foregrounds to vary with position. Spectral index changes from −5.1 to −2.1, depending on the region, have been measured (Kiss *et al* 2003) and also changes with wavelength that indicate the existence of dust at different temperatures, in particular of a cold, extended component. A catalogue of Galactic cold clumps derived with Planck data gives a spectral index between 1.4 and 1.8 (Planck Collaboration 2011b). Even spectral index variations as small as $\Delta\alpha \sim 0.1$ can have a substantial impact on how channels should be combined and on the attainable accuracy (Tegmark 1998), so the errors of the subtraction assuming a certain power spectrum for the dust are serious.

Another technique that does not use a templates is ILC (internal linear combination) (Bennett *et al* 2003). This technique assumes nothing about the particular frequency dependencies or morphologies of the foregrounds and tries to minimise the variance in different regions of the sky with the combination of the available frequencies. There is a degeneracy of solutions, an infinite number of maps can be generated, and there is no way to test whether the maximum likelihood solution is the correct one (Robitaille 2007). Those that have applied this method (e.g., Bennett *et al* 2003) warn against its use for cosmological analysis; it is not effective in removing all residual foregrounds (Eriksen *et al* 2004). The ILC method performs quite badly, especially for dust (Eriksen *et al* 2004), in part because of the variability of the spectral indices. Therefore, ILC maps are not clean enough to allow cosmological conclusions to be derived: '[The] ILC map, which by eye looks almost free of foreground residuals, has been extensively used for scientific purposes—despite the fact that there are strong (and difficult to quantify) residual foregrounds present in the map. (…) the ILC map is indeed highly contaminated by residual foregrounds, and in particular, that the low-ℓ components, which have received the most attention so far, are highly unstable under the ILC cleaning operation' (Eriksen *et al* 2005). The ILC provides satisfactory results only under rather restrictive conditions (Vio and Andreani 2008).

The WI-FIT method ('Wavelet based hIgh resolution Fitting of Internal Templates') (Hansen *et al* 2006) does not require *a priori* templates, but obtains information about the foregrounds by taking differences of temperature maps at different frequencies. However, for its application to presently available maps, it requires the assumption that the spectral indices are constant in space, which, as has just been said, is a very inaccurate approximation. Their assumption that the Galactic emission in each pixel i is proportional to the difference in temperature maps with different frequencies ν, ν' ($(T_i^\nu - T_i^{\nu'}) \propto T_i^\nu$) is in general incorrect because $T_i^{\nu'}$ is not proportional to T_i^ν, the temperature in each pixel being the superposition of many different emissions with different temperatures. Again, we have here the same problem as with the use of templates: the assumption that there is only one temperature along each line of sight, and that the intensity of this emission is describable with a simple average fixed power law multiplying a blackbody emission by an average temperature. Hansen *et al* (2006) have claimed that their method is good because they obtain similar results to those of Bennett *et al* (2003) with different methods, but this may be due to their similar assumptions.

That putative clean maps are not totally clean is clear from the fact there are still some correlations between these clean maps and some surveys Galactic or extra-galactic sources. Foreground contamination residuals are found even for the best available supposedly clean CMBR maps. Some correlation is found between the $\Delta\ell = 4n$ and $n = 1{,}2$ spherical harmonic multipole domain, which is caused by a symmetric signal in the Galactic coordinate system (Naselsky *et al* 2006). The spatial association on scales of 1–2 degrees between interstellar neutral hydrogen, integrated in maps over ranges of 10 km s^{-1} and CMBR maps cleaned of foreground contamination through the ILC methods, is especially significant for the present discussion too (Verschuur 2007). Several extended areas of excess emission at high galactic latitudes ($b > 30°$) are present in both maps. These structures are thought to have typical distances from the Sun of order 100 pc (Verschuur 2007). More recent analyses were carried out with much better data and with more frequencies (for instance, with the PLANCK data; Planck Collaboration 2016a), but the philosophy of extrapolation and correlation is still similar, and the maps which are supposed to be free of contamination still show signs of dust contamination (Axelsson *et al* 2015).

Hence, considering only the Galactic dust component, the methods used to remove it (templates, cross-correlations, assumption of a Galactic power spectrum, MEM, ILC, etc) are all inaccurate and one should not expect to produce maps clean of Galactic dust contamination by applying them. The analysis of galactic latitude dependence of these anisotropies and the fact that the power spectrum is almost independent of the frequency over the range 50–250 GHz may be considered at least as a proof that Galactic dust emission is lower than 10% on the ~1 degree scale (López-Corredoira 2007), possibly higher for lower ℓ multipoles. Even if it is not a dominant contribution, this uncertainty in Galactic contamination may produce important systematic errors in some cosmological parameters. The present error bars calculated for cosmological parameters are significantly underestimated, and the range of possible values is not as small as indicated by the claims of 'precision cosmology'.

Galactic contamination is also important for polarized CMBR light. Synchrotron radiation in the Galaxy is strongly polarized; it is observed in the radio (408 MHz), and has loops with an excess of ~20 μK, probably due to shells of very old supernova remnants (Sarkar 1982), which are not taken into account in the subtraction of foregrounds for polarized CMBR light (Sarkar 2014). Thermal dust also leaves an imprint on polarized light. As a matter of fact, a claim of the discovery of proofs of gravitational waves due to inflation announced with great ballyhoo, based on the BICEP2 detection of a ~0.3 μK B-mode signal, was later shown to be a false result owing to defective analysis of Galactic contamination. The measurements were done in a region of a loop with important radio emission and with some contamination of dust too (Sarkar 2014, Liu *et al* 2014). The anomalous radiation has the spectrum of magnetic dipole radiation.

5.6 Anomalies in the anisotropies

The three lowest multipoles of the anisotropies (dipole, quadrupole, and octopole) are not statistically isotropic. The planes of the quadrupole and the octopole are unexpectedly aligned with a random probability of being purely by chance of <0.2% (de Oliveira-Costa *et al* 2004, Starkman *et al* 2012, Schwarz *et al* 2016). This has been called the 'Axis of Evil'. Furthermore, this axis is surprisingly aligned with the geometry and direction of the motion of the solar system: The plane they define is perpendicular to the ecliptic plane. In addition, some papers claim that this CMBR quadrupole and octopole are more or less aligned with the CMBR parity violation axis and the CMBR kinematic dipole, and it is claimed to have a similar direction to the velocity flows, quasar alignment, anisotropy of cosmic acceleration, handedness of spiral galaxies, and the angular distribution of the fine-structure constant (Zhao and Santos 2015).

A solution to explain some of these anomalies is the presence of a certain amount of contamination in relation with the solar system or its neighbourhood, although part of the effect may also be due to non-uniform sky coverage (Chyzy *et al* 2005). If a hypothetic foreground produced by a cold spot in the Local Supercluster is subtracted from the CMBR data, the amplitude of the quadrupole is substantially increased, and the statistically improbable alignment of the quadrupole with the octopole is substantially weakened, but this does not explain the coincidence of the alignment with the ecliptic (Abramo *et al* 2006).

The systematic errors and contamination of the subtraction of the motion of our Local Group of galaxies may also play an important role in explaining the multiple anomalies related with the alignment of quadrupole, octopole, and other discovered preferent directions (Zhao and Santos 2015). However, the plasma column density in the Local Group Hot Gas is too low to contribute significantly to the observed quadrupole anisotropy in the CMBR (Osone *et al* 2002). Extragalactic point sources and the intergalactic medium may be dominant in the CMBR but only on arcminute scales ($\ell > 2000$) (Toffolatti *et al* 2005, Atrio-Barandela and Mücket 2006). Perhaps the most important argument against a foreground-related explanation of the CMBR anomalies is their frequency independence (Schwarz *et al* 2016).

Anomalous alignments of CMBR data could also be explained by synchrotron radiation emitted in the heliosheath, a broad transitional region between the inner heliosphere and the external environment (Sharpe 2009b, 2009c, 2009a). Radiative and plasma/magnetohydrodynamic processes within the heliosheath could affect the propagation through it of an otherwise isotropic CMBR background. The geometric distortion of the termination shock might be a possible source for the quadrupole. Free–free thermal radiation which involves the picking up of ions at the termination shock region may explain both the blackbody quadrupole and the non-blackbody component of the quadrupole. There is also a prediction for the octopole.

If there were no such contamination and the tension of the low quadrupole, the low octopole, and the alignment of the quadrupole and octopole were real, the tension could be alleviated with a different cosmological model; for instance, $R_h = ct$ cosmology (Melia 2015).

A second important issue is that the angular two-point correlation function at scales of >60 degrees in the regions outside the Galactic cut is approximately zero in all wavebands—or, equivalently, there is a lack of quadrupole CMB signal—and it is discrepant with the best fit ΛCDM inflationary model (Schwarz *et al* 2016): there is a less than 0.1% probability of its being by chance using the larger masks of Planck or WMAP data, which points towards a violation of statistical isotropy (Copi *et al* 2009); see figure 5.3. The absence of wide-angle correlations is best explained by the introduction of a non-zero minimum wave number for the fluctuation power spectrum, and the data rule out a zero minimum wave number (as required by the standard model) in comparison with the best fit at a confidence level exceeding 8σ (Melia and López-Corredoira 2018). According to some authors (e.g., Starkman *et al* 2012, Liu and Li 2013, Melia and López-Corredoira 2018), this is a serious challenge to the standard model.

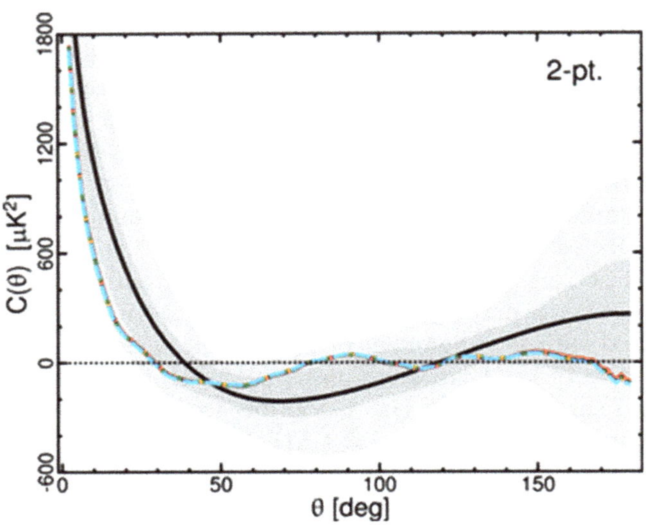

Figure 5.3. Two-point correlation function determined from Planck CMB 2015 temperature maps. The clear colour lines correspond to the data. The black solid line indicates the mean determined from 1000 SMICA simulations. The shaded dark and light grey regions indicate the corresponding 68% and 95% confidence regions, respectively. (Credit: Figure 40/left/top of Planck Collaboration 2016b, reproduced with permission © ESO.)

A third issue is the hemispherical asymmetry (Schwarz *et al* 2016). The plane that maximises the asymmetry is again approximately the ecliptic. It is found that the northern hemisphere in Planck and WMAP data has an extremely low (significant at 3–4σ) variance evaluated on scales of 4–14 degrees compared to what is expected in the standard cosmological model.

Joining the anomalies indicated in the previous paragraphs, we therefore have three significant features at the level of <1% each of happening by chance. As claimed by Schwarz *et al* (2016), one of these three anomalies might be an unlikely statistical fluke, but at least two of these effects should be real statistical flukes, and the combined probability of two anomalies is around $\sim 10^{-5}$. Biasing due to *a posteriori* selection of anomalous effect might somewhat reduce this probability (Melia 2015), but it is still a low number.

Opposite to this, the official WMAP team (Bennett *et al* 2011) did not find a significant detection of hemispherical or dipole power asymmetry across the sky. And they find that the amplitude of the quadrupole is well within the expected 95% CL that there is no anomaly in the lack of large angular CMBR power. They admit, however, that there is a remarkably unexpected alignment of the quadrupole and octopole.

There are still more anomalies in the literature. Isotropy in the ILC maps is ruled out to confidence levels of better than 99.9% (Samal *et al* 2008). There are abnormally high (low) powers in the $\ell = 6$, 12–17 modes (Su and Chu 2008); the probabilities for having anomalous amplitudes in the $\ell = 5, 6, 17$ modes are about 0.1%, 1% and 1% respectively according to the Gaussian conjecture. There is an odd parity correlation in the power spectrum in the $\ell = 2$–22 modes (Kim *et al* 2012,

Schwarz *et al* 2016) with a 99.7% CL; that is, odd and even multipoles ℓ in the power spectrum present unexpectedly different powers according to the standard model; some authors think that this asymmetry is due to contamination in a few regions of the sky (Creswell and Naselsky 2021). Minami and Komatsu (2020) examines the evidence of parity-violating physics with polarization data, which mitigates some possible systematic uncertainties, and report on a measurement of the cosmic birefringence angle of $\beta = 0.35 \pm 0.14$ deg., 2.4-σ away from the expected zero value from the standard model. The first peak in the anisotropies also does not have a blackbody shape and shows a clear hemispherical anisotropy (Jiang *et al* 2010). Recent Planck data combined with other data sets (di Valentino *et al* 2020, 2021) also shown that a model with a simple dark energy term and preserving the flatness of the Universe is excluded at a >99% CL; if the non-flatness of the Universe is confirmed, this is a serious problem for the standard model, since it invalidates an inflation prediction that is fundamental for keeping the validity of the standard cosmological model. Furthermore, a cold spot was detected with a non-negligible significance in recent Planck data (Schwarz *et al* 2016, Planck Collaboration 2016b); however, the official WMAP team (Bennett *et al* 2011) had claimed previously that the cold spots in the WMAP map are statistically consistent with random CMB fluctuations.

Another consideration that could point to the importance of non-cosmological extragalactic contamination over photons of cosmological origin is that the simulations of the gravitational lensing of the microwave background by galaxy clusters at $z < 1$ under any plausible ΛCDM model variation produces far more dispersion in the angular size of the primary acoustic peaks than allowed by observations (Lieu and Mittaz 2005). When all the effects are taken together, it is difficult to understand how CMBR data could reveal no evidence whatsoever of lensing by groups and clusters. Cool spots in the microwave background are too uniform in size to have travelled to us from $z \approx 1100$. There should be a spread of sizes around the average, with some of these cool spots noticeably larger and others noticeably smaller. But this dispersion in sizes is not seen in the data. Too many cool spots have the same size. The observed Sunyaev–Zel'dovich effect caused by clusters accounts only for about 1/4 of the expected decrement (Lieu *et al* 2006). Effects such as a central cooling flow in clusters, the abundance of hot cluster gas, large-scale radial decline in the temperature, uncertainties in the β-model, and the role played by cluster radio sources are too weak to change the estimate (Lieu *et al* 2006). Under standard cosmological model premises, this implies that the cosmological parameters (including the Hubble–Lemaître constant, the amount of dark matter, etc) used predict the original, pre-lensed sizes of the cool and hot spots in the microwave background might be wrong, or that some of these cool spot structures are caused by nearby physical processes and are not really remnants of the creation of the Universe; or perhaps there is some other unknown factor damping the effects of dispersion and focusing. It was speculated that the large-scale curvature of space may not entirely be an initial value problem related to inflation. The absence of gravitational lensing of the CMBR points to the possibility that even effects on light caused by wrinkles in the space of the late (nearby) Universe have been compensated for, beyond some distance scale, by a mechanism that maintains a flat geometry over such scales.

Or high-energy electrons may synchrotron radiate in the intracluster magnetic field of strength $B < \sim 1\ \mu G$ to produce cluster microwave emissions that account for the missing Sunyaev–Zel'dovich effect flux (Lieu *et al* 2010). One is also tempted to interpret this in terms of the importance of the extragalactic non-cosmological contribution in the microwave anisotropies.

Summing up, even if we accepted a cosmological origin for most of the CMBR radiation, putatively clean maps of CMBR anisotropies are most probably not totally free from from solar system, Galactic, or extragalactic contamination, and an accurate way of correcting for all these contributions has still to be devised since we do not have accurate information on the microwave emission of any of these contributions. Furthermore, there are some CMB anomalies that do not seem to be produced by contamination but by some failure in the hypothesis of the cosmological model.

5.7 Other background radiations

Residues in regions away from the Galactic plane ($|b| > 20$ deg.) have been found (Puget *et al* 1996), once the galactic foregrounds (interplanetary and interstellar) and CMBR are removed, in the far infrared associated with the Cosmic Far Infrared Background Radiation (CFIRBR), which was predicted to be due to early galaxies. The contribution of the CFIRBR is one-third of the total flux around 500 micron in the darkest galactic patches. In the optical, an extragalactic background diffuse light has also been detected (Bernstein *et al* 2002). However, other authors (Salvaterra and Ferrara 2006) later showed that the interpretation that the background radiations is due to redshifted radiation from early, very massive, Population III stars must be discarded, as it largely overpredicts the number of J-dropouts and Lyα emitters in ultra-deep field searches. Independently of the detailed physics of Lyα line emission, J-dropouts limit the background excess fraction due to Population III sources to be (at best) $\leqslant 1/24$. As alternative explanations (e.g., miniquasars, decaying neutrinos) can either be rejected or appear unlikely, whereas the reality of the excess is supported by the interpretation of the angular fluctuations, the origin of this component remains puzzling.

5.8 How sure can we be of the standard interpretation of the CMBR?

The standard picture interprets the CMBR in terms of matter–radiation decoupling at $z \approx 1100$ whose anisotropies are shaped by sound waves propagating in the baryon–photon fluid. This scenario works quite well, although not perfectly, given certain anomalies that are found in the comparison of data with predictions. A future measurement of the redshift of some lines associated with the background microwave radiation (Rubiño Martín *et al* 2006) may possibly be an important test to confirm definitively the proposed origin of the CMBR.

A new theory different from ΛCMB with similar number of free parameters (~6) might explain the same data, but the major difficulty is explaining the blackbody shape of CMBR radiation. A local origin in the nearby Universe of the CMBR seems definitively discarded, but other possible scenarios to produce CMBR independently of the cosmological theory remain unexplored. Given that an alternative scenario is not totally discarded, and that definitive direct proofs of the origin of the CMBR at $z \approx 1100$ have not yet appeared, we may remain reasonably sceptical and claim that the standard interpretation is not 100% certain.

References

Abraham R G, Nair P and McCarthy P J *et al* 2007 The gemini deep deep survey. VIII. When did early-type galaxies form? *Astrophys. J.* **669** 184–201

Abramo L R, Sodre L J and Wuensche C A 2006 Anomalies in the low CMB multipoles and extended foregrounds *Phys. Rev.* D **74** 083515

Aguirre A N 2000 The cosmic background radiation in cold big bang *Astrophys. J.* **533** 1–18

Alfonso-Faus A 2010 Sources of cosmic microwave radiation and dark matter identified: millimeter black holes (m.b.h.) (arXiv:1004.2251)

Alpher R A and Herman R 1949 Remarks on the evolution of the expanding universe *Phys. Rev.* **75** 1089–95

Angus G W and Diaferio A 2011 The abundance of galaxy clusters in modified Newtonian dynamics: cosmological simulations with massive neutrinos *Mon. Not. R. Astron. Soc.* **417** 941–9

Assis A K T and Neves M C D 1995 History of the 2.7 K temperature prior to Penzias and Wilson *Apeiron* **2** 79–84 (http://redshift.vif.com/JournalFiles/Pre2001/V02NO3PDF/V02N3ASS.PDF)

Atrio-Barandela F and Mücket J P 2006 The contribution of the intergalactic medium to cosmic microwave background anisotropies *Astrophys. J.* **643** 1–7

Axelsson M, Ihle H T, Scodeller S and Hansen F K 2015 Testing for foreground residuals in the Planck foreground cleaned maps: a new method for designing confidence masks *Astron. Astrophys.* **578** A44

Bennett C L, Hill R S and Hinshaw G *et al* 2003 First-year Wilkinson Microwave Anisotropy Probe (WMAP) observations: foreground emission *Astrophys. J. Supp. Ser.* **148** 97–117

Bennett C L, Hill R S and Hinshaw G *et al* 2011 Seven-year Wilkinson Microwave Anisotropy Probe (WMAP) observations: are there cosmic microwave background anomalies *Astrophys. J. Supp. Ser.* **192** 17

Berger J O and Jefferys W H 1992 The application of robust Bayesian analysis to hypothesis testing and Occam's razor *J. Ital. Stat. Soc.* **1** 17–32

Bernardis P d, Ade P A R and Bock J J *et al* 2000 A flat universe from high-resolution maps of the cosmic microwave background radiation *Nature* **404** 955–9

Bernstein R A, Freedman W L and Madore B F 2002 The first detections of the extragalactic background light at 3000, 5500, and 8000 Å. I. Results *Astrophys. J.* **571** 56–84

Blanchard A, Douspis M, Rowan-Robinson M and Sarkar S 2003 An alternative to the cosmological "concordance model' *Astron. Astrophys.* **412** 35–44

Bond J R and Efstathiou G 1987 The statistics of cosmic background radiation fluctuations *Mon. Not. R. Astron. Soc.* **226** 655–87

Bondi H, Gold T and Hoyle F 1955 *Black giant stars Observatory* **75** 80–1

Born M 1954 On the interpretation of Freundlich's red-shift formula *Proc. Phil. Soc.* A **67** 193–4

Bottino M, Banday A J and Maino D 2008 Foreground analysis of the Wilkinson Microwave Anisotropy Probe 3-yr data with FASTICA *Mon. Not. R. Astron. Soc.* **389** 1190–208

Brynjolfsson A 2004 Redshift of photons penetrating a hot plasma (arXiv:astro-ph/0401420)

Burbidge G R 1958 Nuclear energy generation and dissipation in galaxies *Publ. Astron. Soc. Pac.* **70** 83–9

Burbidge G R 1999 Explosive cosmogony and the quasi-steady state cosmology *Cosmological Parameters and the Evolution of the Universe* ed K Sato (Dordrecht: Kluwer) pp 286–9

Casassus S, Readhead A C S, Pearson T J, Nyman L-A, Shepherd M C and Bronfman L 2004 Anomalous radio emission from dust in the helix *Astrophys. J.* **603** 599–610

Chyzy K T, Novosyadlyj B and Ostrowski M 2005 Gradient and dispersion analyses of the WMAP data (arXiv:astro-ph/0512020)

Ćirković M M and Perović S 2018 Alternative explanations of the cosmic microwave background: a historical and epistemological perspective *Stud. Hist. Phil. Mod. Phys.* **62** 1–18

Clube S V M 1980 The material vacuum *Mon. Not. R. Astron. Soc.* **193** 385–97

Copi C J, Huterer D, Schwarz D J and Starkman G D 2009 No large-angle correlations on the non-galactic microwave sky *Mon. Not. R. Astron. Soc.* **399** 295–303

Cover K S 2009 Sky maps without anisotropies in the cosmic microwave background are a better fit to WMAP's uncalibrated time ordered data than the official sky maps *Europhys. Lett.* **87** 69003

Crawford D F 2011 Observational evidence favors a static universe (part II) *J. Cosmol.* **13** 3947–99

Creswell J and Naselsky P 2021 Asymmetry of the CMB map: local and global anomalies *J. Cosmol. Astropart. Phys.* **2021** 103

Davies R D, Gutiérrez C M and Hopkins J *et al* 1996 Studies of cosmic microwave background structure at dec.=+40 deg—I. The performance of the Tenerife experiments *Mon. Not. R. Astron. Soc* **278** 883–96

de Oliveira-Costa A, Tegmark M and Finkbeiner D P *et al* 2002 A new spin on galactic dust *Astrophys. J.* **567** 363–9

de Oliveira-Costa A, Tegmark M, Gutiérrez C M, Jones A W, Davies R D, Lasenby A N, Rebolo R and Watson R A 1999 Cross-correlation of Tenerife data with galactic templates-evidence for spinning dust? *Astrophys. J. Lett.* **527** L9–12

de Oliveira-Costa A, Tegmark M, Zaldarriaga M and Hamilton A 2004 Significance of the largest scale CMB fluctuations in WMAP *Phys. Rev.* D **69** 063516

di Valentino E, Melchiorri A and Silk J 2020 Planck evidence for a closed universe and a possible crisis for cosmology *Nat. Astron.* **4** 196–203

di Valentino E, Melchiorri A and Silk J 2021 Investigating cosmic discordance *Astrophys. J.* **908** 6

Dicke R H, Peebles P J E, Noll P G and Wilkinson D T 1965 Cosmic black-body radiation *Astrophys. J.* **142** 414–9

Disney M J 2007 Modern cosmology: science or folktale? *Am. Sci.* **95** 383–5

Dodelson S 2003 Coherent phase argument for inflation *Neutrinos, Flavor Physics and Precision Cosmology (AIP Conf. Proc. 689)* ed J F Nieves and R Raymond (Melville, NY: AIP) pp 184–96

Doroshkevich A G and Novikov I D 1964 Mean density of radiation in the metagalaxy and certain problems in relativistic cosmology *Sov. Phys. - Dokl.* **9** 111–3

Draine B T and Lazarian A 1998 Diffuse galactic emission from spinning dust grains *Astrophys. J. Lett.* **494** L19–22

Draine B T and Lazarian A 1999 Magnetic dipole microwave emission from dust grains *Astrophys. J.* **512** 740–54

Eddington A S 1926 *Internal constitution of the stars* (reprinted: 1988) (Cambridge: Cambridge University Press)

Eriksen H K, Banday A J, Górski K M and Lilje P B 2004 On foreground removal from the Wilkinson Microwave Anisotropy Probe data by an internal linear combination method: limitations and implications *Astrophys. J.* **612** 633–46

Eriksen H K, Banday A J, Górski K M and Lilje P B 2005 Astro-ph communication: Simulations of the WMAP internal linear combination sky map (arXiv:astro-ph/0508196)

Fahr H J and Zönnchen J H 2009 The "writing on the cosmic wall": is there a straightforward explanation of the cosmic microwave background? *Ann. Phys.* **521** 699–721

Fernández-Cerezo S, Gutiérrez C M and Rebolo R *et al* 2006 Observations of the cosmic microwave background and galactic foregrounds at 12-17 GHz with the COSMOSOMAS experiment *Mon. Not. R. Astron. Soc.* **370** 15–24

Finkbeiner D P, Davis M and Schlegel D J 1999 Extrapolation of galactic dust emission at 100 microns to cosmic microwave background radiation frequencies using FIRAS *Astrophys. J.* **524** 867–86

Finkbeiner D P, Langston G I and Minter A H 2004 Microwave interstellar medium emission in the green bank galactic plane survey: evidence for spinning dust *Astrophys. J.* **617** 350–9

Finlay-Freundlich E 1954 Red shifts in the spectra of celestial bodies *Proc. Phys. Soc.* A 67 192–3

Fixsen D J, Cheng E S, Gales J M, Mather J C, Shafer R A and Wright E L 1996 The cosmic microwave background spectrum from the full COBE FIRAS data set *Astrophys. J.* **473** 576–87

Gamow G 1953 The expanding universe and the origin of galaxies *Kgl. Dan. Vidensk. Selsk. Mat. Fys. Medd* **27** 3–15 (http://gymarkiv.sdu.dk/MFM/kdvs/mfm%2020-29/mfm-27-10.pdf)

Garrett M A 2002 The FIR/radio correlation of high redshift galaxies in the region of the HDF-N *Astron. Astrophys.* **384** L19–22

Greisen K 1966 End to the cosmic-ray spectrum? *Phys. Rev. Lett.* **16** 748–50

Guillaume C E 1896 *Les Rayons X et la photographie a travers les corps opaques* Gauthier-Villars et Fils, Imprimeurs-Libraires, du Bureau des Longitudes, de L'école Polytechnique, Paris

Gutiérrez de la Cruz C M, Davies R D, Rebolo R, Watson R A, Hancock S and Lasenby A N 1995 Dual-frequency mapping with the Tenerife cosmic microwave background experiments *Mon. Not. R. Astron. Soc.* **442** 10–22

Hanany S, Ade P and Balbi A *et al* 2000 MAXIMA-1: a measurement of the cosmic microwave background anisotropy on angular scales of 10'–5° *Astrophys. J. Lett.* **545** L5–9

Hansen F K, Banday A J, Eriksen H K, Górski K M and Lilje P B 2006 Foreground subtraction of cosmic microwave background maps using WI-FIT (wavelet-based high-resolution fitting of internal templates) *Astrophys. J.* **648** 784–96

Hoyle F, Burbidge G and Narlikar J V 1993 A quasi-steady state cosmological model with creation of matter *Astrophys. J.* **410** 437–57

Hoyle F, Burbidge G and Narlikar J V 1994 Astrophysical deductions from the quasi-steady state cosmology *Mon. Not. R. Astron. Soc.* **267** 1007–19

Hoyle F, Wickramashinghe N C and Reddish V C 1968 Solid hydrogen and the microwave background *Nature* **218** 1124–6

Hu W and Dodelson S 2002 Cosmic microwave background anisotropies *Annu. Rev. Astron. Astrophys.* **40** 171–216

Hu W, Fukugita M, Zaldarriaga M and Tegmark M 2001 Cosmic microwave background observables and their cosmological implications *Astrophys. J.* **549** 669–80

Jefferys H and Berger J 1992 Ockham's razor and Bayesian analysis *Am. Sci.* **80** 64–72

Jiang B-Z, Lieu R, Zhang S-N and Wakker B 2010 Significant foreground unrelated non-acoustic anisotropy on the 1 degree scale in Wilkinson Microwave Anisotropy Probe 5-year observations *Astrophys. J.* **708** 375–80

Jorgensen H E, Kotok E, Naselsky P and Novikov I 1995 Evidence for Sakharov oscillations of initial perturbations in the anisotropy of the cosmic microwave background *Astron. Astrophys.* **294** 639–47

Kashti T and Waxman E 2008 Searching for a correlation between cosmic-ray sources above 10^{19} eV and large scale structure *J. Cosmol. Astropart. Phys.* **2008** 21

Kim D Y, Lasenby A N and Hobson M P 2012 Symmetry and antisymmetry of the CMB anisotropy pattern *Adv. Astron.* **2012** 960509

Kiss C, Ábrahám P, Klaas U, Lemke D, Héraudeau P, del Burgo C and Herbstmeier U 2003 Small-scale structure of the galactic cirrus emission *Astron. Astrophys.* **399** 177–85

Kogut A, Banday A J, Bennett C L, Górski K M, Hinshaw G, Smoot G F and Wright E L 1996 Microwave emission at high galactic latitudes in the four-year DMR sky maps *Astrophys. J.* **464** L5–9

Krishan V 2009 Optical depth of the cosmic microwave background due to scattering and absorption (arXiv:0909.0125)

Larson D, Dunkley J and Hinshaw G *et al* 2011 Seven-year Wilkinson Microwave Anisotropy Probe (WMAP) observations: power spectra and WMAP-derived parameters *Astrophys. J. Suppl. Ser.* **192** 16

Leitch E M, Readhead A C S, Pearson T J, Myers S T, Gulkis S and Lawrence C R 2000 A measurement of anisotropy in the cosmic microwave background on 7'-22' scales *Astrophys. J.* **532** 37–56

Lerner E J 1988 Plasma model of microwave background and primordial elements–an alternative to the big bang *Laser Part. Beams* **6** 457–69

Lerner E J 1990 Radio absorption by the intergalactic medium *Astrophys. J.* **361** 63–8

Lerner E J 1991 *The Big Bang Never Happened: A Startling Refutation of the Dominant Theory of the Origin of the Universe* (Toronto: Random House)

Lerner E J 1993 Confirmation of radio absorption by the intergalactic medium *Astrophys. Space Sci.* **207** 17–26

Lerner E J 1995 Intergalactic radio absorption and the COBE data *Astrophys. Space Sci.* **227** 61–81

Li T-P, Liu H, Song L-M, Xiong S-L and Nie J-Y 2009 Observation number correlation in WMAP data *Mon. Not. R. Astron. Soc.* **398** 47–52

Lieu R and Mittaz J P D 2005 On the absence of gravitational lensing of the cosmic microwave background *Astrophys. J. Supp. Ser.* **628** 583–93

Lieu R, Mittaz J P D and Zhang S-N 2006 The Sunyaev-Zel'dovich effect in a sample of 31 clusters: a comparison between the x-ray predicted and WMAP observed cosmic microwave background temperature decrement *Astrophys. J.* **648** 176–99

Lieu R, Quenby J and Bonamente M 2010 The non-thermal intracluster medium *Astrophys. J.* **721** 1482–91

Lin W and Ishak M 2017 Cosmological discordances II: Hubble constant, Planck and large-scale-structure data sets *Phys. Rev.* D **96** 083532

Liu H and Li T-P 2013 Missing completely of CMB quadrupole in WMAP data *Chin. Sci. Bull.* **58** 1243–9

Liu H, Mertsch P and Sarkar S 2014 Fingerprints of galactic loop I on the cosmic microwave background *Astrophys. J. Lett.* **789** L29

Liu H, Xiong S-L and Li T-P 2011 Diagnosing timing error in WMAP data *Mon. Not. R. Astron. Soc.* **413** L96–100

López-Corredoira M 1999 A conspicuous increase of galactic contamination over CMBR anisotropies at large angular scales *Astron. Astrophys.* **346** 369–82

López-Corredoira M 2007 Some doubts on the validity of the foreground galactic contribution subtraction from microwave anisotropies *J. Astrophys. Astron.* **28** 101–16

López-Corredoira M 2013 Peaks in the CMBR power spectrum II: Physical interpretation for any cosmological scenario *Int. J. Mod. Phys.* D **22** 1350032

López-Corredoira M and Gabrielli A 2013 Peaks in the CMBR power spectrum. I. Mathematical analysis of the associated real space features *Physica* A **392** 474–84

López-Corredoira M, Gutiérrez C M and Génova R 2017 Analysis of the amplitude of the Sunyaev-Zel'dovich effect out to redshift $z = 0.8$ *Astrophys. J.* **840** 62

Lorentz H A 1904 Electromagnetic phenomena in a system moving with any velocity less than that of light *Proc. R. Acad. Amsterdam* **6** 809–30

Mao M Y, Huynh M T, Norris R P, Dickinson M, Frayer M, Helou G and Monkiewick J A 2011 No evidence for evolution in the far-infrared-radio correlation out to $z \sim 2$ in the extended Chandra Deep Field South *Astrophys. J.* **731** 79

Masi S, Ade P A R and Bock J J *et al* 2001 High-latitude galactic dust emission in the BOOMERANG maps *Astrophys. J. Lett.* **553** L93–6

Mather J C, Cheng E S and Cottingham D A *et al* 1994 Measurement of the cosmic microwave background spectrum by the COBE FIRAS instrument *Astrophys. J.* **420** 439–44

McGaugh S S 2004 Confrontation of modified Newtonian dynamics predictions with Wilkinson Microwave Anisotropy Probe first year data *Astrophys. J.* **611** 26–39

Melia F 2015 Cosmological implications of the CMB large-scale structure *Astron. J.* **149** 6

Melia F 2020 Reassessing dust's role in forming the CMB *Eur. Phys. J. Plus* **511** http://dx.doi.org/10.2139/ssrn.3626464

Melia F and López-Corredoira M 2018 Evidence of a truncated spectrum in the angular correlation function of the cosmic microwave background *Astron. Astrophys.* **610** 5

Meyers R 2003 A Brief history of competing ideologies in cosmology and evidence for non-cosmological redshifts as a case for alternative theoretical interpretations in cosmology *PhD thesis* (University of Western Sydney)

Minami Y and Komatsu E 2020 New extraction of the cosmic birefringence from the Planck 2018 polarization data *Phys. Rev. Lett.* **125** 221301

Mukherjee P, Jones A W, Kneissl R and Lasenby A N 2001 On dust-correlated galactic emission in the Tenerife data *Mon. Not. R. Astron. Soc* **320** 224–34

Narlikar J V, Burbidge G and Vishwakarma R G 2007 Cosmology and cosmogony in a cyclic universe *J. Astrophys. Astron.* **28** 67–99

Narlikar J V, Vishwakarma R G, Hajian A, Souradeep T, Burbidge G and Hoyle F 2003 Inhomogeneities in the microwave background radiation interpreted within the framework of the quasi-steady state cosmology *Astrophys. J.* **585** 1–11

Naselsky P D, Novikov I G and Chiang L-Y 2006 Correlations from galactic foregrounds in the first-year Wilkinson Microwave Anisotropy Probe data *Astrophys. J.* **642** 617–24

Navia C E, Augusto C R A and Tsui K H 2007 On the ultra high energy cosmic rays and the origin of the cosmic microwave background radiation (arXiv:0707.1896)

Novikov I 2001 Discovery of CMB, Sakharov oscillations and polarization of the CMB anisotropy *Historical Development of Modern Cosmology (ASP Conf. Ser. 252)* ed V J Martínez, V Trimble and M J Pons-Borderí a (San Francisco, CA: Astronomical Society of the Pacific) pp 43–53

Orenstein B J, Collier J D and Norris R P 2019 The redshift distribution of infrared-faint radio sources *Mon. Not. R. Astron. Soc.* **484** 1021–30

Osone S, Makishima K, Matsuzaki K, Ishizaki Y and Fukazawa Y 2002 Search for hot gas in the local group with ASCA *Publ. Astron. Soc. Japan* **54** 387–92

Paladini R, Montier L, Giard M, Bernard J P, Dame T M, Ito S and Macias-Pérez J F 2007 A broadband study of galactic dust emission *Astron. Astrophys.* **465** 839–54

Pecker J-C, Narlikar J V, Ochsenbein F and Wickramasinghe C 2015 The local contribution to the microwave background radiation Res *Astron. Astrophys.* **15** 461

Peebles P J E 1998 The standard cosmological model *Le Rencontres de Physique de la Vallee d'Aoste: Results and Perspectives in Particle Physics* ed M Greco (Frascati: Poligrafica Laziale s.r.l.) p 39

Peebles P J E and Yu J T 1970 Primeval adiabatic perturbation in an expanding universe *Astrophys. J.* **162** 815–36

Peiris H 2005 First year Wilkinson Microwave Anisotropy Probe results: implications for cosmology and inflation *Contemp. Phys.* **46** 77–91

Penzias A and Wilson R 1965 A measurement of excess antenna temperature at 4080 Mc/s *Astrophys. J.* **142** 419–21

Pietrobon D, Górski K M and Bartlett J *et al* 2012 Analysis of WMAP 7 year temperature data: astrophysics of the galactic haze *Astrophys. J.* **16** 69

Planck Collaboration 2011a Planck early results. X. Statistical analysis of Sunyaev-Zel'dovich scaling relations for x-ray galaxy clusters *Astron. Astrophys.* **536** A10

Planck Collaboration 2011b Planck early results. XXIII. The first all-sky survey of galactic cold clumps *Astron. Astrophys.* **536** A23

Planck Collaboration 2016a Planck 2015 results. XIII. Csomological parameters *Astron. Astrophys.* **594** A13

Planck Collaboration 2016b Planck 2015 results. XVI. Isotropy and statistics of the CMB *Astron. Astrophys.* **594** A16

Puget J L, Abergel A, Bernard J-P, Boulanger F, Burton W B, Desert F-X and Hartmann D 1996 Tentative detection of a cosmic far-infrared background with COBE *Astron. Astrophys.* **308** L5–8

Rana N C 1981 Cosmic thermalization and the microwave background radiation *Mon. Not. R. Astron. Soc.* **197** 1125–37

Regener E 1933 Der energiestrom der ultrastrahlung *Z. Phys.* **80** 666–9

Robitaille P-M 2007 WMAP: a radiological analysis *Prog. Phys.* **1/2007** 3–18 (http://www.ptep-online.com/2007/PP-08-01.PDF)

Roukema B F 2010 On the suspected timing error in Wilkinson Microwave Anisotropy Probe map-making *Astron. Astrophys.* **518** 7

Rubiño Martín J A, Chluba J and Sunyaev R A 2006 Lines in the cosmic microwave background spectrum from the epoch of cosmological hydrogen recombination *Mon. Not. R. Astron. Soc.* **371** 1939–52

Sachs R K and Wolfe A M 1967 Perturbations of a cosmological model and angular variations of the microwave background *Astrophys. J.* **147** 73–90

Salvaterra R and Ferrara A 2006 Where are the sources of the near-infrared background *Mon. Not. R. Astron. Soc.* **367** L11–5

Samal P K, Saha R, Jain P and Ralston J P 2008 Testing isotropy of cosmic microwave background radiation *Mon. Not. R. Astron. Soc.* **385** 1718–28

Sarkar S 1982 Does the galactic synchrotron radio background originate in old supernova remnants *Mon. Not. R. Astron. Soc.* **199** 97–108

Sarkar S 2014 Galactic foregrounds for the CMB *Proc. of Science* (PoS(FFP14)095)

Schlegel D J, Finkbeiner D P and Davis M 1998 Maps of dust infrared emission for use in estimation of reddening and cosmic microwave background radiation foregrounds *Astrophys. J.* **500** 525–53

Schwarz D J, Copi C J, Huterer D and Starkman G D 2016 CMB anomalies after Planck *Class. Quantum Grav.* **33** 184001

Scott D, Silk J and White M 1995 From microwave anisotropies to cosmology *Science* **268** 829–35

Sharpe H N 2009a A heliosheath model for the origin of the CMB quadrupole moment (arXiv:0905.2978)

Sharpe H N 2009b Heliosheath synchrotron radiation as a possible source for the arcade 2 CMB distortions (arXiv:0902.0181)

Sharpe H N 2009c A model for the WMAP anomalous ecliptic plane signal (arXiv:0904.1697)

Shpenkov G P and Kreidik G 2002 Microwave background radiation of hydrogen atoms *Rev. Cienc. Exatas Naturais* **4** 9–18 (https://revistas.unicentro.br/index.php/RECEN/article/view/464/0)

Skordis C and Zlosnik T 2021 A new relativistic theory for modified Newtonian dynamics *Phys. Rev. Lett.* **127** 161302

Soberman R K and Dubin M 2001 Dark matter is baryons (arXiv:astro-ph/0107550)

Soler Gil F J 2012 *Discovery or Construction? Astroparticle Physics and the Search for Physical Reality* (Frankfurt am Main: Peter Lang)

Sorrell W H 2008 The cosmic microwave background radiation in a non-expanding universe *Astrophys. Space Sci.* **317** 59–70

Starkman G D, Copi C J, Huterer D and Schwarz D 2012 The oddly quiet universe: how the CMB challenges cosmology's standard model (arXiv:1201.2459)

Su S-C and Chu M-C 2008 New anomalies in cosmic microwave background anisotropy: violation of the isotropic Gaussian hypothesis in low-ℓ modes (arXiv:0805.1316)

Tegmark M 1998 Removing real-world foregrounds from cosmic microwave background maps *Astrophys. J.* **502** 1–6

Toffolatti L, Negrello M, González-Nuevo J, de Zotti G, Silva L, Granato G L and Argüeso F 2005 Extragalactic source contributions to arcminute-scale cosmic microwave background anisotropies *Astron. Astrophys.* **438** 475–80

Van Flandern T 1993 Is the microwave radiation really from the big bang 'fireball'? *Reflector (Astron. League Newsl.)* **XLV** 4

Vavryčuk V 2018 Universe opacity and CMB *Mon. Not. R. Astron. Soc* **478** 283–301

Verschuur G L 2007 High galactic latitude interstellar neutral hydrogen structure and associated (WMAP) high-frequency continuum emission *Astrophys. J.* **671** 447–57

Vio R and Andreani P 2008 A statistical analysis of the "internal linear combination" method in problems of signal separation as in cosmic microwave background observations *Astron. Astrophys.* **487** 775–80 von Nernst W 1937 Weitere prüfung der annahme lines stationären zustandes im weltall *Z. Phys.* **106** 633–61

Walker M, Ohishi M and Mori M 2002 Microwave anisotropies from the galactic halo (arXiv: astro-ph/0210483)

Watson R A, Rebolo R, Rubiño Martín J A, Hildebrandt S, Gutiérrez C M, Fernández-Cerezo S, Hoyland R J and Battistelli E S 2005 Detection of anomalous microwave emission in the Perseus molecular cloud with the COSMOSOMAS experiment *Astrophys. J. Lett.* **624** L89–92

Zapsepin G T and Kuzmin V A 1966 Upper limit of the spectrum of cosmic rays *Sov. Phys.-JETP Lett.* **4** 78–80

Zhao W and Santos L 2015 Preferred axis in cosmology *Universe* **3** 9–33

IOP Publishing

Fundamental Ideas in Cosmology
Scientific, philosophical and sociological critical perspectives
Martín López-Corredoira

Chapter 6

The abundance of light elements

In the standard model, it is claimed that helium-4, lithium-7, and other light elements were created in the primordial Universe, and the existence of these elements was used as a proof for the necessity of a hot Universe in its first minutes of life. However, only helium-4 has had successful direct confirmation of the predictions, although at the price of requiring a baryon density that raises other problems. The observed abundance of lithium-7 is 3–4 times lower than predicted. The other light elements are affected by uncertainties in the theoretical model or by later creation or destruction associated with stellar nucleosynthesis, cosmic rays, or other astrophysical processes, so they cannot be used to corroborate cosmological predictions. Moreover, there are alternatives to primordial nucleosynthesis to explain the observed abundances, even for helium-4.

The origin of the Big Bang theory by George Gamow (1904–1968) and collaborators (Gamow 1947, Alpher *et al* 1948) was at first an attempt to explain the abundance of the elements, on the assumption that there was no process that could explain the present-day abundances. Initially, they contended that the heavier elements had been formed during an early hot scenario of the initial stage of expansion, since they thought that stars could not achieve temperatures high enough to produce such elements. This idea would be rejected by Fred Hoyle (1915–2001), who showed that heavy elements can indeed be formed in the stars (Hoyle 1946, 1947). Nonetheless, for the light elements, the standard cosmological model would remain as the best hypothesis for their formation and has been one of the most important pillars of modern cosmology since the 1950s.

With higher accuracy of present-day measurements of abundances and the determination of ratios according to the theoretical model, the sturdiness of this pillar is in question, as we shall see in the following sections.

6.1 Basic aspects of primordial nucleosynthesis

According to the standard model, primordial nucleosynthesis formed most of the Universe's helium as the isotope helium-4 (^{4}He), along with small amounts of the

doi:10.1088/978-0-7503-3775-5ch6

hydrogen isotope deuterium (^{2}H), the helium isotope helium-3 (^{3}He), the lithium isotope lithium-7 (^{7}Li), and other unstable radioactive isotopes. This process took place in the interval from roughly 10 seconds to 20 minutes after the beginning of the Universe, which corresponds to the temperature range when the Universe was cool enough for deuterium to survive, but hot and dense enough for fusion reactions to occur at a significant rate (Coc and Vangioni 2017).

The core of Big Bang nucleosynthesis calculations, frequently used in particle astrophysics, is the Boltzmann equation in an expanding Universe. The number density, n, of any particle species evolves as

$$\frac{dn}{dt} = -\langle\sigma v\rangle_a n^2 + \Sigma_i\langle\sigma v\rangle_{p,i} X_i^2 - 3n\dot{a}/a.$$

Here, the first term represents the thermally averaged annihilation cross section (σv), the second term is the sum over thermally averaged production cross sections (X_i is the abundance of the reactants involved in their production), and the final term is the dilution due to the expansion of the Universe. There are several parameters on which these equations depend.

The neutron half-life is longer when the reactions to interconvert neutrons and protons is weaker. Its value is found by particle physicists to be $\tau_N = 879.6 \pm 0.8$ s (Particle Data Group 2020). Neutrons (n) can react with positrons or electron neutrinos to create protons (p) and other products in one of the following reactions:

$$n + e^+ \rightleftharpoons \bar{\nu}_e + p,$$

$$n + \nu_e \rightleftharpoons p + e^-.$$

Hence, the neutron-to-proton ratio within the first seconds of the Universe is well established, provided that our knowledge of the Standard Model of particle physics may be extrapolated to the early Universe.

The number of neutrinos is another important parameter in the equations, since the expansion rate of the Universe is directly proportional to the density of the radiation gas, which in turn is directly proportional to the number of species in the radiation gas. The Standard Model of particle physics predicts the existence of three species, but different numbers have also been proposed to solve the inconsistencies, as we shall see in the following sections.

The baryon density is a third important parameter. The greater the primordial density, the greater the density of baryons (neutrons and protons) and, hence, the higher the production rate of deuterium. Hence, the earlier a critical density of deuterium nuclei forms, the faster is the production of helium achieved. The baryon-to-photon ratio (barion density divided by photon density) is the key parameter determining the abundances of light elements after nucleosynthesis ends. Baryons and light elements can fuse in the following main reactions:

$$p + n \rightarrow {}^2\text{H} + \gamma,$$

$$p + {}^2\text{H} \rightarrow {}^3\text{He} + \gamma,$$

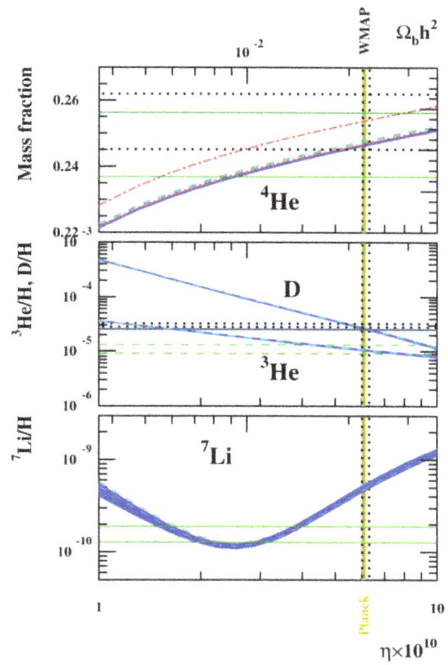

Figure 6.1. Abundances of the cosmological elements as a function of the baryon-to-photon ratio, η, for three neutrino families. The vertical areas correspond to Cosmic Microwave Background baryonic density observations from the WMAP (dotted lines) and Planck (solid lines) satellites. The horizontal areas (hatched and dotted lines) represent the different adopted observational abundances. The dash–dotted lines correspond to He calculated with a different value for the number of neutrino families derived from the Planck mission. (Reproduced with permission from figure 2 of Coc *et al* 2014.)

$$^2\mathrm{H} + {}^2\mathrm{H} \rightarrow {}^3\mathrm{He} + n,$$

$$^2\mathrm{H} + {}^2\mathrm{H} \rightarrow {}^3\mathrm{H} + p,$$

$$^3\mathrm{He} + {}^2\mathrm{H} \rightarrow {}^4\mathrm{He} + p,$$

$$^3\mathrm{H} + {}^2\mathrm{H} \rightarrow {}^4\mathrm{He} + n,$$

along with some other low-probability reactions involving lithium, beryllium, and other light elements. Figures 6.1 and 6.2 show the estimated abundances of the different isotopes produced.

6.2 Helium-4

It is said that the ^{4}He abundance is consistent with that expected minutes after the Big Bang, provided the present Universe has a baryon density (Ω_b) in the range $0.018 < \Omega_b h^2 < 0.022$ (Schramm and Turner 1998). This is more or less correct: helium-4 abundance measurements of $Y_P = 0.256\,5 \pm 0.001\,0(\mathrm{stat.}) \pm 0.005\,0(\mathrm{syst.})$

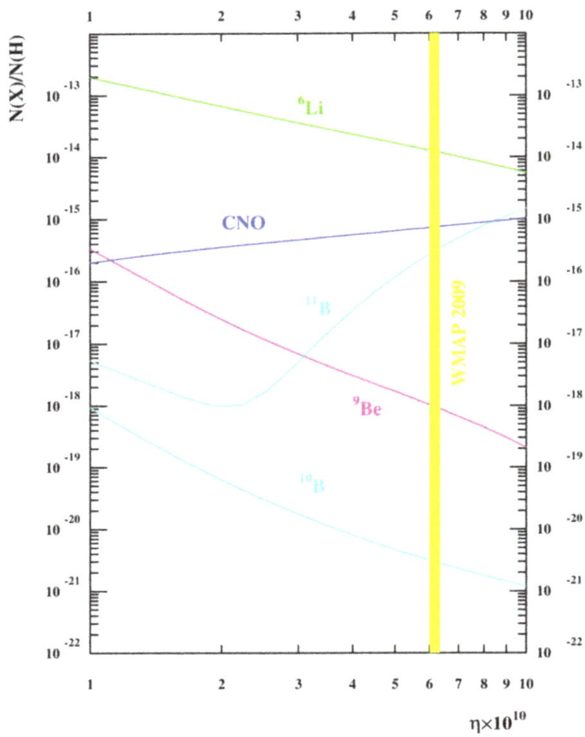

Figure 6.2. Abundances by number of atoms relative to H for ^{6}Li, ^{7}Be, ^{10}B, ^{11}B, C, N, O as a function of η, the baryon-to-photon ratio. We see that these isotopes are extremely rare at the end of primordial nucleosynthesis. (Reproduced by permission of the AAS from figure 14 of Coc *et al* 2012.)

(Izotov and Thuan 2010) are only 1.8σ higher than the prediction of standard Big Bang nucleosynthesis (0.247 09 ± 0.000 17) (Pitrou *et al* 2018). Nonetheless, methods used by Izotov and Thuan (2010) involve eliminating galaxies where many lines of helium are not measured, a process that ignores the lowest values and might produce some bias. In any case, even if the 1.8σ were correct, it could be considered compatible with the standard model within the statistical uncertainties. As a matter of fact, other measurements get Y_P closer to 0.247 (e.g., Aver *et al* 2015, 2021).

There are minor problems with the ^{4}He abundance. For instance, there are galaxies of low metal content, implying minimal stellar contamination, that show a mass fraction of ^{4}He as low as 0.21 (Terlevich *et al* 1992). Even allowing for error bars, this is far too low to match the standard model since we do not know how ^{4}He can be destroyed. Only an ad hoc explanation of inhomogeneities in the primordial set-up conceals primordial nucleosynthesis with this result.

There are also stars in our own Galaxy that have a low helium abundance in their atmospheres (the subdwarf B stars), but it has been shown (Sargent and Searle 1967) that these have also revealed other peculiarities, such as phosphorus and similar elements, which were associated with the chemically peculiar stars of population I. It was believed that these objects do not show a normal helium abundance and so they were not taken into account. Also, anomalously low helium content (lower than primordial helium) is required to fit the luminosities and temperatures of the metal-poor K dwarfs (Casagrande *et al* 2007).

6.3 Lithium

It is also considered that the ^{7}Li abundance should be consistent with what is expected when the baryon density is in the range $0.018 < \Omega_b h^2 < 0.022$ (Schramm and Turner 1998). Here, however, the problems of compatibility are greater. According to many authors, the standard model fails in its predictions of the abundance of this element.

The ^{7}Li abundance is a factor three to four lower than the predictions (Ryan *et al* 2000, Coc *et al* 2012, Famaey and McGaugh 2012). The Li problem remains and is indeed becoming more exacerbated as time goes by; the discrepancy is at the 4.2σ level (from globular cluster stars) to 5.3σ level (from halo field stars) (Cyburt *et al* 2008). The observation of the 'Spite plateau' (the name given to the baseline in the abundance of lithium found in old stars with very low metallicities orbiting the galactic halo) is relevant since it is believed that such stars were formed early on in the Universe from material that had not been significantly modified by other processes. Indeed, it was found that lithium does not exhibit a plateau below [Fe/H] = −3 (Sbordone *et al* 2010). A meltdown of the Spite plateau below that metallicity is established, and there is no depletion that can justify higher values of lithium for lower metallicities. That no star below [Fe/H] = −3 lies above the plateau suggests that they formed at plateau level and experienced subsequent depletion.

There is an extensive literature on proposed solutions to the lithium problem (e.g., Fields 2011, Mathews *et al* 2020). For example, it has been suggested that this discrepancy could be caused by a modification of surface lithium abundances during the stars' lifetimes (Korn *et al* 2006): the lithium is destroyed in old, metal-poor halo stars. This idea has some support in the fact that the lithium abundance is higher in the interstellar medium of the Magellanic Clouds, where the destruction process cannot happen, and it is in agreement with primordial nucleosynthesis predictions (4 times larger than in metal-poor halo stars) (Howk *et al* 2012). Lithium can also be created by stars (Starrfield *et al* 2020). The general impression, however, is that either the problem is not solved (e.g., Mathews *et al* 2020) or, it is solved ad hoc adding complex details of stellar evolution that can destroy or create ^{7}Li, which makes Big Bang predictions impossible to test owing to the degeneracy with many other astrophysical processes. In the best of cases, we should remove ^{7}Li from the numbers predicted by the standard model that fit the observations.

Lithium-6 may also be produced in primordial nucleosynthesis, but with ratios $\sim 10^4$ smaller than lithium-7 (Fields 2011), so it is not so often investigated.

6.4 Deuterium and helium-3

In stellar nucleosynthesis processes, any deuterium that might be present is quickly converted into helium-3. These fusion processes can only destroy, but never produce, deuterium. Precisely for this reason, measurements of primordial deuterium use the oldest objects such as distant quasars (Weiss 2006). Deuterium is an excellent indicator of cosmological parameters because its abundance is highly sensitive to the primordial baryon density and also depends on the number of neutrino species permeating the early Universe (Mossa *et al* 2020). However, the evolution of deuterium and the mechanisms of possible deuterium production from the standard cosmological model are still not properly understood (Vidal-Madjar *et al* 1998, Prodanovic and Fields 2003). The fit with observational data is also hampered by uncertainties in the cross section of the deuterium-burning process, in which a deuterium nucleus and a proton combine to form a helium-3 nucleus, and although new cross section calculations are being improved (Mossa *et al* 2020, Budrikis 2020), the discussion of these numbers is still going on. Hence, again, given the uncertainties with the calculations, we may conclude that deuterium observations cannot be used to validate the standard cosmological model.

Helium-3 is probably the most difficult element for which to measure a primordial value (Weiss 2006). In low-mass stars, it is produced in substantial amounts, while in massive stars, equally substantial amounts of helium-3 are transformed into heavier nuclei. Moreover, the properties of helium-3 and helium-4 nuclei are very similar, so it is practically impossible to detect helium-3 in extragalactic systems. Instead, measurements in HII-clouds in the Milky Way are used. These surveys of HII-clouds indicate that more helium-3 is produced in stars than is destroyed, which gives as an upper limit on the original abundance of helium-3. So far, measurements of helium-3 abundance teach us more about stars than about primordial nucleosynthesis (Weiss 2006).

6.5 Abundances of other elements

Primordial nucleosynthesis produced very few nuclei of elements heavier than lithium. The predicted abundance of isotopes of carbon, nitrogen, and oxygen are expected to be of the order of 10^{-15} times that of H, making them essentially undetectable and negligible (Coc *et al* 2012, 2014, Coc 2016) (see figure 6.2); moreover they cannot be separated from the production of these elements by stellar nucleosynthesis. Apart from hydrogen, helium, and lithium, other light elements have been investigated in connection with primordial nucleosynthesis; for example, beryllium-7, beryllium-9, boron-10, and boron-11. Figure 6.2 shows the relative values of the abundances of these elements.

For beryllium-9, excessive abundances with respect to the predictions of primordial nucleosynthesis are observed in the stars (Casuso and Beckman 1997), but this is justified by the argument that they have no null metallicity, so they are contaminated with stellar nucleosynthesis, so the abundance should be lower in the primordial Universe. It is also argued that our Galaxy produces beryllium, or the accretion of matter (Casuso and Beckman 1997), including non-thermal collisions between cosmic rays and interstellar matter (Vangioni and Cassé 2017).

Another element usually implied by primordial nucleosynthesis is boron, in some of its isotopes ^{10}B or ^{11}B. It has very low abundances, even lower than C, N, O (Coc *et al* 2012, 2014, Vangioni and Cassé 2017), and they are usually dominated by non-thermal collisions between cosmic rays and interstellar matter (Vangioni and Cassé 2017).

An inhomogeneous nucleosynthesis might increase the ratio of primordial beryllium and boron (Boyd and Kajino 1989), but this is an ad hoc solution that does not solve other problems.

6.6 Baryon fraction

In order to get the observed abundances of light elements with primordial nucleosynthesis, a baryon fraction is predicted for the ΛCDM model. However, this number does not fit the observations—at least not directly. Less than 50% of the baryons predicted at low redshift have been found in some way in galaxies or the intergalactic medium (Takei *et al* 2007, Anderson and Bregman 2010, McGaugh *et al* 2010). Where is the rest of it? A second problem with baryons is that galaxies are observed to have a significantly smaller baryon fraction relative to the cosmic average (Anderson and Bregman 2010, McGaugh *et al* 2010). Tentative explanations of these missing baryons are given: a large fraction of the missing baryons may reside in the filaments of the cosmic web (Eckert *et al* 2015, Nicastro *et al* 2018).

To answer the inconsistencies of the nucleosynthesis predictions, some nuclear cosmologists were also proposing that, in the quark–hadron transition stage in the very early Universe (Olive 1991), there could emerge some regions which have more protons and with more neutrons than the near-equality assumed in the thermodynamic equilibrium value of the standard model. Another example of acrobatics in theoretical physics in an effort to explain ad hoc what cannot be explained directly. Nonetheless, the cosmological consequences of a quark–hadron phase transition no longer hold and it is still unclear how baryons (not hadrons) could form at that cosmological transition (Bonometto and Maisini 2016).

6.7 Light element abundance without primordial nucleosynthesis

In the standard model, it is claimed that ^4He and other light elements (deuterium, ^3He, ^7Li, ^9Be) were created in the primordial Universe, and the existence of these elements is used as a proof for the necessity of a hot Universe in the past. However, there are alternatives. These alternatives may have certain caveats to explain all the observations accurately, but neither is the standard cosmological model free of problems, as we have seen.

Helium could be created in several explosions such as those in the (Quasi-)Steady State Theory, or could be synthesised in massive objects evolving in the nuclear regions of galaxies (Burbidge 1971). Massive stars were formed in the first moments of the formation of galaxies, which in few hundred million years would produce the 24% helium abundance now observed, and this would be distributed throughout interstellar space by supernova explosions. Indeed, Audouze *et al* (1985) and Burbidge and Hoyle (1998) have argued that a case can be made for making all the light nuclei in stars. Population III stars are believed to contribute to the observed near-infrared background and heavy element pollution of the intergalactic medium, and could contribute to the primordial He abundance (Salvaterra and Ferrara 2003). Some theoreticians object that in such a case there should be more oxygen and carbon than observed; this could be solved if the least massive stars had not exploded but ejected their outer layers (pure helium) (Lerner 1991 [chapter 6]). Certain rare light isotopes cannot have been produced in this way, but cosmic rays generated by early stars colliding with the background plasma would generate them.

Some light elements such as deuterium, ^{3}He, ^{7}Li, boron, and beryllium, might be totally produced by cosmic rays (Ramaty and Lingenfelter 1969, Meneguzzi *et al* 1971, Walker *et al* 1985); and many of these elements can also be produced by basic plasma physics phenomena (Lerner 1988). Lerner (2020) thinks that the origin of the light elements gives a good fit to the predictions of models with cosmic ray processes or thermonuclear origins.

Even within standard cosmology, there are variations that may compete with the official version. There are some cosmological parameters to play with: neutron decay time, the number of neutrino species, the baryon-to-photon ratio, and others; although the first two are perhaps more or less known. Curiously, a much better fit between primordial nucleosynthesis and observations is given when the number of neutrino species is two instead of the three predicted by the standard model for particles (Hata *et al* 1995).

Therefore, there are alternatives to the Big Bang to explain the production of any of these elements, although, of course, the standard model is the proposal that offers the most complete and elaborate theory to date.

6.8 Light elements, weighty problems

According to Kurucz (1992), the most important aspect to criticise here is the very methodology used for testing the primordial nucleosynthesis. Instead of trying to make predictions that can be tested by observation, cosmologists take observations that can be made, such as measurements of the abundances in Population II material, and try to determine the primordial abundances while knowing *a priori* what answer is needed to fit some particular model. If predictions made before observations were compared with observations, they would almost all be wrong. They then take those derived abundances and adjust the model to match more closely, so that circularity of reasoning is completely guaranteed. In this way, the baryon density derived is too low to account for the subsequent large scale structure of the Universe, and an ad hoc addition of matter must be introduced.

Still, as we have seen, some open questions are being discussed about primordial nucleosynthesis predictions, and whether these can be substituted by alternative models. The question is not as simple as the Big Bang making a unique prediction and the observations confirming it; there is a lot of 'cooking' of the numbers, a lot of astrophysical processes on element production and evolution that are proposed ad hoc in order to fit certain observations that do not fit the *a priori* predictions. But even with these extra ad hoc corrections, there remain problems to be solved.

More remarkably, the successful predictions of primordial nucleosynthesis reduce basically to only one: that for ^{4}He, excluding minor problems with possible solutions. For the other elements, either failures of agreement with the data are more important, or there are many ways including stellar nucleosynthesis or cosmic rays to generate the observed abundance. There are alternative explanations even for ^{4}He abundances. Therefore, one of the oldest pillars of the Big Bang may now be considered as one of its weakest supports. Light element abundances were a motivation to generate the standard cosmological model, but now they cannot be used to sustain the theory.

References

Alpher R A, Bethe H and Gamow G 1948 The origin of chemical elements *Phys. Rev.* **73** 803–4

Anderson M E and Bregman J N 2010 Do hot halos around galaxies contain the missing baryons? *Astrophys. J.* **714** 320–31

Audouze J, Lindley D and Silk J 1985 Big bang photosynthesis and pregalactic nucleosynthesis of light elements *Astrophys. J.* **293** L53–7

Aver E, Berg D A, abd Olive K A, Pogge R W, Salzer J J and Skillman E D 2021 Improving helium abundance determinations with Leo P as a case study *J. Cosmol. Astropart. Phys.* **2021** 27

Aver E, Olive K A and Skillman E D 2015 The effects of He I λ10830 on helium abundance determinations *J. Cosmol. Astropart. Phys.* **2015** 11

Bonometto S A and Maisini R 2016 Baryon number transfer could delay quark–hadron transition in cosmology *Universe* **4** 32

Boyd R and Kajino T 1989 Can be-9 provide a test of cosmological theories? *Astrophys. J. Lett.* **336** L55–8

Budrikis Z 2020 Deuterium experiments reveal conditions in the early universe *Nat. Rev. Phys.* **2** 664

Burbidge G R 1971 Was there really a big bang? *Nature* **233** 36

Burbidge G R and Hoyle F 1998 The origin of helium and the other light elements *Astrophys. J. Lett.* **509** L1–3

Casagrande L, Flynn C, Portinari L, Girardi L and Jiménez R 2007 The helium abundance and $\delta y/\delta z$ in lower main-sequence stars *Mon. Not. R. Astron. Soc.* **382** 1516–40

Casuso E and Beckman J E 1997 Beryllium and boron evolution in the galaxy *Astrophys. J.* **475** 155–62

Coc A 2016 Primordial nucleosynthesis *J. Phys. Conf. Ser.* **665** 012001

Coc A, Goriley S, Xu Y, Saimpert M and Vangioni E 2012 Standard Big Bang nucleosynthesis up to CNO with an improved extended nuclear network *Astrophys. J.* **744** 158

Coc A, Uzan J-P and Vangioni E 2014 Standard big bang nucleosynthesis and primordial CNO abundances after Planck *J. Cosmol. Astropart. Phys.* **2014** 050

Coc A and Vangioni E 2017 Primordial nucleosynthesis *Int. J. Mod. Phys.* E **26** 1741002

Cyburt R H, Fields B D and Olive K A 2008 An update on the big bang nucleosynthesis prediction for ^{7}li: the problem worsens *J. Cosmol. Astropart. Phys.* **2008** 12

Eckert D, Jauzac M and Shan H *et al* 2015 Warm-hot baryons comprise 5-10 per cent of filaments in the cosmic web *Nature* **528** 105–7

Famaey B and McGaugh S 2012 Modified Newtonian dynamics (MOND): observational phenomenology and relativistic extensions *Living Rev. Relativ.* **159** 10

Fields B D 2011 The primordial lithium problem *Annu. Rev. Nucl. Part. Sci* **61** 47–68

Gamow G 1947 *One, Two, Three, Infinity* (New York: Mentor)

Hata N, Scherrer R J, Steigman G, Thomas D, Walker T P, Bludman S and Langacker P 1995 Big bang nucleosynthesis in crisis? *Phys. Rev. Lett.* **75** 3977–80

Howk J C, Lehner N, Fields B D and Mathews G J 2012 Observation of interstellar lithium in the low-metallicity small magellanic cloud *Nature* **489** 121–3

Hoyle F 1946 The synthesis of the elements from hydrogen *Mon. Not. R. Astron. Soc.* **106** 343–83

Hoyle F 1947 On the formation of heavy elements in stars *Proc. Phys. Soc. London* **59** 972–8

Izotov Y I and Thuan T X 2010 The primordial abundance of ^{4}He: Evidence for non-standard big bang nucleosynthesis *Astrophys. J. Lett.* **710** L67–71

Korn A J, Grundahl F, Richard O, Barklem P S, Mashonkina L, Collet R, Piskunov N and Gustafsson B 2006 A probable stellar solution to the cosmological lithium discrepancy *Nature* **442** 657–9

Kurucz R L 1992 Gedanken astrophysics: the universe since recombination *Comments Astrophys* **16** 1–15

Lerner E J 1988 Plasma model of microwave background and primordial elements—an alternative to the big bang *Laser Part. Beams* **6** 457–69

Lerner E J 1991 *The Big Bang Never Happened: A Startling Refutation of the Dominant Theory of the Origin of the Universe* (Toronto: Random House)

Lerner E J 2020 A Reassessment of the Galactic Origin of Light Elements Hypothesis: Data Contradict BBN Predictions and Confirms Generation in Young Galaxies (private communication)

Mathews G J, Kedia A and Sasankan N *et al* 2020 Cosmological solutions to the lithium problem *Proc. of the 15th Int. Symp. on Origin of Matter and Evolution of Galaxies (JSP Conf. Proc. 31)* ed L Kostro, A Posiewnik, J Pykacz and M Zukowski (Japan: The Physical Society of Japan) 011033

McGaugh S S, Schombert J M, de Blok W J G and Zagursky M J 2010 The baryon content of cosmic structures *Astrophys. J. Lett.* **708** L14–7

Meneguzzi M, Audouze J and Reeves H 1971 The production of the elements li, be, b by galactic cosmic rays in space and its relation with stellar observations *Astron. Astrophys.* **15** 337–59

Mossa V, Stöckel K and Cavanna F *et al* 2020 The baryon density of the universe from an improved rate of deuterium burning *Nature* **587** 210–3

Nicastro F, Kaastra J and Krongold Y *et al* 2018 Observations of the missing baryons in the warm-hot intergalactic medium *Nature* **558** 406–9

Olive K A 1991 The quark-hadron transition in cosmology and astrophysics *Science* **251** 1194–9

Particle Data Group 2020 *n* mean life *Prog. Theor. Exp. Phys.* **2020** 083C01

Pitrou C, Coc A, Uzan J-P and Vangioni E 2018 Precision big bang nucleosynthesis with improved helium-4 predictions *Phys. Rep.* **754** 1–66

Prodanovic T and Fields B D 2003 On nonprimordial deuterium production by accelerated particles *Astrophys. J.* **597** 48–56

Ramaty R and Lingenfelter R E 1969 Cosmic-ray deuterium and helium-3 of secondary origin and the residual modulation of cosmic rays *Astrophys. J.* **155** 587–608

Ryan S G, Beers T C, Olive K A, Fields B D and Norris J E 2000 Primordial lithium and big bang nucleosynthesis *Astrophys. J. Lett.* **530** L57–60

Salvaterra R and Ferrara A 2003 Is primordial ^{4}He truly from the big bang *Mon. Not. R. Astron. Soc.* **340** L17–20

Sargent W L W and Searle L 1967 The interpretation of the helium weakness in halo stars *Astrophys. J. Lett.* **150** L33–7

Sbordone L, Bonifacio P and Caffau E *et al* 2010 The metal-poor end of the spite plateau. I. Stellar parameters, metallicities, and lithium abundances *Astron. Astrophys.* **522** 26

Schramm D N and Turner M S 1998 Big-bang nucleosynthesis enters the precision era *Rev. Mod. Phys.* **70** 303–18

Starrfield S, Bose M, Iliadis C, Hix W R, Woodward C E and Wagner R M 2020 Carbon-oxygen classical novae are galactic 7li producers as well as potential supernova Ia progenitors *Astrophys. J.* **895** 70

Takei Y, Henry J P, Finoguenov A, Mitsuda K, Tamura T, Fujimoto R and Briel U G 2007 Warm-hot intergalactic medium associated with the coma cluster *Astrophys. J.* **655** 831–42

Terlevich E, Terlevich R, Skillman E, Stepanian J and Lipovetskii V 1992 The extremely low He abundance of SBS:0335-052 *Elements and the Cosmos* ed M G Edmunds and R Terlevich (Cambridge: Cambridge University Press) pp 21–7

Vangioni E and Cassé M 2017 Cosmic origin of the chemical elements rarety in nuclear astrophysics *Phys. Quasars* **10** 84–97

Vidal-Madjar A, Ferlet R and Lemoine M 1998 Deuterium abundance and cosmology *The Scientific Impact of the Goddard High Resolution Spectrograph (ASP Conf. Series 143)* ed J C Brandt, T B I Ake and C C Petersen (San Francisco, CA: Astronomical Society of the Pacific) pp 3–17

Walker T P, Mathews G J and Viola V E 1985 Astrophysical production rates for Li, Be, and B isotopes from energetic H(1) and He(4) reactions with HeCNO nuclei *Astrophys. J.* **299** 745–51

Weiss A 2006 Elements of the past: big bang nucleosynthesis and observation *Einstein Online* **2** 02/1019 (https://www.einstein-online.info/en/spotlight/bbn_obs/)

Chapter 7

Large-scale-structure and the formation and evolution of galaxies

Here I review some examples of the difficulties that ΛCDM cosmology has in explaining certain observations of galaxies and their large-scale space distributions on scales from Mpc to hundreds of Mpc. Moreover, there are some caveats to that theory's explanation of the distribution of velocities, isotropy, the constraint of the reionization epoch, the too early formation of massive galaxies and black holes, the high metallicities and dust abundances at high z, and others. And even the most successful elements on the formation of structures, such as the observed acoustic waves of Baryon Acoustic Oscillations (BAO), have problems within ΛCDM, and/or they have alternative explanations in other cosmologies.

Cosmology is a comprehensive attempt to explain all observable phenomena in the Universe, and the most direct observations of the cosmos are perhaps its fundamental bricks: the galaxies and how they are distributed giving rise to the large-structure of the Universe. Cosmic microwave background radiation also provides important observational evidence, but, as we have seen in chapter 5, it has to be theoretically interpreted to know what it means. Light elements abundances are a directly measurable, but subject to interpretation in terms of possible contamination or destruction, as seen in chapter 6. The dark elements of chapter 3 are neither directly observable nor interpretable. Neither is expansion directly observed either but is interpreted as the most likely explanation for cosmological redshifts (chapter 4). However, galaxies are galaxies and they are there, open to the gaze of the big eyes of our telescopes, independently of any cosmological theory. Extragalactic astronomy is perhaps the most important element in the realm of observations with respect to which any cosmological theory should be examined.

In the present chapter, as in previous ones, I focus on the problems faced by the standard cosmological model in explaining present observations, without entering

doi:10.1088/978-0-7503-3775-5ch7

too deeply into the theoretical aspects. The tensions are noteworthy in certain topics, such as the reionization epoch accompanying the formation of the first galaxies, discussion of the oldest galaxies, and especially statistical analyses of the large-scale structure of the Universe that has now become an observed reality.

7.1 Reionization epoch

In ΛCDM cosmology, reionization is the process that ionised the matter in the Universe after the dark ages. As the majority of baryonic matter is in the form of hydrogen, reionization usually refers to the reionization of hydrogen gas. This occurred once objects that were energetic enough to reionize neutral hydrogen[1] started to condense in the early Universe.

As these objects formed and radiated energy, the Universe reverted from being neutral to once again being an ionised plasma. At this time of reionization, matter would have been diffused by the expansion of the Universe, and the scattering interactions of photons and electrons would have been much less frequent than before electron–proton recombination. Thus, a Universe full of low density ionised hydrogen would remain transparent, as is the case today.

Before reionization, there was diffuse neutral hydrogen, which should have produced an absorbing trough shortwards of a Lyman-α emission line in Quasi-Stellar-Objects (QSOs)—the Gunn–Peterson effect (e.g., Becker *et al* 2015). A complete Gunn–Peterson trough was firstly discovered at $z = 6.28$ (Becker *et al* 2001), which was interpreted as the Universe approaching the end of the reionization epoch at $z_{\rm eor} \approx 6$ (Becker *et al* 2001, McGreer *et al* 2015). At $z \approx 6$, the amount of absorption increases quickly with redshift, indicating that $z \approx 6$ is close to the end of the reionization epoch. However, the $z = 6.5$ Lyman-alpha lines are not strongly suppressed by a neutral intergalactic medium, which indicates that reionization could be almost complete at $z_{\rm eor} = 6.5$ (Malhotra and Rhoads 2004). In an analysis of gas temperatures inferred from the line widths in near zones observed in several QSO spectra, Raskutti *et al* (2012) found that hydrogen reionization was complete at $z_{\rm eor} > 7.9$ (6.5) at 68(95)% confidence level (CL).

There are galaxies with spectroscopically confirmed redshifts observed even at $z \approx 11$ (Oesch *et al* 2016, Jiang *et al* 2020), which, if real, would mean that the Universe was not totally opaque, and that reionization was already under way at those redshifts. Either the reionization ended at much larger redshift than at $z = 6$ or the range of reionization, Δz_r, is very extensive. There are also cases of observations with lower-redshift ($z \approx 5.5$) troughs in the Lyman-α forest (Davies *et al* 2018). Hence, the epoch of reionization derived from observations are at odds with each other. This is puzzling, and not expected directly from the predictions of the standard model. Inhomogeneous reionization (Becker *et al* 2001, Malloy and Lidz 2015) is a possibility that might explain the apparent disagreement of the different data. This is a possibility, but, as usual, this is an ad hoc correction to the model.

[1] It is noteworthy that predictions of the temperature of hydrogen at $z \approx 17$ did not fit the observations, which show much lower amplitude and temperature in the 21 cm absorption line (Bowman *et al* 2018); yet another in the long list of failed predictions of the ΛCDM model.

There are other ways to get the redshift associated with the end of the reionization epoch (Becker *et al* 2015): ultraviolet luminosity functions, Lyman-α emitter surveys, gamma-ray burst afterglows, redshifted 21-cm line and Cosmic Microwave Background Radiation (CMBR; see chapter 5) anisotropies. With the exception of the CMBR, most of these solutions have high uncertainties.

Measurements of CMBR anisotropies a decade or two ago gave a reionization epoch at higher redshift than those obtained from Gunn–Peterson effect: e.g. a central epoch of reionization $11 < z_{c,r} < 30$ (95% C.L,) from WMAP 1 yr data (Kogut *et al* 2003), $z_{c,r} = 10.5 \pm 1.2$ from WMAP 7 yr data (Jarosik *et al* 2011); these would be incompatible with an end of reionization at $z_{\rm eor} = 6$ if we take typical values derived from CMBR $\Delta z_r = 1.3^{+0.2}_{-0.6}$ (Chowdhury *et al* 2021) for the total extent of that epoch (one would expect roughly $z_{\rm eor} \approx z_{c,r} - \frac{1}{2}\Delta z_r$). However, later values of $z_{c,r}$ using CMBR progressively decreased $z_{c,r}$ until reaching a value more compatible with Gunn–Peterson troughs in QSOs: $z_{c,r} = 8.8^{+1.3}_{-1.2}$ with Planck 2015 data (Planck Collaboration 2015), $z_{c,r} = 7.7 \pm 0.8$ with Planck 2018 data (Planck Collaboration 2018). It is surprising to see these initial apparent tensions smoothed over in such a short time. Why were WMAP values so different from Planck results within their error bars? Could this be an effect of cosmologists forcing their analyses of the data towards consensus and self-suppressing possible discrepancies? I do not know the answer, but in any case, one thing seems clear: the analysis of CMBR data for deriving the reionization epoch is complex enough to permit moving the results up or down since they are quite sensitive to how the modelling is done. Deriving the optical depth and central epoch of reionization requires adopting a basis set for the epoch of reionization history and is indeed quite sensitive to it (Qin *et al* 2020). It is not model independent.

How this reionization is produced is also another matter for discussion. The QSO population is unlikely to provide enough photons to ionise the Universe at $z \approx 6$ (Fan *et al* 2001), even less so at higher redshift, where there is a lower number of QSOs. It is thought that reionization might be due to light from massive stars. But this still poses a problem for reionization at those very high redshifts because the measured star formation density is found to decrease too quickly with increasing redshift (Bunker *et al* 2010, Bouwens *et al* 2011). Another exotic ad hoc proposal to solve the problem uses, for instance, fast accretion shocks formed around the cores of the most massive haloes (Dopita *et al* 2011).

7.2 Formation and evolution of stars and galaxies in the early Universe

Populations of objects at high redshift differ in morphology from their counterparts at low redshifts, luminosity, etc (Peacock 1985). The ratio of mergers of galaxies and the pair fraction is also larger for higher redshifts (Carlberg 1991, Fèvre *et al* 2000). An evolutionary rather than a stationary (such as the one first proposed in the 1940s for the Steady State Model) is observed. Nonetheless, there are important differences between the particular details of the observed evolution and the formation of astrophysical objects vis-à-vis the *a priori* predictions of ΛCDM model.

The formation of galaxies takes a long time, and observations of galaxies at very high redshift do not well match the *a priori* predictions of the standard cosmological model. The discovery of galaxies with spectroscopic redshift $z \approx 11$ (Oesch *et al* 2016, Jiang *et al* 2020), if corroborated, is more of a problem than good news for the progress of our knowledge of the Universe. Distant QSOs are also unique tracers for the study of the formation of the earliest supermassive black holes. There are QSOs at $z = 7.6$ with a black hole of mass greater than 10^9 solar masses (Wang *et al* 2021), when the Universe was 670 million years old according to the standard model, a finding that challenges the theoretical models of supermassive black holes growth.

Furthermore, the standard cosmological model requires stars, QSOs, and non-active galaxies in the early Universe to be primitive, which means mostly metal-free, because it requires many generations of supernovae to build up the metal content in stars. But the evidence shows the existence of even higher than solar metallicities in the earliest QSOs and other galaxies (Becker *et al* 2001, Fan *et al* 2001, Constantin *et al* 2002). The iron-to-magnesium ratio increases at higher redshifts (Iwamuro 2002). And what is even more amazing is that there is no evolution in some line ratios, including iron abundance, between $z = 6.5$ and $z = 0$ (Dietrich *et al* 2003, Freudling *et al* 2003, Maiolino *et al* 2003, Barth *et al* 2003). There is no evolution of CaII absorbers with redshift either (Sardane *et al* 2014 [figure 9]). The amount of dust in high redshift Lyman-alpha-emitting QSOs and other galaxies is much higher than expected (Dunne *et al* 2003, Finkelstein 2010, Laporte *et al* 2017). In view of all this evidence and the tensions arising over the results expected *a priori*, cosmologists supporting the standard model have claimed that star formation began very early and produced metals up to the solar abundance quickly, in a few hundred Myr, when the most massive galaxies assembled substantial amounts of their stellar content rapidly (in 1–2 Gyr) at high z in very intense star formation episodes (Pérez-González *et al* 2008). This idea may find some observational support, such as the discovery of a massive starburst galaxy at $z = 6.34$ (Riechers *et al* 2013), with mass around 3×10^{11} solar masses. A 'maximum starburst' converts the gas into stars at a rate more than 2000 times that of the Milky Way, a rate that is among the highest observed at any epoch. All well and good, but this scenario is at odds with the predictions of the Galaxy formation in hierarchical scenarios of the ΛCDM cosmological model, so the hypothesis which served to explain the large-scale structure successfully is failing here.

According to the standard model, galaxy formation (including the hierarchical formation of galaxies), very massive elliptical galaxies should not have formed in the very early Universe. However, it has been established that such galaxies did indeed already exist in the early Universe, which poses a new challenge. From a comparison with a synthesis model, the colours of some very red galaxies at $z > 2.5$ could be related to the ages of a stellar population in passively evolving galaxies older than 1 Gyr on average (Castro-Rodríguez and López-Corredoira 2012, López-Corredoira *et al* 2017) (see figure 7.1). These turn out to be massive galaxies (stellar masses of $\sim 10^{11}$ $M_\odot$) that where formed on average when the Universe was very young (<1 Gyr), assuming standard cosmological parameters. Red galaxies at $z > 2$ have been analysed (Labbé *et al* 2005) with observations that provide rest-frame

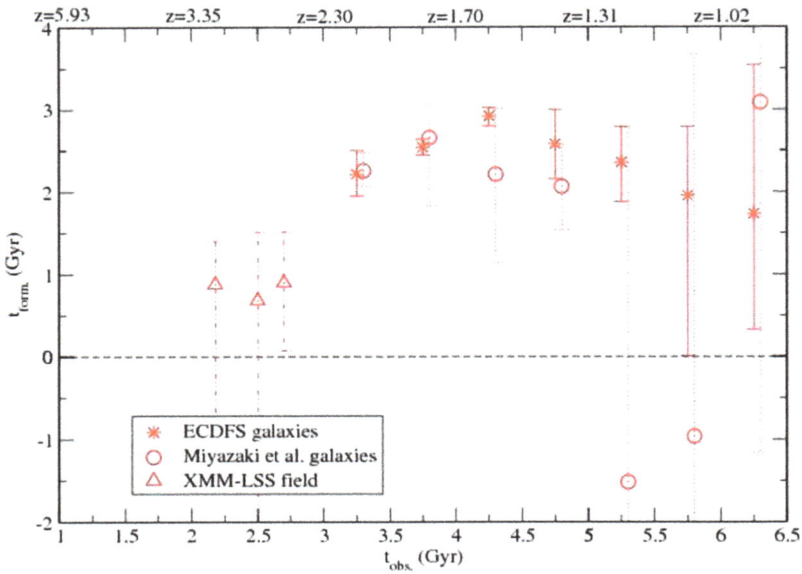

Figure 7.1. Average age (t_{form}) of the Universe at which the stellar populations of the extremely red and massive elliptical galaxies observed at redshift z (equivalent to an age of the Universe t_{obs}) have been formed. (Credit: Figure 6 of Castro-Rodríguez & López-Corredoira (2012), reproduced with permission © ESO.)

ultraviolet to near-infrared photometry, and it has been found that three of the 14 red galaxies were indeed old galaxies with Single Stellar Population best-fit ages 2.6 Gyr at $z = 2.7 \pm 0.4$, 3.5 Gyr at $z = 2.3 \pm 0.3$, and 3.5 Gyr at $z = 2.3 \pm 0.3$ respectively—representative of the average age of the stellar populations—so the oldest stars must be still older. The age of the Universe in the standard model at redshifts of both 2.7 ± 0.4 and 2.3 ± 0.3 is 2.5 ± 0.5 Gyr and 3.0 ± 0.4 Gyr respectively. We are again faced with the case of galaxies almost as old as the Universe. Toft *et al* (2005) fitted synthesis models to the spectra of red compact galaxies with ages 5.5, 3.5, and 1.7 Gyr for redshifts 1.2, 1.9, and 3.4 (ages of the Universe respectively: 5.4, 3.6, 2.0 Gyr), in the limit. Galaxies at very high redshift ($z \sim 4$) have been analysed (Rodighiero *et al* 2007) to find (with optical, near-infrared, and mid-infrared surveys) an important ratio of old ($\sim$1 Gyr or older) and massive ($\sim 10^{11}$ $M_{\odot}$) galaxies. Evidence has been presented with similar techniques for massive and evolved galaxies at redshifts between 5 and 6.5 (Eyles *et al* 2005, Wiklind *et al* 2008). This is what has been called the impossible early galaxy problem, according to which observations find several orders of magnitude more very massive haloes at very much higher redshift than predicted, which implies that these massive galaxies formed at an impossibly early stage (Steinhardt *et al* 2016). Furthermore, the large rotational velocity and large content of cold gas in some high redshift galaxies remain a challenge to efforts to reproduce the standard predictions of formation of galaxies (Neeleman *et al* 2020). All these results are at odds with semi-analytical ΛCDM models, which claim that very massive galaxies were formed much later (Guo *et al* 2011). Nonetheless, possible ad hoc speculative solutions have been proposed in terms of either a 'downsizing' scenario of galaxy formation or a mass function variation with redshift (Manrique and Salvador-Solé 2015).

The age of the oldest objects in the Universe was a matter of great controversy before the introduction of dark energy in the cosmological model. The preferred cosmological models of the early 1990s were either the Einstein–de Sitter ($\Omega = \Omega_m = 1$) or Friedmann–Lemaître–Robertson–Walker model in an open Universe with $\Omega = \Omega_m = 0.2$, and they would give an age of the Universe of 9.3 and 11.8 Gyr respectively (assuming $H_0 = 70$ km s^{-1} Mpc^{-1}), which was significantly shorter than the estimated age of some of the oldest galaxies or stars. However, the introduction of dark energy increased the age of the Universe. Today, ΛCDM gives a value of 13.787 ± 0.020 Gyr (Planck Collaboration 2018)[2], which is roughly equal to the age of the oldest known objects. Nonetheless, although the tension was alleviated, there remain in the literature claims of finding objects significantly older than 13.8 Gyr; e.g. globular clusters, QSOs and stars in the Milky Way with ages over that value (Wang *et al* 2010, Verde *et al* 2013, Ge *et al* 2016). However, either there are only few σ of deviation, or there are uncertainties or systematic errors unaccounted for in the determinations of the ages for individual objects, which makes them a not totally solid argument for supporting an age crisis. In any case, the story of the age crisis is not over yet and future improvements in the determination of age of astrophysical objects may give rise new controversies.

7.3 Large-scale structure

7.3.1 Theories

The formation of large-scale structure (a name given to the distribution of galaxies in space on large scales) is usually used as an argument to support the standard cosmological model. There are sophisticated simulations (e.g., Somerville and Bryan 2019, Aricò *et al* 2020) that show structures similar to those observed. The density distribution of the matter field—derived either analytically (Betancort-Rijo and López-Corredoira 2002) or through simulations (Klypin *et al* 2018)—approximately matches observations according to standard cosmologists. In figure 7.2, I show an example of such simulations produced by the Center for Cosmological Physics at the University of Chicago, and in figure 7.3 (Scodeggio *et al* 2018) we see structures observed in the real Universe, which resemble the simulations.

Nonetheless, one should be cautious when interpreting these kinds of simulations as a definitive proof that the standard cosmological model is correct. Reproducing some observations is a necessary condition to validate a theory but it is not enough, since there may be different ways to reproduce the same facts.

[2] Assuming $H_0 = 67.4$ km s^{-1} Mpc^{-1}, as derived from Cosmic Microwave Background Radiation data (Planck Collaboration 2018). Note, however, that there is some discussion concerning the value of the Hubble–Lemaître constant, with some groups using the local Cepheid–supernova distance ladder in galaxies that give a 10% greater value (see section 4.4) (Riess *et al* 2019), which would give a 10% lower age of the Universe, and this would start to be a more serious age crisis.

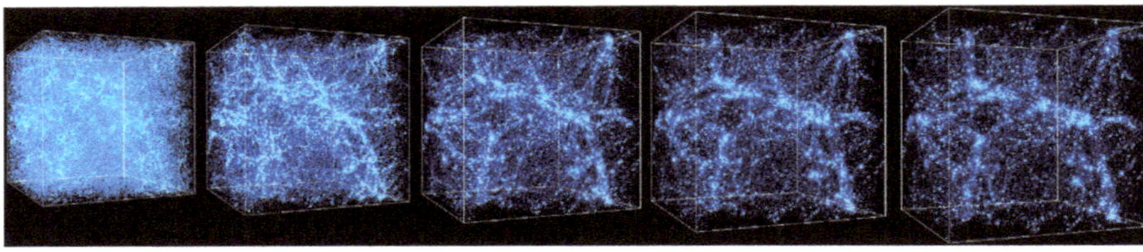

Figure 7.2. Simulation on the formation of clusters and large-scale filaments in the Cold Dark Matter model with dark energy. The frames show the evolution of structures in a 43 Mpc (comoving) box at $z = 10.0, 2.0, 1.0, 0.5, 0.0$. Simulations were performed at the National Center for Supercomputer Applications by Andrey Kravtsov (University of Chicago) and Anatoly Klypin (New Mexico State University). Visualisations by Andrey Kravtsov. Publicly available at: http://cosmicweb.uchicago.edu/filaments.html (version April 2021).

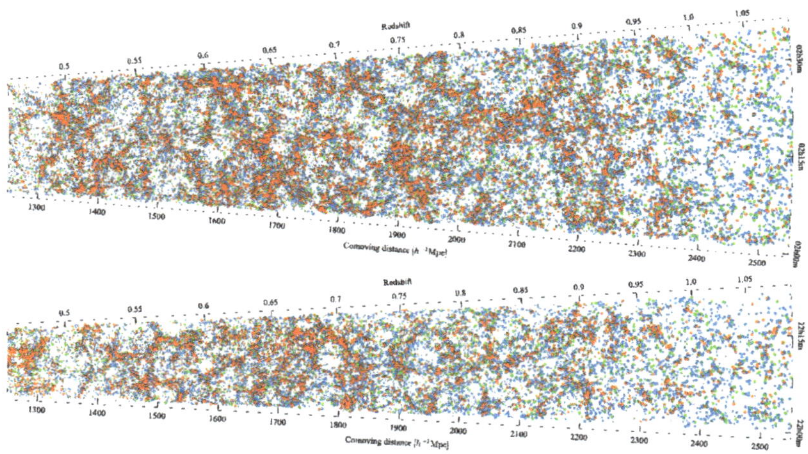

Figure 7.3. Distribution of galaxies at $0.45 < z < 1.1$ as shown by the final VIPERS PDR-2 catalogue in the W1 and W4 fields (top and bottom cone diagram, respectively). Galaxy positions are projected in declination (1.5 degrees for W4, about 1.8 degrees for W1). Each galaxy is represented by a filled circle of size proportional to its *B*-band luminosity and coloured according to its rest-frame $U - V$ colour. (Credit: Figure 15 of Scodeggio *et al* 2018, reproduced with permission © ESO.)

In cosmology, there are other theories that explain the formation of the large-scale structure: radiation given off by primordial galaxies and QSOs (Kurucz 1992); galaxies that beget other galaxies (Burbidge 1999); plasmas within magnetic fields and electric currents that create filaments and other structures (Lerner 1991, Battaner *et al* 1997, Florido and Battaner 1997, Battaner *et al* 1997, Battaner and Florido 1998); the Quasi Steady State Theory (Nayeri *et al* 1999), which reproduces the observed two-point correlation function; zero-active mass (Yennapureddy and Melia 2021); and others.

As usual, the alternative explanations do certainly not work perfectly. Neither does the standard model work perfectly, as we shall see in the following subsections.

7.3.2 Departures from homogeneity and isotropy

The standard theory requires large-scale homogeneity on scales of distances greater than a few tens of Mpc. The distribution of galaxies and clusters of galaxies should be random on very large scales. However, departures from homogeneity at these scales have been observed, indicating that a fractal Universe and regularity of structure are a frequent occurrence. A spatially inhomogeneous matter distribution can be compatible with the Copernican Principle but not with the stronger version of it, the Cosmological Principle, which requires the additional hypothesis of spatial homogeneity (Sylos Labini and Baryshev 2010). Some observations show large-scale inhomogeneities and peculiar velocities typical of a fractal on scales of up to at least 100 Mpc, so the density fluctuations inside the fractal inhomogeneity cell will lead to a strong disturbance of pure Friedmann–Lemaître–Robertson–Walker on scales less than 100 Mpc (Haggerty and Wertz 1972, Fang *et al* 1991, Ribeiro 1992b, 1992a, 1993). Some authors have pushed the limits of statistically significant non-homogeneity up to scales much greater than 100 Mpc (Best 2000, Yadav *et al* 2010, Sylos Labini *et al* 2009, Sylos Labini 2011). This is puzzling because the Hubble–Lemaître expansion law should be a consequence of homogeneity in matter distribution, and a striking linearity of the Hubble–Lemaître law at distances <25 Mpc (Peebles 1998, Ekholm *et al* 2001, Karachentsev *et al* 2002, Teerikorpi *et al* 2005) would suggest homogeneity above these scales.

Most cosmologists also appeal to the highly isotropic character of the CMBR as one of the principal justifications for assuming that the Universe is homogeneous on large scales. However, by itself, the fact that some observer sees isotropic background radiation is inconclusive, for this could also be true in an inhomogeneous Universe (Matravers *et al* 1995).

Another foundational assumption of the standard ΛCDM model is isotropy in the large scale of the structures. This has also been challenged numerous times, significant departure from isotropy having been found with respect to predictions concerning the distribution of different variables of sources: the average redshift for galaxies of a given brightness (corrected for extinction) (Mathewson *et al* 1992, Lindley 1992, Finkbeiner 1992, Hudson *et al* 1999), Hubble–Lemaître expansion flow (Wiltshire *et al* 2012), the distribution of radio galaxies (Bengaly *et al* 2018) or QSOs (Secrest *et al* 2021), the x-ray luminosity–temperature relation of galaxy clusters (Migkas *et al* 2020, 2021), cosmic acceleration (dark energy) traced by Type Ia supernovae (Colin *et al* 2019), extragalactic cosmic rays (The Pierre Auger Collaboration 2017), the fine structure constant (traced by QSO spectra) (King *et al* 2012), etc. Some of these anisotropies might be due to some strong peculiar velocity of the Local Group (see section 7.3.6) or underdensity (see section 7.3.5), but there are other possible causes. Many of these cited papers claim that the cosmological principle and its implementation within ΛCDM cosmology give rise to serious problems. Nonetheless, these observations are still under discussion and there are explanations in standard terms; for instance, the cosmic rays anisotropies may be explained in term of the magnetic field of the Milky Way (Globus *et al* 2019).

7.3.3 Oversized structures

In 3D maps of the large-scale Universe, with points representing galaxies or clusters of galaxies, one may see different structures, either by eye or by applying different statistical criteria. These overdensities might be the fruit of random fluctuations in a given field. In the standard cosmological model, there are small probabilities of large sizes in the density distribution of the matter field. The larger in amplitude and/or size these overdensities are, the smaller are their probabilities. Certainly, there is a lot of structure on the scales of galaxies or clusters of galaxies, and there is even some space correlation for clusters of galaxies, but this correlation is expected to decrease quickly for large distances, so finding large structures with large average densities and sizes much greater than the minimum scale at which the Universe becomes homogeneous should be a quite weird phenomenon. However, the scientific literature contains plenty of cases of megastructures.

Two-dimensional wall-like or sheet-like galaxy features at fairly regular intervals were found to be common (Broadhurst *et al* 1990, Kurki-Suonio 1990), with characteristic scales >140 Mpc.[3] These structures may also be quite thin in one of their dimensions, constituting filaments. Some prominent cases of big walls are the 'Great Wall' with a length of around 240 Mpc (Geller and Huchra 1989), the filaments of clusters Perseus-Pegasus, which is over 400 Mpc long (Batuski and Burns 1985), the Sloan Great Wall, extending over 420 Mpc (Einasto *et al* 2011), the South Pole Wall, also extending over around 420 Mpc (Pomarède *et al* 2020), and the 'Giant Arc', which is a 1.0 Gpc long filament of Mg II absorbing galaxies (López 2021). Maps of some of these structures are shown in figure 7.4. An analysis of the distribution of counts of galaxies versus photometric redshift, $N(z)$, in a grid of deep

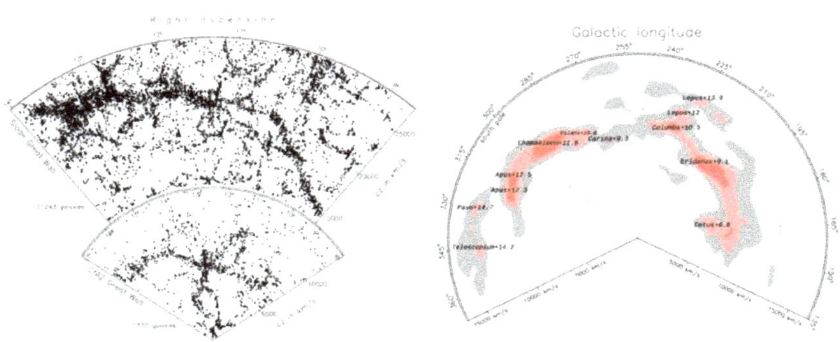

Figure 7.4. A comparison of the extent of the South Pole Wall with two other nearby structures: the CFA Great Wall (bottom left) and the Sloan Great Wall (top left). The South Pole Wall on the right is represented by isodensity contour levels derived from the peculiar velocity field. The linear scales are the same. The South Pole Wall is roughly intermediate in distance from us compared with the other two. (Reproduced by permission of the AAS from figure 3 of Pomarède *et al* 2020.)

[3] Throughout this subsection, a value of $H_0 = 70$ km s^{-1} Mpc^{-1} is adopted. Comoving scales are given; that is, equivalent sizes with the size of the Universe today.

fields reveals the possible existence of super-large structures with a contrast $\Delta N/N \sim 50\%$, and tangential and radial dimensions of about 1000 Mpc (Massey *et al* 2007, Nabokov and Baryshev 2010). However, photometric redshifts are not as reliable as spectroscopic redshifts.

Walls or filaments comprise superclusters (clusters of galaxy clusters) or isolated galaxy clusters. Superclusters may be quite big too, one supercluster alone even constituting a large structure. Many of these huge structures have been observed, such as the eight superclusters with median diameter between 130 and 240 Mpc catalogued by Einasto *et al* (1997), or the local supercluster 'Laniakea' (with a major axis of 160 Mpc) containing the Milky Way (Tully *et al* 2014), the Horologium–Reticulum supercluster (major axis of 170 Mpc) (Fleenor *et al* 2005), the 'Saraswati' supercluster (~200 Mpc) (Bagchi *et al* 2017), 'Caelum' (major axis = 280 Mpc) (Einasto *et al* 2001), superclusters A and B of the BOSS Great Wall at $z = 0.47$ each of diameter of around 250 Mpc (Lietzen *et al* 2016), and a protosupercluster (an overdensity in the process of forming a supercluster) at $z = 2.45$ with a major axis of 150 Mpc (comoving) (Cucciati *et al* 2018).

Large QSO groups with sizes up to 300 Mpc are quite common (Komberg *et al* 1996, Clowes 2001), but there are even larger pieces. The most extreme proposal for a group of this type is U1.27 at $z = 1.27$ (Clowes *et al* 2013), which has an estimated major axis of 1.24 Gpc.

Moreover, there are claims of having found the largest structures ever reported with groups of Gamma Ray Bursts (GRBs) (Horvath *et al* 2014, Balázs *et al* 2015) with a major axis reaching the amazing size of 3.0 Gpc for the Hercules–Corona Borealis Great Wall, but there are few GRBs in their maps, and the significance of the detection of a structure above the noise is low (Balázs 2018, Christian 2020).

Some of these huge structures apparently show excess clustering with respect to the predictions of the standard model, their existence is difficult to reconcile with the assumption of Gaussian initial conditions within typical parameters of the standard model; they even may cast doubt on the validity of the cosmological principle (Sheth and Diaferio 2011, Thomas *et al* 2011, Clowes *et al* 2013, Mészáros 2019, Rozgacheva and Kuvshinova 2020, López 2021). However, some overdensities of very great size may be found as statistical fluctuations, patterns in noise (Kaiser and Peacock 1991, Nadathur 2013, Park *et al* 2012, 2015, Balázs 2018). The discussion is lively and is not yet over. In spite of the impressive size of the huge structures, one should bear in mind that their low density contrast $\frac{\Delta\rho}{\rho_c}$ (ρ_c is the average matter density of the Universe), and that it is possible to find small fluctuations when we explore the whole observable Universe. Nonetheless, given the number of these discoveries, I miss a study that statistically analyses the whole distribution in terms of the predicted structure given by ΛCDM, rather than calculating the probabilities of single structures.

7.3.4 Clusters of galaxies

According to the official version, the simulations resemble the observations on the scale of clusters. There are, however, some papers that draw attention to a problem with profiles of cluster haloes: ΛCDM predicts a low, shallow concentration and density profiles in contrast to observations that indicate a denser high concentration cluster haloes (Perivolaropoulos 2009). Furthermore, according to Perivolaropoulos (2009), ΛCDM predicts a smaller fraction of disc galaxies owing to recent mergers that are expected to disrupt cold rotationally supported discs.

It was also found that the large-scale structure measured by weak gravitational lensing is smoother than that from clustering measurements (Jeffrey *et al* 2021). Something is not being properly understood in this matter.

Another episode within this long list of disagreements between the predictions of the standard model and cosmological observations is the formation of clusters or protoclusters of galaxies at very high redshift associated with epochs that are too early for the measured masses. There are protoclusters of galaxies at redshifts greater than four (Oteo *et al* 2018, Miller *et al* 2018) or even at redshifts higher than six (Overzier 2018). Compact groups are a short-lived phase of group evolution, and we would expect them to be extremely rare at high redshift (Sohn *et al* 2015); instead, we find a significant number of compact groups in the nearby Universe, with well over 100 identified as Hickson groups.

7.3.5 Oversized voids

Another matter for discussion is the existence of big voids, that is, very large volumes of space with high underdensities of galaxies and consequently of mass: $\frac{\Delta\rho}{\rho_c} < 0$.

A local void of 8 Mpc radius is observed to be too empty in comparison to standard cosmological predictions (Tikhonov *et al* 2009, Peebles and Nusser 2010). Indeed, not only is there a local void of a few Mpc around us, but there is a huge volume, called the KBC void, which is a local underdensity with the observed relative density contrast $\delta \equiv 1 - \rho/\rho_c = 0.46 \pm 0.06$ in the range of distances between 40 and 300 Mpc around the Local Group (so it has a maximum diameter of 600 Mpc). Some simulations show that the KBC void causes a 6σ discrepancy with ΛCDM (Keenan *et al* 2013, Haslbauer *et al* 2020). Other surveys at different wavelengths and with different sources (Böhringer *et al* 2015, Whitbourn and Shanks 2016, Wong *et al* 2022) concur that there is a huge underdense region around us, although they give different numbers for the size and the underdensity.

Neither is it the only huge underdense region: there are other similar or bigger structures observed at $z > 0.3$, with sizes of up to 1 Gpc (although not necessarily spherical regions) and $\delta \sim 0.5$ (Shimakawa *et al* 2021 [figure 11]), but only with photometric redshift surveys so far, which might contain certain errors in the distance determination of the sources.

There is yet another minor question: standard cosmology also predicts more dwarf or irregular galaxies in voids than observed (Perivolaropoulos 2009).

For the main topic of this subsection, the size and emptiness of some voids, again, we may express considerations similar to those voiced in subsection 7.3.3. There are plenty of discoveries of huge voids in the literature, associated in many cases with huge walls and filaments around them, which, according to some authors (Tavasoli *et al* 2013, Haslbauer *et al* 2020), are incompatible with the probability of voids expected in the standard cosmological model (e.g., Betancort-Rijo 1990). Some authors are able to explain some of the voids as random fluctuations (e.g., Park *et al* 2012), but many remain without explanation so far.

7.3.6 Streaming velocities

Some anisotropies of the large-scale distribution (see section 7.3.2) might indicate there are strong peculiar motions in the local Universe. As a matter of fact, the Local Group's absolute space velocity inferred from the CMBR dipole anisotropy was measured (Lauer and Postman 1994, Colin *et al* 2017) and it is too high in comparison to the velocities expected for the kinds of structures predicted by the standard model. The standard model interprets this as the existence of a puzzling group flow of galaxies relative to the microwave radiation on scales of at least 130 Mpc. Earlier, the existence of this flow led to the hypothesis of a 'Great Attractor' pulling all these galaxies in its direction. But in newer studies, no infall was found on the other side of this hypothetical feature, possibly because it is difficult to observe given that it is masked behind the plane of the Milky Way. In any case, no evidence for the existence of the Great Attractor has yet been found, and if there turns out not to be any Great Attractor, the large stream motion of the Local Group is problematic for the standard interpretation.

There are other anomalously large bulk flows on scales of 100 Mpc h^{-1} or greater that challenge the ΛCDM model, as well as a large class of competitive models for dark energy and modified gravity (Ayaita *et al* 2009). There is also the problem that the motions of some individual galaxies or clusters seem to be too large. This is the case, for instance, of El Gordo (ACT-CL J0102-4915), an extremely massive galaxy cluster at redshift $z = 0.87$ composed of two subclusters with mass ratio 3.6 merging at a speed $v_{\rm infall} = 2500$ km s^{-1}. Such a fast collision between individually rare massive clusters is unexpected in ΛCDM cosmology at such high z and is ruled out at 6.2σ level (Asencio *et al* 2021).

Results from the measurement of large-scale peculiar velocities of clusters of galaxies using an x-ray cluster catalogue and CMBR data (Kashlinsky *et al* 2009) find a strong and coherent bulk flow on scales out to at least $>300\ h^{-1}$ Mpc. This flow is difficult to explain by gravitational evolution within the framework of the ΛCDM model. The results were confirmed with more recent Planck CMBR data (Atrio-Barandela *et al* 2015), which show that the observed dipole of the kinetic Sunyaev–Zel'dovich effect that gives rise to the dark flow is not contaminated by the thermal Sunyaev–Zel'dovich effect. This dipole was not found with Type Ia supernova peculiar velocities (once the Hubble–Lemaître flow is removed), but the sample is quite inhomogeneous. Possible interpretations are: (i) an intrinsic CMBR dipole in the last scattering surface (but the prediction of inflation gives a much lower dipole), or (ii) the flow is real. Both interpretations are problematic.

7.3.7 BAO peak

Baryon Acoustic Oscillations (BAO) found in the large-scale structure (Eisenstein 2005, Anderson *et al* 2014) are a strong point in favour of the standard model. This is perhaps one of the most successful predictions of the ΛCDM model related to large-scale structure in the last two decades. BAO is observed in the two-point correlation function as a peak at a distance of around 150 Mpc. It is interpreted as fluctuations in the density of the visible baryonic matter of the Universe, caused by acoustic density waves in the primordial plasma of the early Universe. The oscillations that produce the peaks and troughs in the CMBR angular power spectrum can be related to this BAO peak in the large-scale structure. The length of 150 Mpc is given by the maximum distance the acoustic waves could travel in the primordial plasma before the epoch of recombination, which stopped the expansion of the plasma density waves, freezing them into place.

But not everything is rosy concerning this topic. The BAO peak position depends slightly on the environment of galaxies, whether or not in superclusters, a point that is not predicted by the standard model (Roukema *et al* 2015). On the other hand, there are other cosmological models that also explain the BAO peak (e.g., Kiselev and Timofeev 2012, Dolgov *et al* 2014, Melia 2020). It is not surprising that, if we accept that the CMBR power spectrum peaks and the BAO peak are related, we may have similar explanations for both of them in terms of some kind of acoustic wave scenario, and if there are alternative ideas for CMBR anisotropies (see section 5.4) these may also be applicable for BAO explanations.

7.3.8 Different cosmological parameters for CMBR analyses

The dream of present-day standard cosmologists is that their model will fit all kinds of data consistently, and that the cosmological parameters derived from some types of observations are the same as these parameters obtained with an independent method for different types of data. This was indeed the goal of 'precision cosmology'. However, the Universe seems to be more complex than the view seen from our simplistic perspective, and we continuously find ourselves enmeshed in ever-increasing controversies as the data become more precise.

General disagreements over large-scale structure and other cosmological data are also rife. There is an inconsistency between current geometry data sets (Type Ia supernova data, BAO peak, high multipoles of CMBR) and growth data sets (redshift space distortions, weak lensing, CMB lensing, the Sunyaev–Zel'dovich effect, and low multipoles of the CMBR) (Lin and Ishak 2017a), and between CMBR parameters and large-scale structure parameters up to $\approx 3.0\sigma$ (Lin and Ishak 2017b). In particular, a discrepancy of at least 3σ is found between the results of large-scale structure and CMBR anisotropies analyses for such parameters as σ_8 (Asgari *et al* 2021, Jeffrey *et al* 2021), although there are also proposals to reduce this disagreement by studying different reconstructions of the growth history of matter inhomogeneity (Benisty 2021). We also have the Hubble–Lemaître constant controversy (see section 4.4), just to mention the most popular ones, but the list continues to grow.

7.4 Large-scale problems for the standard model

The predictions of the ΛCDM model do not work on galactic scales (see section 3.4), and it is not clear that they are working on supercluster scales, as we have seen in sections 7.3.3 and 7.3.5. On the scale of clusters the doubts are minimal, but not totally absent, as we have seen in section 7.3.4. Hence, there is no universal agreement on any scale regarding the capacity of the standard model to match observations. Moreover, there are certain caveats to explanations of the distribution of velocities, isotropy, the constraint of the reionization epoch, the formation of massive galaxies and black holes much faster than expected, high metallicities and dust abundances at high z, and others. Even the most successful elements, such as the formation of structures and the observed BAO acoustic waves have difficulties in explaining certain details. Of course, no theory is perfect. Let us, however, accept that ΛCDM gives at least a 'rough' picture of the reality: that simulations like the one in figure 7.2 resemble quite well, although not perfectly, the distributions observed as shown in figure 7.3.

Large-scale structure theory is mainly based on two assumptions: the standard cosmological metric and the idea that only gravity—general relativity, which can be approximated to a Newtonian gravity for the purpose of explaining the formation of structures—is responsible for the formation of these structures. It works reasonably well when we do not seek explanations down to the smallest detail, but are there no other possible cosmological scenarios and other possible gravity laws, or other forces different from gravity that can also 'roughly' explain the distributions? And if such alternative models are not as accurate as the standard model is, it is mainly because less manpower and computer time, and much fewer theoreticians have been dedicated dedicated to sweeping away the emerging inconsistencies with new ad hoc assumptions.

References

Anderson L, Aubourg E and Bailey S *et al* 2014 The clustering of galaxies in the SDSS-III baryon oscillation spectroscopic survey: baryon acoustic oscillations in the data releases 10 and 11 galaxy samples *Mon. Not. R. Astron. Soc.* **441** 24–62

Aricò G, Angulo R E, Hernández-Monteagudo C, Contreras S, Zennaro M, Pellejero-Ibañez M and Rosas-Guevara Y 2020 Modelling the large-scale mass density field of the universe as a function of cosmology and baryonic physics *Mon. Not. R. Astron. Soc.* **495** 4800–19

Asencio E, Banik I and Kroupa P 2021 A massive blow for ΛCDM–the high redshift, mass, and collision velocity of the interacting galaxy cluster El Gordo contradicts concordance cosmology *Mon. Not. R Astron. Soc.* **500** 5249–67

Asgari M, Lin C-A and Joachimi B *et al* 2021 Kids-1000 cosmology: cosmic shear constraints and comparison between two point statistics *Astron. Astrophys.* **645** 31

Atrio-Barandela F, Kashlinsky A, Ebeling H, Fixsen D J and Kocevski D 2015 Probing the dark flow signal in WMAP 9-year and Planck cosmic microwave background maps *Astrophys. J.* **810** 143

Ayaita Y, Weber M and Wetterich C 2009 Peculiar velocity anomaly from forces beyond gravity? (arXiv:0908.2903)

Bagchi J, Sankhyayan S, Sarkar P, Raychaudhury S, Jacob J and Dabhade P 2017 Saraswati: an extremely massive ~200 megaparsec scale supercluster *Astrophys. J.* **844** 25

Balázs L G, Bagoly Z, Hakkila J E, Horváth I, Kóbori J, Rácz I I and Tóth L V 2015 A giant ring-like structure at $0.78<z<0.86$ displayed by GRBs *Mon. Not. R. Astron. Soc.* **452** 2236–46

Balázs L G, Lí dia R and Gábor T 2018 Some statistical remarks on the giant GRB ring *Mon. Not. R. Astron. Soc.* **473** 3169–79

Barth A J, Martini P, Nelson C H and Ho L C 2003 Iron emission in the $z = 6.4$ quasar SDSS j114 816.64+525 150.3 *Astrophys. J. Lett.* **594** L95–8

Battaner E and Florido E 1998 Magnetic fields and large scale structure in a hot universe. IV. The egg-carton universe *Astron. Astrophys.* **338** 383–5

Battaner E, Florido E and García-Ruiz J M 1997 Magnetic fields and large scale structure in a hot universe. III. The polyhedric network *Astron. Astrophys.* **327** 8–10

Battaner E, Florido E and Jiménez-Vicente J 1997 Magnetic fields and large scale structure in a hot universe. I. General equations *Astron. Astrophys.* **326** 13–22

Batuski D J and Burns J O 1985 A possible 300 megaparsec filament of clusters of galaxies in Perseus-Pegasus *Astrophys. J.* **299** 5–13

Becker G D, Bolton J S and Lidz A 2015 Reionisation and high-redshift galaxies: the view from quasar absorption lines *Publ. Astron. Soc. Aust.* **32** E045

Becker R H, Fan X and White R L *et al* 2001 Evidence for reionization at $z \sim 6$: detection of a Gunn-Peterson trough in a $z = 6.28$ quasar *Astron. J.* **122** 2850–7

Bengaly C A P, Maartens R and Santos M G 2018 Probing the cosmological principle in the counts of radio galaxies at different frequencies *J. Cosmol. Astropart. Phys.* **2018** 31

Benisty D 2021 Quantifying the s_8 tension with the redshift space distortion data set *Phys. Dark Universe* **31** 100766

Best J S 2000 An examination of the large-scale clustering of the Las Campanas Redshift Survey *Astrophys. J.* **541** 519–26

Betancort-Rijo J 1990 Probabilities of voids *Mon. Not. R. Astron. Soc.* **246** 608–15

Betancort-Rijo J and López-Corredoira M 2002 Probability distribution of density fluctuations in the nonlinear regime *Astrophys. J.* **566** 623–9

Böhringer H, Chon G, Bristow M and Collins C A 2015 The extended ROSAT-ESO flux-limited x-ray galaxy cluster survey (reflex II). V. Exploring a local underdensity in the southern sky *Astron. Astrophys.* **A26** 8

Bouwens R J, Illingworth G D and Labbe I *et al* 2011 A candidate redshift $z \sim 10$ galaxy and rapid changes in that population at an age of 500 Myr *Nature* **469** 504–7

Bowman J D, Rogers A E E, Monsalve R A, Mozdzen T J and Mahesh N 2018 An absorption profile centred at 78 megahertz in the sky-averaged spectrum *Nature* **555** 67–70

Broadhurst T J, Ellis R S, Koo D C and Szalay A S 1990 Large-scale distribution of galaxies at the galactic poles *Nature* **343** 726–8

Bunker A J, Wilkins S and Ellis R S *et al* 2010 The contribution of high-redshift galaxies to cosmic reionization: new results from deep WFC3 imaging of the Hubble ultra deep field *Mon. Not. R. Astron. Soc.* **409** 855–66

Burbidge G R 1999 Explosive cosmogony and the quasi-steady state cosmology *Cosmological Parameters and the Evolution of the Universe* ed K Sato (Dordrecht: Kluwer) pp 286–9

Carlberg R G 1991 A limit on the cosmological constant *Astrophys. J.* **375** 429–31

Castro-Rodríguez N and López-Corredoira M 2012 The age of extremely red and massive galaxies at very high redshift *Astron. Astrophys.* **A31** 6

Chowdhury T R, Mukherjee S and Paul S 2021 Cosmic microwave background constraints on a physical model of reionization *Mon. Not. R. Astron. Soc.* **501** L7–11

Christian S 2020 Re-examining the evidence of the Hercules-Corona Borealis Great Wall *Mon. Not. R. Astron. Soc.* **495** 4291–6

Clowes R G 2001 Large quasar groups—a short review *The New Era of Wide Field Astronomy (ASP Conf. Ser. 232)* ed R Clowes, A Adamson and G Bromage (San Francisco, CA: Astronomical Society of the Pacific) pp 108–13

Clowes R G, Harris K A, Raghunathan S, Campusano L E, Söchting I K and Graham M J 2013 A structure in the early universe at $z \sim 1.3$ that exceeds the homogeneity scale of the r-w concordance cosmology *Mon. Not. R. Astron. Soc.* **429** 2910–6

Colin J, Mohayaee R, Rameez M and Sarkar S 2017 High-redshift radio galaxies and divergence from the CMB dipole *Mon. Not. R. Astron. Soc.* **471** 1045–55

Colin J, Mohayaee R, Rameez M and Sarkar S 2019 Evidence for anisotropy of cosmic acceleration *Astron. Astrophys.* **631** L13–8

Constantin A, Shields J C, Hamann F, Foltz C B and Chaffee F H 2002 Emission-line properties of $z > 4$ quasars *Astrophys. J.* **565** 50–62

Cucciati O, Lemaux B C and Zamorani G *et al* 2018 The progeny of a cosmic titan: a massive multi-component proto-supercluster in formation at $z = 2.45$ in VUDS *Astron. Astrophys.* **A49** 21

Davies F B, Becker G D and Furlanetto S R 2018 Determining the nature of late Gunn-Peterson troughs with galaxy surveys *Astrophys. J.* **155** 12

Dietrich M, Hamann F, Appenzeller I and Vertergaard M 2003 Fe II/Mg II emission-line ratio in high-redshift quasars *Astrophys. J.* **596** 817–29

Dolgov A, Halenka V and Tkachev I 2014 Power-law cosmology, SN Ia, and BAO *J. Cosmol. Astropart. Phys.* **10** 47

Dopita M A, Krauss L M, Sutherland R S, Kobayashi C and Lineweaver C H 2011 Re-ionizing the universe without stars *Astrophys. Space Sci.* **335** 345–52

Dunne L, Eales S, Ivison R, Morgan H and Edmunds M 2003 Type II supernovae as a significant source of interstellar dust *Nature* **424** 285–7

Einasto M, Einasto J, Tago E, Müller V and Andernach H 2001 Optical and x-ray clusters as tracers of the supercluster-void network. I. Superclusters of Abell and x-ray clusters *Astron. J.* **122** 2222–42

Einasto M, Liivamägi L J and Tempel E *et al* 2011 The Sloan Great Wall. Morphology and galaxy content *Astrophys. J.* **736** 51

Einasto M, Tago E, Jaaniste J, Einasto J and Andernach H 1997 The supercluster-void network. I. The supercluster catalogue and large-scale distribution *Astron. Astrophys. Suppl. Ser* **123** 119–33

Eisenstein D J 2005 Dark energy and cosmic sound *New Astron. Rev.* **49** 360–5

Ekholm T, Baryshev Y, Teerikorpi P, Hanski M O and Paturel G 2001 On the quiescence of the Hubble flow in the vicinity of the local group. A study using galaxies with distances from the Cepheid PL-relation *Astron. Astrophys.* **368** L17–20

Eyles L P, Bunker A J, Stanway E R, Lacy M, Ellis R S and Doherty M 2005 Spitzer imaging of i'-drop galaxies: old stars at $z \approx 6$ *Mon. Not. R. Astron. Soc.* **364** 443–54

Fan X, Narayanan V K and Lupton R H *et al* 2001 A survey of $z > 5.8$ quasars in the Sloan Digital Sky survey. I. Discovery of three new quasars and the spatial density of luminous quasars at $z \sim 6$ *Astron. J.* **122** 2833–49

Feng L L, Ruffini R and Mo H J 1991 The cellular structure of the universe and cosmological tests *Astron. Astrophys.* **243** 283–94

Fèvre O L, Abraham R and Lilly S J *et al* 2000 Hubble space telescope imaging of the CFRS and LDSS redshift surveys–IV. Influence of mergers in the evolution of faint field galaxies from $z \sim 1$ *Mon. Not. R. Astron. Soc.* **311** 565–75

Finkbeiner A 1992 Mapping the river in the sky *Science* **257** 1208–10

Finkelstein S L 2010 Searching for the first galaxies *New Horizons in Astronomy: Frank N. Bash Symp. 2009 (ASP Conf. Ser., Vol. 432)* L M Stanford, J D Green, H Hai and Y Mao (San Francisco, CA: Astronomical Society of the Pacific) pp 33–48

Fleenor M C, Rose J A, Christiansen W A, Hunstead R W, Johnston-Hollitt M, Drinkwater M J and Saunders W 2005 Large-scale velocity structures in the Horologium-Reticulum supercluster *Astron. J.* **130** 957–67

Florido E and Battaner E 1997 Magnetic fields and large-scale structure in a hot universe. II. Magnetic flux tubes and filamentary structure *Astron. Astrophys.* **327** 1–7

Freudling W, Corbin M R and Korista K T 2003 Iron emission in $z \sim 6$ QSOs *Astrophys. J. Lett.* **587** L67–70

Ge Z S, Bi S L, Chen Y Q, Li T D, Zhao J K, Liu K, Ferguson J W and Wu Y Q 2016 Ages of 70 dwarfs of three populations in the solar neighborhood: considering O and C abundances in stellar models *Astrophys. J.* **833** 161

Geller M J and Huchra J P 1989 Mapoing the universe *Science* **246** 897–903

Globus N, Piran N, Hoffman Y, Carlesi E and Pomarède D 2019 Cosmic-ray anisotropy from large scale structure and the effect of magnetic horizons *Mon. Not. R. Astron. Soc.* **484** 4167–73

Guo Q, White S and Boylan-Kolchin M *et al* 2011 From dwarf spheroidals to cD galaxies: simulating the galaxy population in a ΛCDM cosmology *Mon. Not. R. Astron. Soc.* **413** 101–31

Haggerty M J and Wertz J R 1972 On the redshift-magnitude relation in hierarchical cosmologies *Mon. Not. R. Astron. Soc.* **155** 495–503

Haslbauer M, Banik I and Kroupa P 2020 The KBC void and Hubble tension contradict ΛCDM on a Gpc scale—Milgromian dynamics as a possible solution *Mon. Not. R. Astron. Soc.* **499** 2845–83

Horvath I, Hakkila J and Bagoly Z 2014 Possible structure in the GRB sky distribution at redshift two *Astron. Astrophys.* **L12** 4

Hudson M J, Smith R J, Lucey J R, Schelegel D J and Davies R L A 1999 Large-scale bulk flow of galaxy clusters *Astrophys. J. Lett.* **512** L79–82

Iwamuro F, Motohara K, Maihara T, Kimura M, Yoshii Y and Doi M 2002 Fe II/Mg II emission-line ratios of QSOs within $0<z<5.3$ *Astrophys. J.* **565** 63–77

Jarosik N, Bennett C L and Dunkley J *et al* 2011 Seven-year Wilkinson Microwave Anisotropy Probe (WMAP) observations: sky maps, systematic errors, and basic results *Astrophys. J. Supp. Ser.* **192** 14

Jeffrey N, Gatti M and Chang C *et al* 2021 Dark energy survey year 3 results: curved-sky weak lensing mass map reconstruction *Mon. Not. R. Astron. Soc.* **505** 4626–45

Jiang L, Kashikawa N and Wang S *et al* 2020 Evidence for GN-z11 as a luminous galaxy at redshift 10.957 *Nat. Astron.* **5** 256–61

Kaiser N and Peacock J A 1991 Power-spectrum analysis of one-dimensional redshift surveys *Astrophys. J.* **379** 482–506

Karachentsev I D, Sharina M E and Makarov D I *et al* 2002 The very local Hubble flow *Astron. Astrophys.* **389** 812–24

Kashlinsky A, Atrio-Barandela F, Kocevski D and Ebeling H A 2009 Measurement of large-scale peculiar velocities of clusters of galaxies: results and cosmological implications *Astrophys. J. Lett.* **686** L49–52

Keenan R C, Barger A J and Cowie L L 2013 Evidence for a ∼300 megaparsec scale under-density in the local galaxy distribution *Astrophys. J.* **775** 62

King J A, Webb J K, Murphy M T, Flambaum V V, Carswell R F, Bainbridge M B, Wilczynska M R and Koch F E 2012 Spatial variation in the fine-structure constant–new results from VLT/UVES *Mon. Not. R. Astron. Soc.* **422** 3370–414

Kiselev V V and Timofeev S A 2012 Cosmological extrapolation of modified Newtonian dynamics *Class. Quantum Grav.* **29** 065015

Klypin A, Prada F, Betancort-Rijo J and Albareti F D 2018 Density distribution of the cosmological matter field *Mon. Not. R. Astron. Soc.* **481** 4588–601

Kogut A, Spergel D N and Barnes C *et al* 2003 First-year Wilkinson Microwave Anisotropy Probe (WMAP) observations: temperature-polarization correlation *Astrophys. J. Supp. Ser.* **148** 161–73

Komberg B V, Kravtsov A V and Lukash V N 1996 The search and investigation of large quasar groups *Mon. Not. R. Astron. Soc.* **282** 713–22

Kurki-Suonio H 1990 Galactic beads on a cosmic string *Sci. News* **137** 287

Kurucz R L 1992 Gedanken astrophysics. The universe since recombination *Comments Astrophys.* **16** 1–15

Labbé I, Huang J and Franx M *et al* 2005 IRAC mid-infrared imaging of the Hubble deep field-south: star formation histories and stellar masses of red galaxies at $z > 2$ *Astrophys. J.* **624** L81–4

Laporte N, Ellis R S and Boone F *et al* 2017 Dust in the reionization era: Alma observations of a $z = 8.38$ gravitationally lensed galaxy *Astrophys. J.* **837** L21

Lauer T R and Postman M 1994 The motion of the local group with respect to the 15,000 kilometer per second Abell cluster inertial frame *Astrophys. J.* **425** 418–38

Lerner E J 1991 *The Big Bang Never Happened: A Startling Refutation of the Dominant Theory of the Origin of the Universe* (Toronto: Random House)

Lietzen H, Tempel E and Liivamägi L J *et al* 2016 Discovery of a massive supercluster system at z 0.47 *Astron. Astrophys.* **588** L4

Lin W and Ishak M 2017a Cosmological discordances: a new measure, marginalization effects, and application to geometry versus growth current data set *Phys. Rev.* D **96** 023532

Lin W and Ishak M 2017b Cosmological discordances II: Hubble constant, Planck and large-scale-structure data sets *Phys. Rev.* D **96** 083532

Lindley D 1992 Not so great attractor? *Nature* **356** 657

López A 2021 A giant arc on the sky. Oral presentation in: 238th meeting of the American Astron. Soc. virtual, 7–9 June 2021

López-Corredoira M, Vazdekis A, Gutiérrez C M and Castro-Rodríguez N 2017 Stellar content of extremely red quiescent galaxies at $z > 2$ *Astron. Astrophys.* **600** 13

Maiolino R, Juarez Y, Mujica R, Nagar N and Oliva E 2003 Early star formation traced by the highest redshift quasars *Astrophys. J. Lett.* **596** L155–8

Malhotra S and Rhoads J 2004 Luminosity functions of lyα emitters at redshifts z=6.5 and z=5.7 : Evidence against reionization at z<=6.5 *Astrophys. J. Lett.* **617** L5–L8

Malloy M and Lidz A 2015 How to search for islands of neutral hydrogen in the $z \sim 5.5$ IGM *Astrophys. J.* **799** 18

Manrique A and Salvador-Solé E 2015 An improved treatment of cosmological intergalactic medium evolution *Astrophys. J.* **803** 103
Massey R, Rhodes J and Ellis R *et al* 2007 Dark matter maps reveal cosmic scaffolding *Nature* **445** 286–90
Mathewson D S, Ford V L and Buchhorn M 1992 No back-side infall into the great attractor *Astrophys. J. Lett.* **389** L5–8
Matravers D R, Ellis G F R and Stoeger W R 1995 Complementary approaches to cosmology—relating theory and observations *Q. J. R. Astron. Soc.* **36** 29–45
McGreer I D, Mesinger A and D'Odorico V 2015 Model-independent evidence in favour of an end to reionization by $z \approx 6$ *Mon. Not. R. Astron. Soc.* **447** 499–505
Melia F 2020 Reassessing dust's role in forming the CMB *Eur. Phys. J. Plus* **135** 511
Mészáros A 2019 An oppositeness in the cosmology: distribution of the gamma ray bursts and the cosmological principle *Astron. Nachr.* **340** 564–9
Migkas K, Pacaud F, Schellenberger G, Erler J, Nguyen-Dang N T, Reiprich T H, Ramos-Ceja M E and Lovisari L 2021 Cosmological implications of the anisotropy of ten galaxy cluster scaling relations *Astron. Astrophys.* **649** 38
Migkas K, Schellenberger G, Reiprich T H, Pacaud F, Ramos-Ceja M E and Lovisari L 2020 Probing cosmic isotropy with a new x-ray galaxy cluster sample through the $L_X - T$ scaling relation *Astron. Astrophys.* **636** A15
Miller T B, Chapman S C and Aravena M *et al* 2018 A massive core for a cluster of galaxies at a redshift of 4.3 *Nature* **556** 469–72
Nabokov N V and Baryshev Y 2010 Method for analyzing the spatial distribution of galaxies on gigaparsec scales. II. Application to a grid of the HUDF-FDF-COSMOS-HDF surveys *Astrophysics* **53** 101–11
Nadathur S 2013 Seeing patterns in noise: gigaparsec-scale 'structures' that do not violate homogeneity *Mon. Not. R. Astron. Soc.* **434** 398–406
Nayeri A, Engineer S, Narlikar J V and Hoyle F 1999 Structure formation in the quasi-steady state cosmology: a toy model *Astrophys. J.* **525** 10–6
Neeleman M, Xavier Prochaska J, Kanekar N and Rafelski M 2020 A cold, massive, rotating disk galaxy 1.5 billion years after the big bang *Nature* **581** 269–72
Oesch P A, Brammer G and van Dokkum P G *et al* 2016 A remarkably luminous galaxy at $z = 11.1$ measured with Hubble space telescope Grism spectroscopy *Astrophys. J.* **819** 129
Oteo I, Ivison R J and Dunne L *et al* 2018 An extreme protocluster of luminous dusty starbursts in the early universe *Astrophys. J.* **856** 72
Overzier R A 2018 The realm of the galaxy protoclusters. A review *Astron. Astrophys. Rev.* **24** 14
Park C, Choi Y-Y, Kim J, Gott J R III, Kim S S and Kim K-S 2012 The challenge of the largest structures in the universe to cosmology *Astrophys. J. Lett.* **759** L7
Park C, Song H, Einasto M, Lietzen H and Heinamaki P 2015 Large SDSS quasar groups and their statistical significance *J. Korean Astron. Soc.* **48** 75–82
Peacock J A 1985 The high-redshift evolution of radio galaxies and quasars *Mon. Not. R. Astron. Soc.* **217** 601–31
Peebles P J E 1998 The standard cosmological model *Le Rencontres de Physique de la Vallee d'Aoste: Results and Perspectives in Particle Physics* ed M Greco (Frascati: Poligrafica Laziale s.r.l.) p 39
Peebles P J E and Nusser A 2010 Nearby galaxies as pointers to a better theory of cosmic evolution *Nature* **465** 565–9

Pérez-González P G, Rieke G H and Villar V *et al* 2008 The stellar mass assembly of galaxies from z = 0 to z = 4: analysis of a sample selected in the rest-frame near-infrared with Spitzer *Astrophys. J.* **675** 234–61

Perivolaropoulos L 2009 Six puzzles for LCDM cosmology *The Problems of Modern Cosmology* ed P M Lavrov (Tomsk: Tomsk State Pedagogical University Press) pp 245–54

Planck Collaboration 2015 Planck 2015 results. XIII. Cosmological parameters *Astron. Astrophys.* **594** A13

Planck Collaboration 2018 Planck 2018 results. VI. Cosmological parameters *Astron. Astrophys.* **641** A6

Pomarède D, Tully R B, Graziani R, Courtois H M, Hoffman Y and Lezmy J 2020 Cosmicflows-3: the south pole wall *Astrophys. J.* **897** 133

Qin Y, Poulin V, Mesinger A, Greig B, Murray S and Park J 2020 Reionization inference from the CMB optical depth and e-mode polarization power spectra *Mon. Not. R. Astron. Soc.* **499** 550–8

Raskutti S, Bolton J S, Wyithe J S B and Becker G D 2012 Thermal constraints on the reionization of hydrogen by population II stellar sources *Mon. Not. R. Astron. Soc.* **421** 1969–81

Ribeiro M B 1992a On modeling a relativistic hierarchical (fractal) cosmology by Tolman's spacetime. I–Theory *Astrophys. J.* **388** 1–8

Ribeiro M B 1992b On modeling a relativistic hierarchical (fractal) cosmology by Tolman's spacetime. II–Analysis of the Einstein-De Sitter model *Astrophys. J.* **395** 29–33

Ribeiro M B 1993 On modeling a relativistic hierarchical (fractal) cosmology by Tolman's spacetime. III. Numerical results *Astrophys. J.* **415** 469–85

Riechers D A, Bradford C M and Clements D L *et al* 2013 A dust-obscured massive maximum-starburst galaxy at a redshift of 6.34 *Nature* **496** 329–33

Riess A G, Casertano S, Yuan W, Macri L M and Scolnic D 2019 Large magellanic cloud cepheid standards provide a 1% foundation for the determination of the Hubble constant and stronger evidence for physics beyond ΛCDM *Astrophys. J.* **876** 85

Rodighiero G, Cimatti A, Franceschini A, Brusa M, Fritz J and Bolzonella M 2007 Unveiling the oldest and most massive galaxies at very high redshift *Astron. Astrophys.* **470** 21–37

Roukema B F, Buchert T, Ostrowski J J and France M J 2015 Evidence for an environment-dependent shift in the baryon acoustic oscillation peak *Mon. Not. R. Astron. Soc.* **448** 1660–73

Rozgacheva I K and Kuvshinova I B 2020 Filaments of the large scale structure and primary scalar and vector gravitational perturbations *Astrophysics* **63** 56–65

Sardane G M, Turnshek D A and Rao S M 2014 Ca II absorbers in the Sloan Digital Sky survey: statistics *Mon. Not. R. Astron. Soc.* **444** 1747–58

Scodeggio M, Guzzo L and Garilli B *et al* 2018 The VIMOS Public Extragalactic Redshift Survey (VIPERS). Full spectroscopic data and auxiliary information release (PDR-2) *Astron. Astrophys.* **609** A84

Secrest N J, Hausegger S v, Rameez M, Mohayaee R, Sarkar S and Colin J 2021 A test of the cosmological principle with quasars *Astrophys. J. Lett.* **908** L51

Sheth R K and Diaferio A 2011 How unusual are the Shapley supercluster and the Sloan Great Wall? *Mon. Not. R. Astron. Soc.* **417** 2938–49

Shimakawa R, Higuchi Y and Shirasaki M *et al* 2021 Subaru hyper Suprime-Cam excavates colossal over- and under-dense structures over 360 $\deg^2$ out to $z = 1$ *Mon. Not. R. Astron. Soc.* **503** 3896–912

Sohn J, Hwang H S, Geller M J, Diaferio A, Rines K J, Lee M G and Lee G-H 2015 Compact groups of galaxies with complete spectroscopic redshifts in the local universe *J. Korean Astron. Soc.* **48** 381–98

Somerville R S and Bryan G L 2019 Large-scale structure with cold dark matter *Nat. Astron.* **3** 1058–9

Steinhardt C L, Capak P, Masters D and Speagle J S 2016 The impossible early galaxy problem *Astrophys. J.* **824** 21

Sylos Labini F 2011 Inhomogeneities in the universe *Quantum Grav.* **28** 164003

Sylos Labini F and Baryshev Y V 2010 Testing the Copernican and cosmological principles in the local universe with galaxy surveys *J. Cosmol. Astropart. Phys.* **2010** 021

Sylos Labini F, Vasilyev N L and Baryshev Y V 2009 Breaking the self-averaging properties of spatial galaxy fluctuations in the Sloan Digital Sky survey–data release six *Astron. Astrophys.* **508** 17–43

Tavasoli S, Vasei K and Mohayaee R 2013 The challenge of large and empty voids in the SDSS DR7 redshift survey *Astron. Astrophys.* **553** A15

Teerikorpi P, Chernin A and Baryshev Y 2005 The quiescent Hubble flow, local dark energy tests, and pairwise velocity dispersion in a $\omega = 1$ universe *Astron. Astrophys.* **440** 791–7

The Pierre Auger Collaboration 2017 Observation of a large-scale anisotropy in the arrival directions of cosmic rays above 8×10^{18} eV *Science* **357** 1266–70

Thomas S A, Abdalla F B and Lahav O 2011 Excess clustering on large scales in the MegaZ DR7 photometric redshift survey *Phys. Rev. Lett.* **106** 241301

Tikhonov A V, Gottlöber S, Yepes G and Hoffman Y 2009 The sizes of minivoids in the local universe: an argument in favour of a warm dark matter model *Mon. Not. R. Astron. Soc.* **399** 1611–21

Toft S, van Dokkum P, Franx M, Thompson R I, Illingworth G D, Bouwens R J and Kriek M 2005 Distant red galaxies in the Hubble ultra deep field *Astrophys. J.* **624** L9–12

Tully R B, Courtois H, Hoffman Y and Pomarède D 2014 The Laniakea supercluster of galaxies *Nature* **513** 71–3

Verde L, Protopapas P and Jiménez R 2013 Planck and the local universe: quantifying the tension *Phys. Dark Universe* **2** 166–75

Wang F, Yang J and Fan X *et al* 2021 A luminous quasar at redshift 7.642 *Astrophys. J. Lett.* **907** 7

Wang S, Li X-D and Li M 2010 Revisit of cosmic age problem *Phys. Rev.* **D82** 9

Whitbourn J R and Shanks T 2016 The galaxy luminosity function and the local hole *Mon. Not. R. Astron. Soc.* **459** 496–507

Wiklind T, Dickinson M, Ferguson H C, Giavalisco M, Mobasher B, Grogin N A and Panagia N 2008 A population of massive and evolved galaxies at $z > \sim 5$ *Astrophys. J.* **686** 781–806

Wiltshire D L, Smale P R, Mattsson T and Watkins R 2012 Hubble flow variance and the cosmic rest frame *Phys. Rev.* **D88**

Wong J H W, Shanks T and Metcalfe N 2022 The local hole: a galaxy under-density covering 90% of sky to ~200 Mpc *Mon. Not. R. Astron. Soc.* **511** 5742–55

Yadav J K, Bagla J S and Khandai N 2010 Fractal dimension as a measure of the scale of homogeneity *Mon. Not. R. Astron. Soc.* **405** 2009–15

Yennapureddy M K and Melia F 2021 Structure formation and the matter power-spectrum in the $r_h = ct$ universe *Phys. Dark Universe* **31** 100752

Chapter 8

Sociological factors that hinder the development of alternative cosmological models

As we have seen in previous chapters, there are scientific arguments to both support and reject the standard cosmological model. Nevertheless, the reason for the theory's success does not depend only on physics and astronomy. Social factors also intervene. Most cosmologists do not usually work within the framework of alternative cosmologies because they feel that these are not at present as competitive as the standard model. It is true that they are not so developed, but that is because cosmologists do not work on them. This vicious circle is to a great extent due to a sociological phenomenon known as the 'snowball effect', in which resources are distributed to the most successful theory at a given time; the effect acts as a potential in a field that attracts cosmologists, causing funds, research positions, prestige, telescope time, publication in top journals, citations, conferences, and other resources to be dedicated almost exclusively to standard cosmology. This approach might be appropriate in cases where a branch of science has an unquestionably solid basis, but it all too often becomes a dogmatic defence orthodoxy. This dogmatism would perhaps be better replaced by a more pluralistic approach to research in cosmology.

There are many individuals, scientists and non-scientists alike, who believe that science is an open process in which the best ideas are quickly recognised and accepted, while the erroneous ideas are immediately discarded. They think that a researcher could be working hard in a laboratory or observatory, or developing some theoretical idea and, on making a revolutionary discovery, would cry, 'Eureka! Eureka!' as astonished colleagues welcome the new idea, immediately recognising its merits.

Such a vision is naïve and is far from the way in which ideas are usually accepted—even less so when talking about cosmology these days.

Cosmology today has become almost closed to discussion of theoretical models that stray too far from the standard model, not because the model is perfect—we

doi:10.1088/978-0-7503-3775-5ch8

have seen throughout chapters 3–7 that there are many controversies and observations not properly understood with the ΛCDM model—or because the fundamental ideas of alternative models have been definitively rejected, but for other, non-scientific, reasons. Small variations of the standard model (see section 2.1) are given an ear, but not the most outrageous proposals that challenge the basic dogmas, starting with the expansion of the Universe. Some authors have even considered the term 'alternative cosmology' as a pseudoscience on a par with 'alternative medicine' (Ćirković and Perović 2018 [footnote 33]). Cosmologists do not usually work within the framework of exotic alternative cosmologies and they do not pay attention to them because they feel these are not at present as competitive as the standard model. They are certainly not so well developed for the simple reason that cosmologists do not work on them. It is a vicious circle. Furthermore, the machinery of science is becoming more and more collective, moving away from individual scientific adventures. The scientist is more like a small cog in a big machine than a free intellectual. Most scientists are busy with administrative and routine tasks within a predetermined project, trying to survive within the system, so they do not have much time to explore new ideas, much less think about the ideas of people that have scant recognition.

The fact that most cosmologists pay no attention to alternative ideas and dedicate their research time exclusively to the standard model is to a great extent owed to sociological considerations. It is therefore worth devoting some thought to the social structures in cosmological science, as I now do in this chapter and the one that follows. I am not saying that the Universe is a social construct (a typical postmodern idea). Quite the contrary: the Universe exists independently of our human affairs and some of its properties are derivable with our scientific analyses. However, human and social factors play an important role in the selection of theories, and this is perhaps the most important factor in the case of cosmology.

8.1 Deduction and induction in modern cosmology

8.1.1 The methodology of science

Basically, there are two different methodologies for studying Nature, both inherited from different ways of thinking in ancient Greece: the rationalist–deductive method and the empirical–inductive method (e.g., Markie 2012).

8.1.1.1 The rationalist–deductive method:

The rationalist–deductive method was devised by Pythagoras (c. 570–c. 490 BC) and Plato (c. 427–347 BC), according to whom the pure relations of numbers in arithmetic and geometry are the immutable reality behind changing appearances in the world of the senses. We cannot reach the truth through observation with the senses, they argued, but only through pure reason, which may investigate the abstract mathematical forms that govern the world. In this mode of thinking, there is a predominance of abstract theories, and mathematical modelling predominates over experimental and observational results. Some historians think that the thought of a world ruled by numbers is related with the introduction of coins in the marketplace in the society were Pythagoras was living. Anything could be reduced to abstract numbers: the value of a pot, a jar of oil, a plot of land, or a slave, could be expressed by an exact number of coins, as could the wealth and worth of any citizen.

Plato also emphasised that true knowledge comes through reason alone, thus diminishing the role of the senses and the earthly realm as a whole. In his dialogue *Timaeus*, Plato formulates a cosmology with a creator moulding matter into approximations of these ideal shapes to create a Universe ruled by eternal mathematical laws, laws that humans can deduce only through reason. These eternal mathematical laws are the true reality while the changeable Universe we see is a mere appearance; the observation of nature is thus unreliable.

There are good cases of success using a rationalist–deductive approach. An example within modern science is Einstein's (1879–1955) general relativity, which was posited from aesthetic and/or rational principles in a time in which observational data did not require a new gravity theory. In fact, observational tests proved this theory successful. Present-day physics and cosmology are partially Pythagorean when a theory is created before the observations. It is also common among modern Pythagoreans to approve of statements such as the search for beauty in a mathematical construction describing physical reality, or the Divine plan by which the creator designed the Universe (Zee 1986). The physicist–mathematician tries to achieve something close to a mystical approach, tries to read into the Mind of God. Also, analogously with religion, this extremely theoretical physics and cosmology can only be understood by a priestly elite able to think in four or more dimensions or in terms of similar abstractions.

8.1.1.2 The empiricist–inductive method:

As opposed to the preceding method, this one points out that nature should be known through observations and extrapolations of them. This is Anaxagoras' (500–428 BC) method of knowing nature, a view of an infinite world in which there is no difference between the heavens and the Earth, all operate on the same principles and the method of knowing nature was based on observations and extrapolations of those observations to parts of the Universe distant in space and time. Aristotle (385–322 BC) uses both inductive and deductive methods, and he says that 'the mathematical method is not the method of the physicists, because nature, perhaps all, involves matter' (*Metaphysics*, book II). To be sure, mathematics is useful for physicists, in spite of what was said by Aristotle, and this has been clear since Galileo Galilei (1564–1642) established the bases of the scientific method, but I agree with the Greek philosopher that matter is not the same thing as mathematical entities. Matter is not numbers, or geometry, or arithmetic, or the analysis of functions. Matter (or, better, matter–energy) is a component of the physical Universe, and this is what constitutes the reality of nature as studied by the physical sciences.

The empiricism of Galileo Galilei might be an example in modern science, in the sense that observation and experimentation are a requisite prior to theoretization, although all scientists, even Galileo, are also partly Pythagorean; for example, when he rejected Johannes Kepler's (1571–1630) hypothesis of elliptical planetary orbits by thinking that only perfect circular motion could prevail in the Heavens (Galilei 1967).

And all Pythagoreans are in part empiricists too; for example, in the case of Johannes Kepler, Pythagorean in his origins (Field 1988), when he embraced the empiricist deductive line of reasoning in his *Astronomia nova* (1609) in his careful and respectful use of Tycho Brahe's (1546–1601) measurements of the motions of Mars, which led to the first two of his three laws and marked the strongest observationally-based argument for heliocentrism before Isaac Newton (1643–1727). These are extreme positions which cannot usually be found in their pure form, but it is clear that, in some researchers, one of the two trends dominates. Apart from a few exceptions, the empirical–inductive method is more frequent in all branches of science, except pure mathematics.

8.1.2 The methodology of cosmology

In the early stages of modern cosmology both positions, inductive and deductive, were adopted (Gale 1993, 2002). Richard C Tolman (1881–1948), for instance, admitted both positions to be of great merit in cosmology. Edward A Milne (1896–1950) was in favour of a rationalist–deductive science, although some of his points of view may remind one of an empirical approach when he emphasises that space itself has no existence, and that using static space is probably more physically correct, as in ordinary physics. Considering the expansion of matter as motion in this space, he said in one of his papers:

> If the curvature of space cannot be determined, if it is essentially unobservable, then it should be rejected. The time has come when we should remember William of Occam's maxim: '*Entia non sunt multiplicanda proeter necessitatem*'. (Milne 1933)

The physicist and philosopher Herbert Dingle (1890–1978) made an aggressive attack against the rationalist–deductive method in favour of the empiricist–inductive method (Dingle 1937), with terms such as 'paralysis of reason', 'intoxication of the fancy', 'Universe mania', 'delusions', 'traitors', 'treachery'. Howard P Robertson (1903–1961), Arthur S Eddington (1882–1944), and Willem de Sitter (1872–1934) were also in favour of an empiricist–inductive science, although Eddington was also accused by Dingle of inventing scientific hypotheses by out of thin air rather than by strict immersion in observations and observational data. One may hesitate to consider him a empiricist–inductivist. Eddington used to state (ironically): 'not to put overmuch confidence in the observational results… until they are confirmed by theory'.

Cosmology is better derived empirically by first taking the data without preconceived ideas and then interpreting them from all possible theoretical viewpoints. There are certainly always prejudices and intuitions in our minds that push us towards certain avenues of research, but at least we should openly consider all the theoretical possibilities that might explain the data, rather than taking only one (standard) theory and always trying to squeeze the data into it in one way or another. In the words of Sherlock Holmes: 'It is a capital mistake to theorize before you have all the evidence' (Conan Doyle's *A Study in Scarlet*). Holmes further opines: 'before one has data, one begins to twist facts to suit theories instead of theories to suit data' (Conan Doyle's *A Scandal in Bohemia*).

When a feature of a model is ascertained through imposition rather than by experimental or observational confirmation, it is unscientific by virtue of being merely based on personal preference. In other words, a certainty achieved that way becomes a dogma (Ribeiro and Videira 1998)—a common occurrence in science, particularly in the discipline of astrophysics (Öpik 1977). A rather dogmatic attitude has come to prevail in modern cosmology, so we must be wary of the danger of believing too strongly in ideas that are not confirmed by observations. Without such confirmation we render ourselves unable to distinguish between science and metaphysics, as many scientists have noted (Tolman 1934, de Vaucouleurs 1970, MacCallum 1987, Wesson 1987, Rothman and Ellis 1987, Matravers *et al* 1995, Krasiński 1997 [part IX]).

The success of Einstein theory of gravitation with only two observational confirmations in favour of it in 1919—the bending of light in close proximity to the Sun and Einstein's accurate calculation of the perihelion precession of Mercury's orbit—were claims in favour of the purely deductive reasoning (Lerner 1991 [chapter 4]). Einstein's theory of general relativity has now been widely confirmed by experience, but that was far from being the case in the 1920s. This deductive approach formed a precedent for the way in which the science of cosmology is pursued today. As written by Einstein (1934), the physical experience of the experimenter could not lift him into the regions of higher abstraction and science to be guided by purely mathematical, formal considerations in his search for a theory. Pure thought can grasp reality, said Einstein. He also introduced the notion of a homogeneous Universe, an idea that is still, against all empirical evidence, present in cosmology today purely on the basis of simplicity and aesthetics rather than observation. We may therefore reasonably argue that modern cosmology was born with Einstein around 1917.

According to Lerner (1991) and Bonometto (2001), cosmologists and particle theorists defending the standard model use the deductive method, mathematically deducing how the Universe must be, while the method of those who propose new ideas is based on learning from observations. Big Bang defenders seek to derive the present, historically formed Universe from a hypothetical perfection in the past. They begin with a mathematical theory with aesthetic motivations to deduce from that theory how the Universe must have begun and work forward from that beginning to the present-day cosmos. Bonometto (2001) calls this 'post-modern' research.

Some astrophysicists closer to observation than to theory complain about the lack of an empirical approach in cosmology. Gérard de Vaucouleurs (1918–1995), for example, known for his extragalactic surveys and Hubble–Lemaître constant measurements, wrote: '[There are] parallelisms between modern cosmology and medieval scholasticism. … Above all I am concerned by an apparent loss of contact with empirical evidence and observational facts, and, worse, by a deliberate refusal on the part of some theorists to accept such results when they appear to be in conflict with some of the present oversimplified and therefore intellectually appealing theories of the Universe' (de Vaucouleurs 1970). The quantity of data for observational cosmology today is certainly much greater than in 1970 (although there were then ample data: the CMBR, redshifts of galaxies, abundances of light elements, etc); however, I think de Vaucouleurs criticisms are still valid nowadays: cosmology has not changed its methodology a great deal.

To sum up, although the inductive method usually still stands as the ideal approach, in practice cosmology is more dominated by the deductive method, which establishes what is the truth about the Universe before observing it. It then tries to adapt prejudgement fact through consecutive chains of ad hoc theoretical corrections.

8.2 Cosmological models and free parameters: new epicycles?

The number of independent measurements relevant to current cosmology and the number of free parameters of the theory might roughly be of the same order (Disney 2007, 2011): In the 1950s, the 'Big Bang' was a theory with three or four free parameters to fit the few quantities of observational cosmology (basically, the Hubble–Lemaître constant and helium abundance), and the increase in cosmological information from observations, with the CMBR anisotropies and others, has been accompanied by an increase in the number of free parameters and patches (dark matter, dark energy, inflation, initial conditions, etc) in the models to fit these new numbers, until today when the theory has around 20 free parameters, apart from the initial conditions and other boundary conditions introduced in simulations to reproduce certain structures of the Universe, or astrophysical processes such as star formation, stellar nucleosynthesis, and others.

A similar situation also prevails in particle physics (Unzicker and Jones 2013). The standard particle model, for example, has around twenty parameters; it has lost its simplicity, predictive capacity, and unity (Smolin 2008, Iradier 2009, Unzicker and Jones 2013). The origin of many of these parameters remains unknown and, while no significant evidence for a failure of the theory has emerged so far, it may be incomplete because it does not account for facts such as the oscillations of the neutrino; it accounts for the electromagnetic, weak, and strong interactions, but fails to unify this last with the two former and is also understood to be incomplete because it does not account for gravitation or dark energy; neither does it contain any viable dark matter particle (Yao *et al* 2006). Something similar happens with cosmology. In total, it is stated that particle physics and cosmology have in total more than thirty independent fundamental constants (Tegmark *et al* 1998).

In cosmology, for instance, the number of independent measurements in Cosmic Microwave Background Radiation (CMBR; see chapter 5) anisotropies is not high. While its power spectrum shows repeated information in the form of multiple peaks and oscillations, its Fourier transform, the angular correlation function, offers a more compact presentation that condenses all the information of the multiple peaks into a localised real space feature. Oscillations in the power spectrum arise when there is a discontinuity in a given derivative of the angular correlation function at a given angular distance (López-Corredoira and Gabrielli 2013). This allows a physical interpretation of these mathematical properties of CMBR anisotropies in terms of matter distribution in the fluid generating the radiation. A power spectrum with oscillations is a rather normal characteristic expected from any fluid with clouds of overdensities that emit/absorb radiation or interact gravitationally with photons, and with a finite range of sizes and distances for those clouds (López-Corredoira 2013a). The standard cosmological interpretation of 'acoustic' peaks, from the hypothesis of primaeval adiabatic perturbations in an expanding Universe (Peebles and Yu 1970), is just a particular case; peaks in the power spectrum might be generated in scenarios that have nothing to do with oscillations owing to gravitational compression in a fluid. The CMBR angular correlation function can be fitted by a generic function with a total of $\approx$6 free parameters. Saying that the power spectrum/angular correlation function contains hundreds or thousands of independent parameters for a given resolution is incorrect because the different values of C_ℓ are not independent in the same sense that hundreds of observations of the position and velocity of a planet do not indicate hundreds of independent parameters, the information of the orbit of a planet being reduced to six Keplerian parameters.

Today, apart from minor problems brought about an ambition to arrive at a 'precision' cosmology, there is near consensus in the approximate values of the cosmological parameters. There have, however, been many major historical disputes concerning the measurement of these parameters. For example, historical analyses (Croft and Dailey 2015, Faerber and López-Corredoira 2020) show that, before 1995, there has been a great dispersion around two values (50 and 100 km s^{-1} Mpc^{-1}) of the Hubble–Lemaître constant, whereas immediately after 1995 almost all values clustered very close to a preferred value of 70 km s^{-1} Mpc^{-1} with small errors given by the HST Key Project. Measurements of Ω_Λ, a quantity that was considered null before the 1990s, have now settled at 0.7, and since 1995 presents a dispersion much lower than expected statistically from the error bars (Croft and Dailey 2015), which means either that the error bars were overestimated, or that there is a bias in the publication of results towards the preferred value. Other examples could be given.

Nonetheless, in the process of refining the standard model, the controversy has abated at the cost of adding more and more ad hoc corrections and new parameters, which may be metaphorically referred to as Ptolemaic-style 'epicycles'. The development of modern cosmology bears a strong similarity to the development of the Ptolemaic epicyclic theory, a historical example brought up many times by many different authors when topics such as dark matter, dark energy, or inflation (see chapter 3) are brought up.

Alfonso the Wise (Alfonso X of Castile, 1221–1284), said of the Ptolemaic system: 'If the Lord Almighty had consulted me before embarking upon Creation, I should have recommended something simpler'. A similar judgement may be uttered today concerning the standard cosmological model: it is too complicated to be considered an elegant truth, while at the same time not complicated enough to accommodate all the observational evidence. In the future, therefore, we may well see the cosmologists adding new epicycles in an effort to 'save the phenomena'.

Even renowned defenders of the Big Bang may admit the similarity between the dark elements of ΛCDM cosmology and the epicycles, deferents, and quadrants of the Ptolemaic model. They occasionally mention it, although half-heartedly, so as not to be confused with those heterodox researchers that loudly claim that dark matter is like an epicycle in a geocentric model. I remember—because I took notes of it, since my memory is not so good—for instance, Edward W ('Rocky') Kolb (1951–) in a presentation wondering whether the actual cosmological model is a modern version of epicycles. I remember the brave claims of Kolb, saying ironically in his talk: 'We don't know anything about the reason for Λ, the dark matter, the particles of inflation. … Welcome to the golden age of cosmology!'. However, this honest sentence does not appear in the published proceedings (Kolb 2001).

The situation is different for cosmological models different from ΛCDM. In this race to build more and more epicycles, the Big Bang model is allowed to make ad hoc corrections and add more and more free parameters to the theory in order to solve the problems that get in its way; in contrast the alternative models are summarily rejected when gaps in knowledge or inconsistencies arise, and most cosmologists do not heed their ad hoc corrections. One may wonder why different theories are accepted or rejected with such different criteria:

> If physicists do not understand the what of their theories, they'll introduce a new particle. If they don't understand the when, then it must have happened right after the Big Bang. If they don't understand the where, then of course it took place in an extra dimension. And if they don't understand the how, they will postulate a new interaction. If they don't understand the how much, a symmetry breaking will soon appear. If they don't understand anything, they will propose strings and branes. And if they lose interest in all understanding, there is always the strong anthropic principle. (Alexander Unzicker (1965–), physicist, in chapter 17 of *Bankrupting Physics* (Unzicker and Jones 2013))

> Given a theory A self-called orthodox or standard, and a non-orthodox or non-standard theory B. If the observations achieve what was predicted by the theory A and not by the theory B, this implies a large success to the theory A, something which must be divulged immediately to the all-important mass media. This means that there are no doubts that theory A is the right one. Theory B is wrong; one must forget this theory and, therefore, any further research directed to it must be blocked (putting obstacles in the way of publication, and giving no time for telescopes, etc).
>
> If the observations achieve what was predicted by theory B rather than by theory A, this means nothing. Science is very complex and before taking a position we must think further about the matter and make further tests. It is probable that the observer of such had a failure at some point; further observations are needed (and it will be difficult to make further observations because we are not going to allow the use of telescopes to re-test such a stupid theory as theory B). Who knows! Perhaps the observed thing is due to effect 'so-and-so', of course; perhaps they have not corrected the data from this effect, about which we know nothing. Everything is so complex. We must be sure before we can say something about which theory is correct. Furthermore,

by adding some new aspects in the theory A surely it can also predict the observations, and, since we have an army of theoreticians ready to put in patches and discover new effects, in less than three months we will have a new theory A (albeit with some changes) which will agree the data. In any case, while in troubled waters, and as long as we do not clarify the question, theory A remains. (López-Corredoira 2008)

8.3 Social dynamics of an *N*-cosmologist system

Many cosmologists are used to performing simulations of astronomical objects, aimed at understanding the dynamics of the whole Universe. *N*-body simulations are popular, especially among experts on large-scale structure (see section 7.3). Clusters of galaxies, galaxies, or even smaller sized masses are point-like objects in the vast space simulated in mega-computers. Human beings are studied by sociologists in a similar way by looking at global trends and statistical numbers, rather than paying attention to individuals. Human behaviour is in general more complex than stellar or galactic events, but one may derive some basic laws from observing their behaviour. We cannot carry out a simulation of the evolution of humanity (or maybe we can, I am not sure how far sociology has gone in this aspect), but we can extract some rules from mere observation.

Cosmologists, along with researchers in other disciplines, are human and most of them are motivated by the same concerns as the rest of humanity. The dynamics of our present-day society are driven by certain forces that may be identified in terms of economic or other power structures.

8.3.1 The prestige of orthodoxy

Cosmology has developed into a powerful enterprise in recent decades, with greater representation in both absolute and relative terms throughout the disciplines of physics and astronomy. It is interesting to note the huge increase in scientific publications related to cosmology, in both absolute and relative terms, in physics and astronomy (Sanromà 2007)[1]. Consequently, greater prestige and status is now granted to cosmologists than some decades ago, and ambitions are higher for social status.

Alternative cosmological models are not rejected because they are not potentially competitive, but because workers promoting them encounter great difficulties in advancing in their research against the mainstream. A small number of scientists cannot compete with the huge mass of cosmologists dedicated to polishing and refining the standard theory, which is endowed with status, power, and prestige among members of the scientific community, who dedicate large portions of their

[1] Counting the number of papers at ADS-NASA (https://ui.adsabs.harvard.edu/) that contain the word 'Cosmology' in the abstract or as keywords, we find in absolute terms and relative to the total number of papers published: 19 (0.014%), 125 (0.052%), 248 (0.069%), 5418 (0.32%), 40 287 (1.1%), 162 200 (1.8%) respectively for the periods 1901–20, 1921–40, 1941–60, 1961–80, 1981–2000, 2001–20. In the area of astronomy alone, excluding other areas of physics, the ratio of papers with the word 'Cosmology' in the abstract during 2001–20 is 11.9%.

time and effort to the administration and politics of science[2], in a struggle to be leaders of projects. In stark contrast, professional researchers who choose to challenge dominant ideas experience obstacles and ostracism, a phenomenon that has been studied by many sociologists (e.g., Campanario and Martin 2004). It is well-known within the astrophysical community that dissenters from orthodoxy pay a high price in terms of their career prospects.

The present-day methodology of research in cosmology does not favour the exploration of new ideas. Alternative models try to compete with the standard model, but cumulative inertia gives clear social advantage to the standard model. This advantage determines whether or not researchers may continue to explore these alternative ideas. If the number of researchers and funds is lower than a minimum value in a group (critical mass), that group will most probably disappear. In this respect, the whole is more than the sum of the parts (Bunge 2004, Sawyer 2005).

The standard theory in cosmology became dominant because it could explain more phenomena than the alternative ideas, but it is possible that partial successes have propitiated a general view that is misguided and does not let other ideas advance that might be closer to a more correct description of the Universe in some aspects. False hypotheses, once established, become difficult to refute; under such conditions science may therefore be not self-correcting (Ioannidis 2012, Park *et al* 2014), as is commonly believed.

> … views of cosmology at any epoch are largely determined by the ideas of a few strong individuals, rather than by an objective appraisal of the information available. Today, a similar situation prevails, and it is further complicated because we are often trying to use classes of objects to test cosmological theories long before we understand the physics of them. (Geoffrey Burbidge (1925–2010), heterodox cosmologist and editor[3], in his *Nature*'s paper 'Was there really a Big Bang?' (Burbidge 1971))

The status of orthodoxy is granted to the defence of a consensus around what is defined as standard model at any given time rather than to individuals. Even astrophysicists of great prestige who have spent their whole professional lives arguing in favour of orthodoxy in cosmology can fall from grace if they at a given moment depart of the consensus, even over minor questions. For instance, Allan R Sandage (1926–2010) and Gustav A Tammann (1932–2019) were marginalised by the system after 1999, when the community decided that a value of around 71 km/s/Mpc for the Hubble–Lemaître constant was preferable to the value of 56–58 km s^{-1} Mpc^{-1} given by Sandage and Tammann (Burbidge 2006).

Simon White (1951–), a reputed orthodox cosmologist, thinks that, in terms of the definition of the present-day status quo within cosmology, high energy physicists are

---

[2] Elsewhere, I refer to such activities as astropolitics; see López-Corredoira (2008), López Corredoira (2013b [chapters 3, 6]).

[3] Geoff Burbidge was a physicist with an important influence in the political decisions on astrophysics. He was editor of the highest impact factor journal of astrophysics, the *Annual Review of Astronomy and Astrophysics*.

occupying an important space. He also thinks that the status quo is fundamentalist, in pursuit of a single truth, rather than universalist and creative in taking interest in all aspects of the cosmos (White 2007). High energy physicists and astronomers have converged in support of astronomical projects to constrain dark energy, which—according to White—is damaging to astronomy, since dark energy is only one of many options, and not necessarily the one where significant progress is most likely to result from such a major investment of resources.

Philosophers have also perceived how power structures are pushing research in a single direction while preventing other possible ideas to emerge. The philosopher of science Mario Bunge (1919–2020) points out a number of causes that are destroying science in the form of recipes to kill it (Bunge 1997): eliminating all heterodox researchers, restricting freedom of research, subjecting ideas to strict control, rewarding mediocre researchers and punishing those who show any originality, favouring applied over basic research, putting the direction of science and technology in the hands of a public manager (a lawyer, a politician, or at most an engineer or a doctor) with no scientific experience, burdening researchers with administrative tasks so that they have little time to concentrate on science, etc. These recipes apply well to cosmology.

Andrew Holster, a philosopher, and Devon D Brewer, an independent researcher in health and social sciences, make similar observations referring to science nowadays:

> As the system is organised now, intellectuals are excluded, and only those who have developed power networks, who have mastered organisational propaganda, who understand how to bully and manipulate the others in the power networks, can succeed there. (Holster 2016)

> Progress in science depends on researchers who challenge others' ideas and evidence. Doubters are scientific heroes, but researchers generally treat them as villains. Quite simply, those scientists who denounce and discourage doubters undermine the pursuit of truth gravely. (Brewer 2020 [chapter 61])

> If you express unpopular or dissenting ideas in a scientific field afflicted with defensive attitudes, it is likely you will experience some degree of social conflict with or isolation from other researchers as a result. Although open debate about scientific matters is a scientific ideal, in practice, the differences aired in the scientific arena often spill over to social interactions outside of the debate. (Brewer 2020 [chapter 62])

8.3.2 Funding of cosmology

Money distribution acts as a potential in a field that attracts particles around it, and the flows of funding have a very great impact on how science is produced.

> The snowball effect arising from the social dynamics of research funding drove more researchers into the Standard Cosmology fold and contributed to the drying out of alternative ideas (Narlikar and Padmanabhan 2001)

Geoffrey R Burbidge, another of the creators of the quasi-steady-state together with Fred Hoyle (1915–2001) and Jayant V Narlikar (1938–), has an even poorer opinion:

> Let me start on a somewhat pessimistic note. We all know that new ideas and revolutions in science in general come from the younger generation, who look critically at the contemporary schemes, and having absorbed the new evidence, overthrow the old views. This, in general, is the way that science advances. However, in modern astronomy and cosmology, at present, this is emphatically not the case. Over the last decade or more, the vast majority of the younger astronomers have been conformists in the extreme, passionately believing what their leaders have told them, particularly in cosmology. In the modern era the reasons for this are even stronger than they were in the past. To obtain an academic position, to obtain tenure, to be successful in obtaining research funds, and to obtain observing time on major telescopes, it is necessary to conform. (Burbidge 1997)

The general opinion among leading cosmologists about this lack of funding for students who want to spend their time investigating alternative ideas, as said by Geoffrey Burbidge, is partly echoed in a view expounded by Malcolm Longair (1941–):

> Geoffrey [Burbidge] urged us to take these problems seriously. I am sure he is correct, but I have to say that I would not put a research student onto these problems—I would be jeopardizing their future careers too much. I am also reluctant to devote time to them myself, partly because there always seem to be many more interesting problems to tackle. (Longair 1997)

It is not strange that Burbidge and Narlikar were frustrated in their odyssey in steering an alternative cosmological model in a valid direction of research and they link the lack of the social success of their theory with how social dynamics works, but their thoughts and observations look beyond their particular defence of an alternative cosmological model. Here is a similar opinion from a researcher, Abraham ('Avi') Loeb (1962–), who is not particularly heterodox in the field of cosmology, although he is famous for 'exotic' proposals in other areas:

> It is common practice among young astrophysicists these days to invest research time conservatively in mainstream ideas that have already been explored extensively in the literature. This tendency is driven by peer pressure and job market prospects, and is occasionally encouraged by senior researchers. Although the same phenomenon existed in past decades, it is alarmingly

> more prevalent today because a growing fraction of observational and theoretical projects are pursued in large groups with rigid research agendas. In addition, the emergence of a 'standard model' in cosmology (albeit with unknown dark components) offers secure 'bonds' for a safe investment of research time. (Loeb 2010)

Then there is this other observation by the always lucid Mike Disney (1937–), an astrophysicist not particularly attached to any alternative model:

> … after publishing a previous sceptical article on this topic (Disney 2000), I received hundreds of e-mails from professional astronomers saying 'Thank God somebody is saying these things at last—but don't quote me'. Then again many younger astronomers will privately admit that they don't believe a word of Lambda-CDM, 'But if I don't acknowledge it in my grant and observing proposals then I don't succeed.' (Disney 2011)

The snowball effect referred by Narlikar is also called Matthew effect, after the sociologist Robert K Merton (1910–2003), who gave it this name from the Gospel of St. Matthew (25:29) (Merton 1968): 'Unto every one that hath shall be given, and he shall have abundance: but from him that hath not shall be taken away even that which he hath.' This dictum is to a certain extent present in the social dynamics of cosmology, as well as in other speculative areas of science (López Corredoira 2013b [section 3.8]). It is a feedback ball: the more successful the standard theory is, the more money and scientists are dedicated to work on it, and therefore the higher the number of observations that can be explained with more parameters and ingredients (dark matter, dark energy, inflation, etc) introduced ad hoc, and that cause the theory to be considered more successful.

Balls do not take several generations to stop them and give an opportunity to other ideas to initiate their rolling down. The renowned physicist Max Planck (1858–1947) said that a new scientific truth does not triumph by convincing its opponents and making them see the light; it succeeds only because its opponents die off and a new generation grows up that is familiar with it (Planck 1949).

The biologist José Luis Pérez Velázquez, writing about how science works, expresses the following view:

> Funding institutions do not like to take risks, do not appreciate uncertainty, and thus will almost never fund truly creative, innovative research, because, as a consequence of being creative, this research is risky: we do not know what the results will be. Real research is therefore inherently uncertain, it consists in investigating the unknown; if we knew the results that will be observed it would not be called research—someone said once—and therefore the conclusion seems to be that granting agencies do not fund research. What do they fund? (…) mainly projects that aim at confirmation and clarification of already known findings. And the more trivial, easy to produce 'positive' results the details are, the more likely the project will be funded. Kuhn's assessment of

'development-by-accumulation' of accepted facts has been the standard since the middle of the past century, and possibly before that. It is no coincidence that Gerhard Fröhlich is credited with the words: 'Most scientific publications are utterly redundant, mere qualitative 'productivity' ' (Pérez Velázquez 2019 [chapter 3])

In a political context, the Pulitzer winner writer Upton Sinclair (1878–1968) quipped:

> It is difficult to get a man to understand something when his salary depends on his not understanding it. (Singer 1935)

As the philosopher Paul K Feyerabend (1924–1994) put it,

> … the best single entity to get a modern scientist away from what his 'scientific conscience' tells him to pursue is still the dollar. (Feyerabend 1975 [chapter 4])

8.3.3 Telescope time

Money can be important in the development of a scientific idea when hiring new researchers to work on it, but not only that. There are, in fact, many cases in which heterodox researchers self-fund their ideas, at least for the development of theoretical ideas. Nonetheless, getting telescope time to carry out some kinds of special observations beyond the publicly available data is also another part of the task. And it is well-known that heterodox cosmologists rarely get telescope time when they try to work on ideas that stray from the consensus precisely because the committees charged with the allocation of telescope time are in the hands of researchers holding standard views, who form the majority of the research community. Since, as in democracies, the majority decides and the standard cosmology has a much greater number of adherents, the probability of a heterodox researcher gaining access to expensive telescope time is not high.

Halton ('Chip') Arp (1927–2013) was denied access to telescopes for many decades owing to his polemical statements (Burbidge 1988). Later, he would even be fired from his permanent position as astronomer in the US for those views and had to migrate to another country to get a new job. This is one well-known case, but I am sure there are many more cases like this among professional astronomers who decide to spend their career pursuing some line of research contrary to standard views. Of course, one can think that Arp was wrong[4], but the question here is not whether one is wrong or right, since science is not composed of gods knowing the absolute truth deciding who is right and who is wrong. The question is, when a reputed professional astronomer like Arp, with a long career dedicated to astronomy

[4] In a range from 0 to 5 for degree of crankiness on the web page http://www.crank.net/, Chip Arp was given a 3.

and several hundreds of scientific publications, decides to do research in a certain line of research supported by a minority, this is usually ostracised, which denotes that all is not well in science in general and cosmology in particular.

Narlikar (2000) remarks that at the beginning of the 20th century there was venture funding for new ideas in science. For instance, when Hubble and Eddington were asked what they expected to find with the new 200 inch telescope, they answered: 'if we knew the answer, there would be no purpose in building it'. Nowadays, however, access to large telescopes is awarded after astronomers pass a peer-review process in which they state clearly what they expect to find with the observing time requested (Narlikar 2000).

We may certainly agree that telescopes are very expensive instruments. There are thousands of astronomers around the world that want to make use of the largest ones and there is not available time for everybody, so one has to select the 'best science' proposals to distribute that time. This leitmotiv of selecting the 'best science' is indeed the origin of the perversion of the scientific methodology into a social dynamic in standard cosmology and other areas of scientific research: instead of allowing a free circulation of ideas among professional researchers until those ideas that do not even satisfy their creators die a natural death, ideas are killed shortly after they are born, not allowing them to be developed while they are still at an embryonic stage. The distribution of funds and telescope time play here a fundamental role as social forces pushing scientists towards consensus.

Not only is telescope time awarded preferentially to teams with orthodox views in cosmology, but these teams' use of these expensive instruments obliges them to churn out impressive reports for the mass media aimed at the general public as if to say, 'See how your taxes have helped us to produce this fantastic science'. Ever present is the insidious temptation to exaggerate the importance of results.

An instrument capable of cosmologically interesting observations costs millions of euros. The gargantuan amounts of money necessary to carry out professional astronomy is not something new[5], but recent decades have seen an increase of several orders of magnitude in funding. Future telescopes of very large size like the future European E-ELT in Chile even reach the stratospheric cost of over a billion euros. Including operating costs and the expected duration of the telescope, observing on the E-ELT telescope will cost of the order of 10 euros/second (around 300 hundred thousand euros per night of observation). It is easy to image that time allocation committees will be very restrictive in granting use of so costly a telescope, and that they will read with scant attention any kinds of proposals that propose to test of alternative ideas to the Big Bang.

[5] Already in the 19th century, the costs of astronomy were very high. For instance, the organisation and management of the vast Harvard Observatory Surveys of the apparent magnitudes and spectra of very large numbers of stars by Edward Pickering is a good example (Longair 2001): over the years when the great surveys were undertaken, Pickering obtained several hundred thousands of dollars from the Henry Draper Fund, 400 000 dollars from the Paine Fund in 1886, 230 000 dollars from the Boyden Fund in 1887 and 50 000 dollars from Bruce Fund. It is estimated that Pickering's personal contribution to the project amounted to about 100 000 dollars.

The future James Webb Space Telescope (JWST) is yet an order of magnitude more expensive than the E-ELT: around 10 billion dollars, plus maintenance. According to Billings (2010), the JWST has devoured resources meant for other major projects, none of which can begin serious development until the spending spree is over. The philosopher Bjørn Ekeberg (1978–) recently said of this:

> … the purpose of the JWST, notwithstanding the individual motivations of the scientists within the enterprise, is to produce research, to engage scientific labor and create conditions for more capital–intensive projects in return. … As Bergson realised more than one century ago, the objective of the scientific mind is not truth but action—what is at stake is how the production of new truths allows a new reach, expansion and mobilisation.

> … the right question is not whether its metaphysical enframing of the Universe is true or false—whether its explanatory power is absent or all-powerful. Rather, the question is whether, and to whom, the experiment [JWST] is relevant. Will its eventual findings actually matter to anyone but the scientists themselves? (Ekeberg 2019)

8.3.4 Rejection of papers and lack of citation

Although lack of telescope time to test certain ideas is an obstacle, it does not constitute a total hindrance since heterodox cosmologists can work on their ideas regardless. With the huge amounts of publicly available data, one may do a lot of astrophysics without the need to set foot in a telescope control room.

A greater obstacle is the process of publishing scientific results, since the most important journals are usually reluctant to publish research that goes against Big Bang theory: either the referees reject the papers with no possibility of reply, or the editors of the journals take it upon themselves to discard some submitted papers without even sending them to be assessed by a referee if the paper in question seriously challenges the status quo in cosmology. There are exceptions of this behaviour, but they are rare. And there are proposals to circumvent such vetoes[6], but they are usually ignored or found to be too difficult to apply, since leading scientists consider the peer-review process of journals to be sacrosanct. The only other possibilities open to heterodox (professional or non-professional) cosmologists is to disseminate their articles by publishing them either in very minor journals that

[6] Halton Arp wrote me by e-mail in 2006 about one proposal to denounce this situation and to look for solutions: 'As for my attempt to initiate reform I have posted the editor's rejection at the end of astro-ph/0605453. My idea was that the reader could judge for himself whether rejection was obviously biased. Perhaps if we all started including referee's comments, when warranted, on our astro-ph papers maybe we could finally get some editors with integrity.' However, as I explain in the next subsection, neither is the astro-ph section of arXiv.org free from censorship.

are usually ignored by professional cosmologists or in sundry Internet forums in the vast ocean of amateur ideas.

Not all alternative ideas in cosmology encounter the same level of rejection. Ideas that we might classify as variations on the standard model (see section 2.1; e.g., MOND, $R_h = ct$, …) have a better chance of acceptance than other, more extreme positions, such as those claiming the possibility of a static Universe, which are usually rejected outright after reading the title or abstract. In recent years, after the introduction of dark energy in cosmology, the tolerance range has become narrower and narrower. This highly restrictive editorial choice of articles deemed worth of publication creates the false sensation that there is overwhelming evidence in support of ΛCDM and promotes to the impression that the alternative ideas are either dead or maintained by a few rank outsiders.

An even greater obstacle is the lack of attention that the few accepted publications receive. Most cosmologists simply look aside when they find a paper arguing against the standard model or proposing new ideas. It is an understandable reaction. Professional astrophysicists are fed up with so many different models of the Universe and prefer to stay with the standard one, which they can work with and discuss with other colleagues. Garth A Barber, who proposed Self Creation Cosmology, once said:

> After I had my own alternative gravitational/cosmological theory published I was somewhat puzzled why there was not more of a reaction to it, either positive or negative; since then I have realised that there are so many alternatives, all claiming to be 'The Theory' that solves everything, and most of them completely 'crackpot', that it is not surprising that there is such resistance and indifference to new ideas. The way forward is to fully understand the standard models and critically and informatively to ask pertinent questions. Most of the scientific establishment would also agree with this strategy. (altcosmology forum, 2005-1-18; http://cosmology.info/essays/models_marmet.html [version April 2021])

Again, there are some exceptions, especially when a large group of people is dedicated to one line of research such as MOND. But even in this relatively accepted theory of modified gravitation, there are examples of complaints about the lack of citations of MOND by the leaders of cosmological research. Here is David Merritt on the subject:

> One wonders whether this systemic 'failure to cite' is deliberate, or whether it reflects a general ignorance on the part of standard-model cosmologists of the published record … Second: Articles in the standard-model literature that do mention Milgrom's theory rarely credit the theory with more than a single successful prediction. Rarely, if ever, does one find an acknowledgement that Milgrom's theory has produced a string of confirmed novel predictions—at least a half dozen, and perhaps twice that number …. And the standard-model literature is full of statements implying that Milgrom's postulates are the

> modern-day equivalent of Ptolemy's epicycles and equants .... Third: As documented throughout this book, standard-model cosmologists routinely ignore the distinction between prediction and accommodation. That is: they assume that a post hoc explanation of a fact has the same evidentiary status, with regard to the standard cosmological model, that successful prediction of the fact has with regard to Milgrom's theory. (Merritt 2020 [chapter 9])

There is much to discuss concerning the fairness, or unfairness, of the referee system to select the best science in general[7]. In the particular field of cosmology, I think the peer-review system of the journals is a serious hindrance to progress in the area, more than in other fields of research, because cosmology has a higher degree of speculation and discussion of its fundamental ideas is more important than in other areas where a solid theoretical foundation has already been laid.

8.3.5 Censorship at arXiv.org

Apart from the refereed journals, another important tool for communicating scientific results in physics is the preprint server arXiv.org. Its lack of competitors gives it a monopolistic role in physics publishing. Other preprint servers are usually ignored by most astrophysicists. The majority of the papers published in journals are posted on this preprint server (Henneken *et al* 2006), and it is a first port of call for astrophysicists. The situation has become such that papers not posted on arXiv.org will receive scant dissemination within the community, particularly when the papers are not published in a reputed refereed journal, which is often the case for non-mainstream positions.

The development of arXiv.org, first at Los Alamos National Laboratory and later at Cornell University, was a wonderful example of freedom of expression between 1992 to 2004 that provided everybody with an open forum in which to post their ideas. There was a small fraction of papers with 'exotic' ideas, but they were very few (5% or less), so they did not disturb the flow of information. However, after 2004 there was a change in policy and those responsible for the site decided to block the posting of certain contributions. In 2004, a system was introduced in which in order to post something on the site support was requested from a colleague with experience in the field. The methods of the system would become more subtle in the following years, forbidding some scientists from giving support when arXiv moderators noted that they had allowed the publication of very challenging heterodox ideas. Committees were created for the purpose of rejecting papers without any obligation to read them without being accompanied by a referee's report: the committees just read the title and the abstract and, if they did not like the content (and normally they do not like anything that has not been accepted in a refereed journal and smells of heterodoxy, such as denial of the expansion of the Universe or discussions concerning alternative interpretations of the CMBR), they

[7] Talking about science in general, I have expressed a number of view about the question of publications and referees in López Corredoira (2013b [section 3.3]).

shunt the paper, which formerly would have been placed on 'astro-ph.CO', widely read by astrophysicists, to 'physics.gen-ph', which is hardly read by anybody. In some cases, they remove the contribution totally, without further explanation (e.g., Castro Perelman 2008). When asked for an explanation for a rejection, they usually reply with set phrases: 'arXiv reserves the right to reclassify or reject any submission. We are not obligated to provide substantive reasons for every rejection, and usually the moderators do not provide more than a sentence or two, often in a form not appropriate for author viewing'.

8.3.6 Conferences

Last, but not least, is the conference circuit for spreading scientific results. Here it is common to find people with outrageous ideas, yes. Nonetheless, they do not occupy the best places (invited reviews, plenary lectures, etc) and, if they are allowed to present a short oral communication or a poster, the attendees pay almost no attention to them. The physicist Robert B Leighton (1919–97), talking about conferences, once said that scientists seldom listen to other scientists, and if they are dozing to lectures, they may be preoccupied by their own thoughts (Leighton 1971). Conferences are not places to do science by promoting new ideas, but rather sites where people meet colleagues over a beer to discuss the social issues of science, or form collaborations among groups of people with similar ideas. A heterodox cosmologist at a conference with a vast majority of orthodox cosmologists, who are also usually the organisers, will receive very little or no attention at all. Precisely because of that, the organisers try to avoid the formation of big groups with heterodox ideas, because they know that an individual isolated will have little effect, whereas a group of scientists with common ideas may have a considerable impact.

As far back as 1982, the most—or one of the most—important conference of that year on cosmology was held at the Vatican. It was planned as an event to discuss all of the fundamental ideas and facts around cosmology at that time, but it turned out to be completely dedicated to Big Bang cosmology and its proponents, in which only the most important orthodox cosmologists were invited. The most prominent leaders among professional cosmologists with different ideas, such as Fred Hoyle, Víktor Ambartsumián (1908–1996), and Geoffrey Burbidge, were not invited (Burbidge 1997). Forty years later, it is unrealistic to expect that big events concerning the business of cosmology would even consider any attention being given to researchers with alternative ideas beyond variations around the same dogmas. Moreover, with the exception of the superficial attention drawn by MOND, not even the denial of the existence of non-baryonic dark matter is given a look-in.

Some alternative theories may be marginal and dead, not because irrefutable scientific arguments against them have been given, but rather precisely because of this kind of attitude in the organisation of social scientific events (journals, meetings, etc). Alternative theories die because they are being killed off by the same people who declare them to be dead. Most of the scientists who claim that these theories are dead or marginal have never read a paper on these ideas and are merely repeating what they have heard from some colleague. Indeed, what the censors probably mean is that the leaders of alternative ideas in cosmology are passing away before they gain any recognition and their work is not handed on to new generations, so it is understood that there is now no living sacred cow to respect, with the result that the community decides to declare that the alternative theories are dead.

It is sometimes suggested to invite a renowned orthodox cosmologist to tell us why an alternative cosmological model is wrong, without giving the model's proposers the chance to defend them. This is equivalent to organising a meeting on different religious ideas and inviting only Christians to participate, including an address from, say, the Pope telling us why Hinduism is a false doctrine.

Heterodox cosmologists also organise their own conferences, where freedom to challenge the standard model is considered a plus. These events, however, are not usually supported by funds from any organisation within astrophysics, they have small number of participants, they lack impact among other cosmologists and the mass media, and no important collaborations can emerge from these meeting for two reasons: (1) each heterodox cosmologist thinks that his or her model is the right one and is not interested in other ideas; and (2) there is no money to hire students or postdocs, or to organise future science. Rather, these conferences of heterodox cosmologists are venues of illusions in a parallel Universe, far from the social reality that moves the funds for investment in cosmology. These 'alternative' conferences are certainly more relaxed than the stressful 'normal' ones in that participants can exchange ideas, discuss, debate, and think creatively in an atmosphere of true passion for science. It is reminiscent of the golden era of physics when a few people would meet for long hours in friendly discussion of new ideas; such a serene atmosphere is far removed from the typical scientific meetings organised nowadays in which hundreds of participants have a few minutes each to rush through expositions of dull work, often lacking in novelty (much less posing any sort of challenge to accepted ideas), or they are led by a few orthodox organisers to secure support for their own credos.

It would be a bonus if some of the ideas discussed at the conferences of heterodox scientists were shown to be correct. This would mean that it is still possible to do top rank physics, rather than the usual second or third class science. Unfortunately, regardless of whether these ideas are correct or not, the scientific community is probably not yet ready to change its mind and abandon its dogmas, so we cannot know at present how much truth they contain.

8.3.7 Groupthink

How might we best explain the behaviour of the scientific community as described in the previous subsections?

The psychologist Irving Lester Janis (1918–1990) described the symptoms of groupthink as (Janis 1972, Dolsenhe 2011):

1. An illusion of invulnerability, shared by most or all the members of the group, which creates excessive optimism and encourages the taking of extreme risks;

2. An unquestioned belief in the group's inherent morality, allowing its members to ignore the ethical or moral consequences of their decisions;
3. Collective efforts at rationalisation in order to discount warnings or other information that might lead the members to reconsider their assumptions before they recommit themselves to their past policy decisions;
4. Stereotyped views of enemy leaders as too deviant to warrant genuine attempts to negotiate, or as too weak and stupid to counter risky attempts made at defeating their purposes;
5. Self-censorship of deviations from the apparent group consensus, reflecting each member's inclination to minimise to himself the importance of his doubts and counterarguments;
6. A shared illusion of unanimity concerning judgements conforming to the majority view (partly resulting from self-censorship of deviations, augmented by the false assumption that silence means consent);
7. Direct pressure on any member who expresses strong arguments against any of the group's stereotypes, illusions, or commitments, making clear that this type of dissent is contrary to what is expected of all loyal members; and finally,
8. The emergence of self-appointed mindguards—members who protect the group from adverse information that might shatter their shared complacency about the effectiveness and morality of their decisions.

The concept of groupthink is applicable to present-day cosmology (Sanromà 2007). Orthodox cosmology has an important element of groupthink, of following a leader's opinion. Any opinion, however outrageous, can be accepted if it is supported by the leading cosmologist, and in this sense Big Bang theory, even if it is a very speculative set of hypotheses, still finds a place in the psychology of the wider community of scientists.

The psychological profile of a researcher can produce leanings towards either orthodoxy or heterodoxy. According to Luminet (2008), there are two kinds of scientists: the craftsmen who do normal science (95% in string theory) and the imaginative artists of revolutionary science, with an added element of narcissism. Cosmologists tend to fall into one of the two extreme categories, although there are gradations of grey between groupthink and individualistic social forces.

The analysis by Sunstein (2019) of 'Conformity'[8] (highly relevant in our social media age) applies to social dynamics in the sciences. Conformity dynamics is particularly pronounced when it comes to dealing with very difficult problems. Social experiments in a multitude of contexts clearly show that when the problem before us is hard to solve, as in cosmology, people tend to follow the crowd, tending to defer to those who are perceived as authorities on the matter. A key mechanism in this collective effect is the so-called informational cascades, where people primarily rely on the signals conveyed by others rather than on independent information.

[8] See the book review 'Big Bang Conformity?' by Bjørn Ekeberg, at the blog https://www.drbjorn.com/post/big-bang-conformity

'Once this happens, the subsequent statements or actions of few or many others add no new information. They are just following their predecessors.' (Sunstein 2019). This cascade is very hard to stop, as we know from social networks on the Internet. A reputational cascade develops along with and reinforces an informational one. At this stage it simply becomes too risky to go against core consensus.

We could cite here the *Novum Organum* by the English philosopher Francis Bacon (1561–1626):

> The human understanding when it has once adopted an opinion (either as being the received opinion or as being agreeable to itself) draws all things else to support and agree with it. And though there be a greater number and weight of instances to be found on the other side, yet these it either neglects and despises, or else by some distinction sets aside and rejects; in order that by this great and pernicious predetermination the authority of its former conclusions may remain inviolate. (chapter XLVI)

Psychologists and sociologists now refer to the same idea as 'confirmation bias':

> The human organism tends to seek, embellish, and emphasize experiences that support rather than challenge already held beliefs. As an information processor, it filters experiences in a confirmatory manner–highlighting those that are consistent with its conceptual biases and ignoring or discrediting those that … [are] … not. (Mahoney and DeMonbreun 1977 [p 20])

8.4 Optimism and conservatism

8.4.1 Epistemological optimism

Human forces, in both sociological or psychological terms, play an important role in the development of cosmological ideas. However, not everything is a social construct (as some postmodernists claim) or a construct of our minds: the CMBR, the redshift of galaxies, etc are hard and concrete facts and they also have weight in the credibility of the standard model. It is no less truth that, in the debate during the 1950s between the Big Bang and the Steady State, both theories competed on an almost equal footing, with the result that later, after 1965, the first one was the winner because it was able to explain certain observations that the second failed to do *a priori*. So there was a justification to favour Big Bang theory, at least in the 1960s, but later developments of the standard cosmological model were a cascade of failures and ad hoc corrections with the intention to 'save' the theory, rather than allow other voices to debate at the same level.

One may think optimistically that there is a correlation between investment and truth, such that the best science closest to the real description of nature arises from an appropriate distribution of money and institutional support of science. An epistemological optimism encourages the belief that successful theories are successful because they reflect reality.

Between 1970 and 2000, many cosmologists (e.g., Sciama 1971, Weinberg 1977, Peebles *et al* 1991) had already made this case and taken the stand that the hot Big Bang model had attained sufficient maturity to be able to serve as the broadly correct theory of the Universe. Whatever difficulties might arise with the model, they declared, could be sorted out. According to them, no other views different from the standard model were worth considering. Michael S Turner (1949–) even considered that we now know the Universe not only qualitatively but also quantitatively; according to him, we had already (in 1999) reached the stage in which the cosmological parameters were very well constrained (Turner 1999).

The philosopher of science Jesús Mosteirín (1941–2017) said that the epistemologically relevant question is not whether a certain cosmological model is definitive, absolutely true, and is a perfectly accurate representation of the Universe (Mosteirín 1989). No model is that, he said. The relevant question is: which of the present-day available models is the best, which one explains best the observed facts and makes testable predictions that can afterwards be verified. These criteria, however, do not apply in their totality to predictions in cosmology, where there is an absence of fair comparison of models competing with the same conditions. Rather, the standard model is the best one because many patches have been placed on it. Mosteirín (1989) said that scientists do not have any prejudices against accepting alternative cosmologies. He also said:

> [T]here are no working alternatives to the standard big bang cosmological model (or family of models). This fact is not due to the will of the scientists who created the model, still less to the prejudices of the scientific establishment. On the contrary, it is almost exclusively due to the strong observational constraints which reality puts on the activity of model-making. The standard big bang cosmological model is the model no one wanted, but which recalcitrant experience forced everyone to accept, at least for the moment being. (Mosteirín 1989)

This kind of statements are somewhat naïve and denotes an excessive confidence in the fair application of scientific methodology. All the available alternative models may well be wrong, but this does not mean that they have been rejected fairly; and neither does it mean that the standard model is maintained for fair reasons. This epistemological optimism might be justified in certain branches of science but not in areas such as cosmology, which has an important element of speculation.

There are many historical examples that contradict such epistemological optimism. Wisdom and power, as the saying goes, do not always go together.

8.4.2 Conservatism and consensus cosmology

In this subsection[9], I will explain the sociological context of conservatism implicit in this wide consensus concerning the standard model of cosmology, a conservatism

[9] Abridged version of section 5.3 of my book *The Twilight of the Scientific Age* (López Corredoira 2013b).

that is common in many fields of research with well-established bases, or indeed in others with great gaps in knowledge but which present an appearance of solidity and depth.

In general, science is a conservative system based on tradition, on cumulative knowledge, and on the patient accumulation of wisdom that filters out the bad information and distils the good. Science does not function like the arts, in which spontaneity and novelty are more important than tradition. Neither is science a philosophical discipline, in which pluralism is encouraged, with tens or hundreds of ideologies living alongside one another.

Conservatism and tradition in general are also good for society. Our culture is a cumulative process in which we draw profit from our civilisation's past. We cannot and should not destroy all forms of society and begin again from nothing. It is important to keep most of what we have learned from the past, the works of the past, our languages, and so on. In the political and economic order, conservatism also has an important part to play in maintaining stability and equilibrium.

Nonetheless, some degree of progressiveness is also necessary. A purely conservative position, lacking in all progress, would lead our society to a state of stagnation. The philosopher Friedrich Hegel (1770–1831) used the metaphor of stagnant water forming bogs to indicate that a society needs, from time to time, to rise in revolt against the established order. Keeping one form of politics for a long time accords too much power to certain sectors of society and leads to a corruption of the values that originated that form of politics. Humanity must, from time to time, invent new forms, and social revolutions and conflicts are unavoidable in this process.

We could make similar observations with regard to science, although there is a very important important difference in relation to politics: there are no absolute truths in politics; in nature, however, there clearly are absolute truths, and science tries to understand them. Science has achieved small truths which will possibly last forever, and no revolution will be able to deny them. Only in some fields where the truth is not so immediately clear—for instance, the origin of life, cosmology, the final components of matter, etc—is there the possibility for some kind of revolution, and these are the very places where conservative scientists try to defend their power against the possible revolutions of progressive ideas.

The consensus science of large groups of scientists has come to occupy a privileged position, pushing aside the individual ideas of autonomous scientists. The snowball effect leads to a science of consensus in which the important thing is convincing the mass of scientists through mass media sources dedicated to them: reviews, conferences, etc. And, as pointed out by Luminet (2008), a consensus in the scientific community is transformed into orthodoxy and ends up becoming counter-revolutionary.

This counter-revolution constitutes a mechanism of self-defence created by the autonomous structure of knowledge in the building of theories. According to Iradier (2009) in his philosophical approach to science, all disciplines or specialisations were constituted and directed by a process of removing questions. Once a theory is established by the system, its fundamentals are not investigated any further.

The geological stratification of theories is a great obstacle, says Iradier (2009): While a theory of science as an exploration without restrictions or prejudices and as a progressive approach to the truth is given publicity, research on the fundamentals of science is despised, and tricks are played against the people who have alternative ideas to these fundamentals. The first thing to be established becomes the last thing that is allowed to be revised.

This is unavoidable. The evolution of science towards a tyranny of knowledge is unavoidable, because of the logic of the construction of knowledge itself—both because knowledge becomes more and more solid, and also because the dominant groups within the dominant ideas in science become ever more powerful. The distinction between power and truth in science is difficult to disentangle.

8.5 Pluralism

The Austrian physicist Ludwig Boltzmann (1844–1906) defended the idea of pluralism in science (Hiebert 1980, Ribeiro and Videira 1998); that is, maintaining discussion of the explanation of certain phenomena from the points of view of different theories, but this is not the most common approach. Other social ideas within scientific power hierarchies usually dominate the development of a science. This is in part good, because science achieves consensus and avoids sterile discussions that do not lead anywhere, as often occurs in many humanistic approaches, but the positive side arises only in cases where there really is a close match between a scientific theory and reality. In cases where wrong theories are chosen as standard views to the exclusion of others, or in areas like cosmology where no total truth might be reached, the negative effect of dogmatism is the stagnation of intellectual activity over the generations, a misdirection of time and resources in the wrong direction, which will come to be remembered in future generations as a blunder. A pluralistic approach of wrong ideas has at least the advantage of an open-minded attitude, and is one which deserves much greater respect.

Philosophers of science usually bet on a science with multiple theories concerning the same events, as we must remember that the truth is not always on the side of those with the most persuasive arguments. Here are some quotations from two of the most influential thinkers in the philosophy of science from the second half of the 20th century, Karl Popper (1902–1994) and Paul K Feyerabend (1924–1994):

> [T]he orthodoxy produced by intellectual fashions, specialization, and the appeal to authorities is the death of knowledge, … the growth of knowledge depends entirely on disagreement. (Popper 1994 [p x])

> Intolerant dogmatism, however, is one of the main obstacles to science. Indeed, we should not only keep alternative theories alive by discussing them, but we should systematically look for new alternatives. And we should be worried whenever there are no alternatives—whenever a dominant theory

> becomes too exclusive. The danger to progress in science is much increased if the theory in question obtains something like a monopoly. (Popper 1994 [p 16])

> [T]o use alternatives only when refutations have already discredited the orthodox theory puts the cart before the horse. (Feyerabend 1975 [chapter 2])

> Knowledge so conceived is not a series of selfconsistent theories that converges towards an ideal view; it is not a gradual approach to the truth. It is rather an ever increasing ocean of mutually incompatible alternatives, each single theory, each fairy-tale, each myth that is part of the collection forcing the others into greater articulation and all of them contributing, via this process of competition, to the development of our consciousness. (Feyerabend 1975 [chapter 2])

Feyerabend (1975) also adds that pluralism is the only method with humanistic approach; all the others following a tyrannical or fanatic path. I find this approach by Feyerabend, however, too relativist in general, in which all the hypotheses in science and other areas have the same value. I do not agree that, certain ideas such as flat Earth proposals, geocentric cosmogony, or creationist claims that the Earth's age is six thousand years should be discussed on the same level as accepted ideas of a quasi-spherical Earth, heliocentrism, and the scientific arguments supporting a much older age of the Earth. There are many theories in science that have so solid basis as to be considered definitive, such that they no longer require continuous discussion. Feyerabend (1975) also usually mentions 'alternative' Chinese medicine as a clear case of a human approach that should be considered as being on a par with western medicine. I do not agree with that either, but let's concede that the case for this view is at least worthy of debate.

However, there are areas, such as cosmology, where speculation still prevails and where many inconsistencies continue to arise; in such cases the pluralism proposed by Feyerabend and other philosophers of science is appropriate. Alan Heiblum (1982–) is one of several present-day philosophers arguing in favour of pluralism in cosmology[10]. I disagree with considering 'alternative cosmology' as a pseudoscience comparable to 'alternative medicine', as claimed by Ćirković and Perović (2018 [footnote 33]). Alternative cosmologies are in the same category as standard cosmology and contain the same kind of speculative elements, although so far at a less developed stage. Some scientists (e.g., Salpeter 2005) claim that some non-orthodox proposals are based on fallacies, either due to subconscious cheating or the 'file drawer effect' (filing away unfavourable results instead of publishing), but this may be present in some degree among orthodox researchers too. As cited in one of the above quotations, we do not need to wait for the total discrediting of the ΛCDM

[10] See video corresponding to his talk at Instituto de Astrofísica de Canarias 'Pluralism in contemporary cosmology' (2015): http://research.iac.es/iactalks/esp/talks/view/770.

model to be thinking about other options. Pluralism consists in a number of different theories competing with each other in an effort to explain the certain phenomena, not in identifying the winning horse before the race has commenced.

Some researchers in cosmology also think that the best solution is to adopt a theoretical pluralism:

> Inasmuch as no cosmological model can ever be capable of holding ultimate knowledge about the Universe, then it follows that no cosmology can hold eternal truths. Then, the larger the number of cosmologies available to cosmologists, the higher the chances of we obtaining better representations of the Universe. (Ribeiro and Videira 1998)

As explained throughout this chapter, that is not the approach dominating cosmological research today, but it is the one that mostly reflects the current scientific spirit.

References

Billings L 2010 Space science: The telescope that ate astronomy *Nature* **457** 1028–30

Bonometto S A 2001 Modern and post-modern cosmology *Historical Development of Modern Cosmology (ASP Conf. Ser. 252)* ed V J Martínez, V Trimble and M J Pons-Bordería (San Francisco, CA: Astronomical Society of the Pacific) pp 219–36

Brewer D D 2020 Essentials of Scientific Research *A Practical Guide* (Seattle, WA: Evidence Guides)

Bunge M 1997 *Ciencia, técnica y desarrollo* (Buenos Aires: Sudamericana)

Bunge M 2004 How does it work? the search for explanatory mechanisms *Phil. Soc. Sci* **34** 182–210

Burbidge G R 1971 Was there really a big bang? *Nature* **233** 36

Burbidge G R 1988 Review of "quasars, redshifts and controversies *Sky Telesc.* **75** 38–43

Burbidge G R 1997 New vistas in cosmology and cosmogony *The Universe at large. Key Issues in Astronomy and Cosmology* ed G Münch, A Mampaso and F Sánchez (Cambridge: Cambridge University Press) pp 64–109

Burbidge G R 2006 The state of cosmology *Current Issues in Cosmology* ed J-C Pecker and J V Narlikar (Cambridge: Cambridge University Press) pp 3–16

Campanario J M and Martin B 2004 Challenging dominant physics paradigms *J. Sci. Explor* **18** 421–38 (https://documents.uow.edu.au/~bmartin/pubs/04jse.html)

Castro Perelman C 2008 My struggle with Ginsparg () and the road to Cyberia: a scientific-gulag in cyberspace *Against the Tide. A Critical Review by Scientists of How Physics and Astronomy Get Done* ed M López-Corredoira and C Castro Perelman (Boca Raton, FL: Universal Publishers) pp 59–76

Ćirković M M and Perović S 2018 Alternative explanations of the cosmic microwave background: a historical and epistemological perspective *Stud. Hist. Phil. Mod. Phys.* **62** 1–18

Croft R A C and Dailey M 2015 On the measurements of cosmological parameters *Q. Phys. Rev.* **1** 1–14 (arXiv:1112.3108)

de Vaucouleurs G 1970 The case for a hierarchical cosmology *Science* **167** 1203–13

Dingle H 1937 Modern aristotelianism *Nature* **139** 784–6

Disney M J 2000 The case against cosmology *Gen. Relativ. Gravit.* **32** 1125–34
Disney M J 2007 Modern cosmology: science or folktale? *Am. Sci.* **95** 383–5
Disney M J 2011 Doubts about big bang cosmology *Aspects of Today's Cosmology* ed A Alfonso-Faus (Rijeka: InTech) pp 123–32
Dolsenhe O 2011 *A Critique of Science: How Incoherent Leaders Purged Metaphysics of Mind and God* (Raleigh, NC: Lulu)
Einstein A 1934 *Essays in Physics* (New York: Philosophical Library)
Ekeberg B 2019 *Metaphysical Experiments. Physics and the Invention of the Universe* (Minneapolis, MN: University of Minnesota Press)
Faerber T and López-Corredoira M 2020 A chi-squared analysis of the measurements of two cosmological parameters over time *Universe* **6** 114
Feyerabend P K 1975 *Against Method* (London: New Left Books)
Field J V 1988 *Kepler's Geometrical Cosmos* (London: Athlone)
Gale G 1993 Philosophical aspects of the origin of modern cosmology *Encyclopedia of Cosmology* ed N S Hetherington (New York: Garland Publishing) pp 481–95
Gale G 2002 (revised 2019) Cosmology: methodological debates in the 1930s and 1940s *Stanford Encyclopedia of Philosophy* ed E N Zalta (Standford: The Metaphysics Research Lab) https://plato.stanford.edu/entries/cosmology-30s/
Galilei G 1967 *Dialogue Concerning Two Chief World Systems–Ptolemaic & Copernican. English Translation* 2nd edn (Berkeley, CA: University of California Press)
Henneken E A, Kurtz M J, Eichhorn G, Accomazzi A, Grant C, Thompson D and Murray S S 2006 Effect of e-printing on citation rates in astronomy and physics *J. Electron. Publ* **9** https://doi.org/10.3998/3336451.0009.202
Hiebert E N 1980 Boltzmann's conception of theory construction: the promotion of pluralism, provisionalism, and pragmatic realism *Probabilistic Thinking, Thermodynamics and the Interaction of the History and Philosophy of Science (Synthese Library: Studies in Epistemology, Logic, Methodology, and Philosophy of Science, vol 146)* ed J Hintikka, D Gruender and E Gruender (Dordrecht: Springer) pp 175–98
Holster A 2016 *The Death of Science: A Companion Study to Martín López Corredoira's "The Twilight of the Scientific Age* (Boca Raton, FL: Universal Publishers)
Ioannidis J P A 2012 Why science is not necessarily self-correcting Perspect *Psychol. Sci.* **7** 645–54
Iradier M 2009 *La ciencia en coordenadas* (Cádiz: Hurqualya)
Janis I L 1972 *Groupthink: Psychological Studies of Policy Decision and Fiascos* (Boston, MA: Houghton Mifflin)
Kolb E W 2001 Cosmology at the start of a new century *Historical Development of Modern Cosmology (ASP Conf. Ser. 252)* ed V J Martínez, V Trimble and M J Pons-Bordería (San Francisco, CA: Astronomical Society of the Pacific) pp 295–307
Krasiński A 1997 *Inhomogeneous Cosmological Models* (Cambridge: Cambridge University Press)
Leighton R B 1971 Panel discussions: is there any hope? *Phys. Today* **24** 30
Lerner E J 1991 *The Big Bang Never Happened: A Startling Refutation of the Dominant Theory of the Origin of the Universe* (Toronto: Random House)
Loeb A 2010 Taking "the road not taken": on the benefits of diversifying your academic portfolio (arXiv:1008.1586)
Longair M S 1997 Reflections on the key problems *The Universe at Large. Key Issues in Astronomy and Cosmology* ed G Münch, A Mampaso and F Sánchez (Cambridge: Cambridge University Press) p 422

Longair M S 2001 The technology of cosmology *Historical Development of Modern Cosmology (ASP Conf. Ser. 252)* ed V J Martínez, V Trimble and M J Pons-Bordería (San Francisco, CA: Astronomical Society of the Pacific) pp 55–74

López-Corredoira M 2008 What do astrophysics and the world's oldest profession have in common? *Against the Tide. A Critical Review by Scientists of How Physics and Astronomy Get Done* ed M López-Corredoira and C Castro Perelman (Boca Raton, FL: Universal Publishers) pp 145–77

López-Corredoira M 2013a Peaks in the CMBR power spectrum II: Physical interpretation for any cosmological scenario *Int. J. Mod. Phys.* D **22** 1350032

López Corredoira M 2013b *The Twilight of the Scientific Age* (Boca Raton, FL: BrownWalker Press)

López-Corredoira M and Gabrielli A 2013 Peaks in the CMBR power spectrum. I. Mathematical analysis of the associated real space features *Physica* A **392** 474–84

Luminet J-P 2008 Is science nearing its limits? Summarizing dialogue *Is Science Nearing Its Limits?* ed G Steiner, E L Rui Vilar and F ao Calouste Gulbenkian *et al* (Manchester: Carcanet Press) pp 180–200

MacCallum M A H 1987 Strengths and weaknesses of cosmological big-bang theory *Theory and Observational Limits in Cosmology (Proc. Vatican Obs. Conf.)* ed W R Stoeger (Vatican: Specola Vaticana) pp 121–41

Mahoney M J and DeMonbreun B G 1977 Psychology of the scientist: an analysis of problem-solving bias *Cogn. Ther. Res* **1** 229–38

Markie P 2012 Rationalism vs. empiricism *The Stanford Encyclopedia of Philosophy (Summer 2012 Edition)* ed E N Zalta (Stanford, CA: The Metaphysics Research Lab.)

Matravers D R, Ellis G F R and Stoeger W R 1995 Complementary approaches to cosmology–relating theory and observations *Q. J. R. Astron. Soc.* **36** 29–45

Merritt D 2020 *A Philosophical Approach to MOND. Assessing the Milgromian Research Program in Cosmology* (Cambridge: Cambridge University Press)

Merton R K 1968 The Matthew effect in science *Science* **159** 56–63

Milne E A 1933 World-structure and the expansion of the universe *Z. Astrophys.* **6** 1–95

Mosteirín J 1989 A round-trip ticket from philosophy to cosmology *Foundation of Big Bang Cosmology* ed F Walter Meyerstein (Singapore: World Scientific) pp 289–314

Narlikar J V 2000 Venture funding for new ideas *Nature* **404** 707

Narlikar J V and Padmanabhan T 2001 Standard cosmology and alternatives: a critical appraisal *Annu. Rev. Astron. Astrophys.* **39** 211–48

Öpik E J 1977 About dogma in science, and other recollections of an astronomer *Annu. Rev. Astron. Astrophys.* **15** 1–17

Park I-U, Peacey M W and Munafó M R 2014 Modelling the effects of subjective and objective decision making in scientific peer review *Nature* **506** 93–6

Peebles P J E, Schramm D N, Turner E L and Kron R G 1991 The case for the relativistic hot big bang cosmology *Nature* **352** 769–76

Peebles P J E and Yu J T 1970 Primeval adiabatic perturbation in an expanding universe *Astrophys. J.* **162** 815–36

Pérez Velázquez J L 2019 *The Rise of the Scientist-Bureaucrat* (Cham: Springer)

Planck M 1949 *Scientific Autobiography and Other Papers* (London: Williams & Norgate)

Popper K 1994 *The Myth of a Framework: In Defence of Science and Rationality* (London: Routledge)

Ribeiro M B and Videira A A P 1998 Dogmatism and theoretical pluralism in modern cosmology *Apeiron* **5** 227–34

Rothman T and Ellis G F R 1987 *Astronomy* **15** 6–21

Salpeter E E 2005 Fallacies in astronomy and medicine *Rep. Prog. Phys.* **68** 2747–72

Sanromà M 2007 El big bang: història i sociologia d'una teoria científica *Master's thesis* Universidad Oberta de Catalunya

Sawyer R K 2005 *Social Emergence: Societies as Complex Systems* (Cambridge: Cambridge University Press)

Sciama D W 1971 *Modern Cosmology* (Cambridge: Cambridge University Press)

Singer D W 1935 *Candidate for Governor: And How I Got Licked* (New York: Farrar, Strauss and Giroux)

Smolin L 2008 The Trouble with Physics: The Rise of String Theory *The Fall of Science, and What Comes Next* (Boston, MA: Houghton Mifflin Harcourt)

Sunstein C R 2019 *Conformity: The Power of Social Influences* (New York: New York University Press)

Tegmark M, Aguirre A, Rees M J and Wilzek F 1998 Dimensionless constants, cosmology, and other dark matters *Phys. Rev.* D **73** 023505

Tolman R C 1934 Effect of inhomogeneity on cosmological models *Proc. Natl Acad. Sci. USA* **20** 169–76

Turner M S 1999 Cosmological parameters *Second Int. Workshop on Particle Physics and the Early Universe (COSMO-98)* ed D O Caldwell (Melville, NY: AIP) pp 113–28

Unzicker A and Jones S 2013 *Bankrupting Physics. How Today's Top Scientists Are Gambling Away Their Credibility* (New York: Palgrave)

Weinberg S 1977 The First Three Minutes *A Modern View of the Origin of the Universe* (New York: Basic Books)

Wesson P S 1987 Problems with the way in which cosmology is done *Theory and Observational Limits in Cosmology (Proc. Vatican Obs. Conf.)* ed W R Stoeger (Vatican: Specola Vaticana) pp 559–63

White S D M 2007 Fundamentalist physics: why dark energy is bad for astronomy *Rep. Prog. Phys.* **70** 883–97

Yao W M, Amsler C and Asner D *et al* 2006 Review of particle physics *J. Phys. G: Nucl. Part. Phys.* **33** 1–1232

Zee A 1986 *Fearful Symmetry: The Search for Beauty in Modern Physics* (New York: McMillan)

Chapter 9

Cosmology and culture

Cosmology does not belong only to professional physicists and astronomers. It is part of our western culture, and many different sectors of our society have been influenced by the new ideas of the cosmos, and vice versa: The (philosophical, religious, political, etc) ideas of our society have influenced cosmologists in the creation of their science. We may wonder to which degree cosmology has been more affected by human factors (psychological, sociological, ideologies/culture, etc) than most other sciences. Here we analyse some of the topics relevant to these questions. The discussion includes such themes as creation versus eternal universes, the anthropic principle, and fine tuning and multiverse, the influence of the Scholastic Middle Age *Primum Mobile* in the creation of modern cosmology, Anglo-Saxon political power and cultural colonisation, and analytical philosophy versus continental philosophy.

Cosmology is a question that does not concern only specialists. Ideas in cosmology are unlike those in other areas of physics or astronomy; they are not reduced to technical issues such as the analysis of stellar evolution or the decay of fundamental particles. Rather, cosmology is interconnected with many other developments of our culture, often linked to some *Weltanschauungen*, where philosophical, religious, political, or other kinds of ideas influence the development of our view of the cosmos. Cosmology expresses a culture's sense of its existence with respect to the wholeness of the Universe.

The different strata of our society are not independent from one another. We cannot therefore expect cosmology to be independent of the dominant religion, politics, economic model, etc of a society. We cannot expect for instance that a cosmology developed in our neo-liberal democratic secular states—and we may include here countries that do not strictly follow this definition, but that imitate the dominant western way of doing science—will produce the same view of the cosmos that might be produced for instance in Mao Tse Tung's (1893–1976) China, which was the opposite of western democratic values, or in theocratic Islamic Republics.

doi:10.1088/978-0-7503-3775-5ch9

All of these three examples can produce the same nuclear weapons using the same physics, as that is a purely technical issue lacking of any ideology; cosmology, however, is is quite different from most other branches of science.

Therefore, reducing the analysis of cosmology to mere scientific arguments would leave us partially blind to the reality of this cultural expression. In this chapter, I briefly review some connections of cosmology with other areas. Beyond the discussion of professional cosmologists on the subtle details of the standard and alternative models, there is the consideration of how society also participates in this collective creation. Georges Lemaître (1894–1966) and George Gamow (1904–1968) may have laid the foundations of the scientific arguments for Big Bang cosmology, and many other scientists contributed to its further development, but the substratum of these ideas are a general part of our culture. The principles of cosmological research go beyond science, and we must observe their human underpinnings.

9.1 The influence of religion

A remarkable factor carrying weight in the determination of the dominant scenario in cosmology is the ideology of researchers, religious ideas having some relevance in this aspect. The association between cosmology and religion is indeed a very old one, says Kragh (2007b), but there are in my opinion older themes that have remained persistent.

9.1.1 Creation versus eternal Universe in western culture

In *Timaeus*, Plato (c. 427–347 BC) says that time was created simultaneously with the Universe. This idea was introduced into Christianity from the third century A.D., after reconciling Christianity with existing Roman society (Roberts 1924, Lerner 1991 [chapter 2]). Augustine of Hippo (354–430) later introduced certain Platonic ideas into Christianity, such as the untrustworthiness of the senses and the instantaneous creation of the Universe from nothing. A Universe of infinite space and time is exclusive to the Deity and thus prohibited for the material Universe. Saying that the Universe is unlimited is to obscure a crucial difference between God and Nature, and thus advocate pantheism. This idea is still actually contended (e.g., Jaki 1988).

Time was created with the Universe, says Augustine in his *Confessions*. Before Heaven and Earth there was no time, so it is pointless wondering what was done before the creation because there was no 'that time' simply because there was no time. This very much brings to mind the contents of the standard cosmology. There is also another coincidence between Augustine's philosophy and the Big Bang claim: 'Time itself was made by God' declared Augustine to undercut the premise of a challenger who had asked him, 'What did God do before He made Heaven and Earth?' (Grünbaum 1993).

The question of the creation in Big Bang theory is solved with quantum theory (Grünbaum 1993), rather than through the direct action of God. Science is not explicitly defending religious dogmas, but there is some similarity between the action of spontaneous creation in both cosmovisions. According to the present-day standard model, this spontaneity is described as fundamental fluctuations in empty space that lead to the emergence of energy. Indeed, the very distinction between space and time becomes mushy, as does the notion of an individual singular instant of time. However, the union of particle physics and cosmology is not totally achieved yet; no quantum cosmology exists yet.

The Indian heterodox cosmologist Jayant Narlikar (1938–) once said: 'To the western world, the concept of a universe without a beginning and thus lacking a 'creation epoch' was strange at best and philosophically unacceptable' (Narlikar 2001). This is not entirely correct. While ideas of creation have dominated religion and other philosophical notions, eternity has also been present in some views. The debate about the eternity of the Universe is implicit in Aristotle's (385–322 BC) works and according to some interpretations of Plato (Meyers 2003).

Later, during the Middle Ages in Europe, Nicholas of Cusa (also known as Nicolaus Cusanus, or simply Cusanus; 1401–1464), a German bishop (Sigmund 1963), returned in his *On Learned Ignorance* (ca. 1450) to the idea of an infinite, which is everywhere the same, with its centre everywhere and its periphery nowhere; with no beginning or end of time. There can be no 'Theory of Everything', asserted Cusanus, since all learning is finite in the midst of infinite reality. Knowledge can only be a series of better approximations, unifying larger realms of experience. The astronomer, geographer, and mathematician Paolo dal Pozo Toscanelli (1397–1482) disseminated Cusanus' new cosmology and linked it to the emerging observational sciences. One of his pupils was Leonardo da Vinci (1452–1519).

Giordano Bruno (1548–1600) also challenged the idea of *ex nihilo* creation, arguing that the Universe must be unlimited in both space and time, without beginning or end (Singer 1950). Thus, we cannot say that western philosophy was exclusively fixated on the idea of creation. Nonetheless, we may consider that a non-eternal Universe was a dominant idea, in particular during 19th and 20th century, and the influence of Christian dogma is undeniable. As Narlikar might concur, if we compare the Christian Europe and America with the Hinduism in India, there is no doubt that the latter one is more dominated by the idea of an eternal Universe, given the culture in which it is embedded, but Europe, given its variety of classical philosophies, is not influenced by a unique dogma.

9.1.2 Some authors connecting religion and modern cosmology

Because of this historical background and the coincidences of elements of the standard model with certain religious credos, some authors think that nowadays the Big Bang is simply the scientific version of Genesis, and that the Big Bang idea, being a synthesis of astrophysics and the dogma of an *ex nihilo* creation (e.g., Jastrow 1978), is attractive to many people in the same way. Indeed, in 1951 (when the Big Bang was not yet a dominant standard theory), Pope Pius XII (1876–1958) asserted that the Big Bang supports the doctrine of *ex nihilo* creation (Pius XII (Pope) 1952). He wrote in an address to the Pontifical Academy of Sciences:

> In fact, it seems that present-day science, with one sweeping step back across millions of centuries, has succeeded in bearing witness to that primordial 'Fiat Lux' (Let there be light) uttered at the moment when, along with matter, there burst forth from nothing a sea of light and radiation, while the particles of the chemical elements split and formed into millions of galaxies.... Hence, creation took place in time, therefore, there is a creator, therefore, God exists!

Georges Lemaître (1894–1966), who was present at this papal address in 1951, could not hide his disappointment at these words of the Pope: They supported Fred Hoyle's (1915–2001) contention that the Big Bang aspiration and aversion to the Steady State Theory had religious roots; Hoyle wondered mockingly what sort of class of scientific theory was the Big Bang, having been created by a priest and given the papal seal of approval. However, even if Lemaître tried to separate science and religion, he was a man deeply convinced of the imprint of God in the new cosmology that he was creating. The father of the expansion model studied in a seminary for the priesthood and was raised a member of the Catholic hierarchy. He also became the Director of the Pontifical Academy of Science (Lerner 1991 [chapter 4]). The original typed manuscript of the paper (Lemaître 1931), conserved in the archives of the University of Leuven (Belgium), finished with a paragraph (which would be later removed in the published version) in which he expressed the faith in the *Deus absconditus* of biblical resonances: 'I think that every one who believes in a supreme being supporting every being and every acting, believes also that God as essentially hidden and may be glad to see how present physics provides a veil hiding the creation.'

The Pope John Paul II (1920–2005) also touched on the topic (John Paul II (Pope) 1981), although with more ambiguity than Pius XII, and in a somewhat self-contradictory manner. On the one hand, he talks about the independence of religion and science, apologising on behalf of the Church for the persecution of Galileo Galilei (1564–1642) and paraphrasing Galileo by saying that the Bible does not wish to teach how heaven was made but how one goes to heaven. He says that religion must not judge another cosmological theory. On the other hand, he favourably quotes from Pius XII's earlier speech and contends that the question of the beginning of the Universe is not one that can be solved by science alone; above all, it requires knowledge that comes from revelation.

Many scientists have also used the ideas of modern cosmology in order to make theological claims. There are many who talk about the Big Bang leading to a proof of God's existence (e.g., Davies 1983). There are also debates for and against the idea (e.g., Soler Gil and López Corredoira 2008). George F Smoot (1945–), when the discovery of the anisotropies of the Cosmic Microwave Background Radiation was announced, claimed that for a religious person this was like looking at the face of God (Wright 1993). We must also bear in mind that the United States, at present the country leading cosmological research, is dominated by a much higher proportion of followers of the Christian religion than other wealthy countries.

Christianity is not the only religion that has advocated this association of concepts. There also seems to be great acceptance of the standard cosmology in other monotheistic religions. The Israeli physicist and cosmologist Moshe Carmeli (1933–2007) says that not only does the Big Bang scenario agree with the idea of creation described in the Bible, but also with the scenario of creation in six days (Carmeli 2000). The Muslim astronomer Kamel Ben Salem (1956–) analyses the Quranic description of phenomena linked to the evolution of the Universe (Ben Salem 2005).

These three religions also have in common that they only admit one God, in the same way that cosmologists nowadays are not pluralist and admit only one theory as the correct one. One may wonder how the cosmological research would be if, instead of having an origin in western civilisation, it had been developed in some other culture; Japan, for example, is more respectful of pluralistic views.

The opposite trend is also observed. Among heterodox scientists and sceptics (myself included) there has been, and continues to be, a higher ratio of atheists and agnostics. It is known, for instance, that Fred Hoyle was not a believer. And there are cases of practices in communist countries that favoured non-standard cosmologies. For instance, in the People's Republic of China until the 1970s, there was a degree of censorship affecting the circulation of ideas relating to the Big Bang (Hu 2004, Kragh 2007a [pp 199–200]).

Of course, there are also many atheists who accept the Big Bang and vice versa, but there have been correlations between religious dogmas and preferred cosmological scenarios during various stages in development of these ideas, and these correlations are not fortuitous. Nonetheless, from what I observe among my colleagues—whether Christians or atheists—they are mostly pro-Big Bang. Once the standard paradigm is in a dominant position, religious ideas, it seems, no longer exert such a strong influence, other sociological factors seeming to be more important.

9.2 God, multiverse, or neither

Look at the cosmos:[1] its galaxies with myriads of stars, some with planets. Is not it a jaw-dropper? Look at our planet Earth, with all the variety of life it contains and with the presence of the human species, which has taken the development of civilisations to incredible limits. Is it not astounding? How is it possible that all this is here, before our eyes? Priests, philosophers, and scientists all ask the same question, all of us are involved in reflecting on the existential question. This old fascination—a mixture of science, religion, and existential philosophy—is what has driven what we know today as the anthropic principle, or fine tuning. We continue to ask ourselves, as we did centuries and millennia ago, how it is possible to have a Universe such as the one we observe and not one that is sterile and incapable of producing such complex structures as those that surround us from the chaos of the primordial cosmic soup.

[1] Part of this section is an English translation of my Spanish text López-Corredoira (2015).

9.2.1 Anthropic principles

The anthropic principle is the name given to any consideration of the structure of the Universe, the values of the constants of nature, or the laws of nature that has a bearing upon the existence of life (The Editors of Encyclopaedia Britannica 2013). It is argued that scientific observation of the Universe would not even be possible if the laws of the Universe had been incompatible with the development of sentient life. In other words, the constants and laws of the Universe are fine tuned to be able to produce life.

There are different formulations of the anthropic principle, which can be divided into 'weak' and 'strong', depending on the types of cosmological claims. The weak anthropic principle states that the fine tuning is the result of selection bias. Within this framework, the notion of the multiverse emerges, for which there is a statistical population of universes to select from. The strong anthropic principle, as originally proposed by the physicists John D Barrow (1952–2020) and Frank J Tipler (1947–), states that the Universe is in some sense compelled eventually to have conscious and intelligent life emerging within it (Barrow and Tipler 1986).

9.2.2 God of the gaps

We may imagine the ancient inhabitants of the Earth, of some tribe somewhere on the planet, observing the movement of the stars above their heads at night, or observing the regularity of certain natural phenomena in the forest. What might they have thought? That there is some god or gods controlling all this? What if some unexpected phenomenon appeared (for example, a lightning bolt falling from the sky on some house in the village) that the disturbed the regularity, different from the usual run of things? Perhaps they would have thought that some god was angry. In the face of such unpredictable and out of the usual order of events, one would say that there are reasons for such phenomena, but those reasons would be in the mind of a god whose designs one was trying to understand; for example, that the god was angry. Studying these primitive religions, or from anthropological approaches in tribes far away from western civilisation, one can understand certain structures of the human way of thinking, of its psychology, of the way our minds work.

Since immemorial time, man has assigned to the gods the roles of creators with powers of intervention on created matter when no other explanation was available to explain the observed facts. It seems that the eagerness to have an orderly representation of the events around us is part of our existence. We cannot understand many things, but what we do not understand we attribute to a thinking and wilful structure that we assume acts for certain reasons. This may be the root, or one of the roots, of animistic ideas (that the reasons for facts observed in Nature reside in the will and intelligence of supreme beings) and therefore of all religions, or even explanations in terms of aliens, as has become the fashion these days.

In the example of lightning, it is no longer common in our civilisation for someone to attribute the existence of lightning to a divine will, but rather in material terms of electrical discharges between the clouds and the Earth of different electrostatic potential. However, there are certain residues of this in the modern age.

Thus, Isaac Newton (1643–1727), when he was unable to find an explanation for the stability of the orbits of the planets, deduced that it was the hand of God that intervened by tapping the planets from time to time to put them in place so that they would not stray from their orbits. Pierre Simon Laplace (1749–1827) would later demonstrate that a purely physical explanation can be given.

God of the gaps is a type of theological perspective in which holes in scientific knowledge are wielded as evidence or proof of God's existence (see figure 9.1). The term was coined by Christian theologians not to discredit theism, but to point out the fallacy of relying on teleological arguments for the existence of God. It refers to a tendency derived from theism that states that what can be explained by human reason is beyond divine action and, God's action therefore being confined to the gaps that science cannot explain.

The term 'God of the gaps' is often used to describe the decline of religious explanations of physical phenomena in the light of increasingly comprehensive

Figure 9.1. How to connect physics and religion by filling gaps in knowledge through divine action. Copyright ©2021 by Sidney Harris. ScienceCartoonsPlus.com.

scientific explanations. As science has found explanations through different sciences, the need for a god to explain such phenomena has progressively diminished because it has been left with ever narrowing gaps in knowledge. This line of reasoning usually concludes that, as science explains more and more, the space for theistic or divine explanations has become smaller and smaller, and the hypothesis of the existence of God becomes unnecessary.

The term was originally used by Revd. Prof. Henry Drummond (1851–1897) in his Lowell Lectures on the origins of man. Drummond rebukes those Christians who use the things that science cannot explain to embrace all of Nature as the work of an immanent God, the God of evolution, who is infinitely greater than the miracle-worker of ancient theology.

In the 20th century, the term was used by Dietrich Bonhoeffer (1906–1945) in letters written in a prison under Nazi rule during World War II. And it came to attention in 1955 of Charles Alfred Coulson (1910–1974), when in his book *Science and Christian Belief* (Coulson 1955) he stated that there is no 'God of the gaps' at the strategic points where science fails; for such gaps have the inevitable habit of becoming vanishingly small. Used frequently since then in the fields of theology and philosophy, the concept gained more attention with the publication in 1971 of an article by the physicist Richard H Bube (1927–2018), who again attributed the contemporary crisis of faith in part to the loss of ability of such an argument as the explanatory power of science increased. Bube (1971) argued that Darwin's *Origin of Species* was the death toll of the god of the gaps. Bube (1971) also argued that the god of the gaps is not the same as the God of the Bible, implying that he was not making an argument against the existence of God per se.

9.2.3 Fine-tuning

When used to justify the existence of an intelligent designer according to a strong anthropic principle, the issue of fine tuning is a particular case of the 'God of the gaps' argument. We are told that the Universe is finely tuned so that life can exist in it; we do not understand the reason, so we fill the gap in our knowledge with some kind of religious idea. 'Tuning' has certain reminiscences of the idea of a Watchmaker who regulates the mechanisms of the Universe-machine in order that the desired ends be generated. This kind of intelligent design might be comparable to saying that dinosaurs became extinct to make way for human dominance of the planet. Such final causes are shocking to the modern scientific mindset.

There are variations on the theme, and some cosmologists have pointed out that the fine tuning of the constants of physics as necessary condition for life to arise is instead a proof of the existence of infinite number of parallel universes, each with a different set of physical constants. In this hypothesis, the fact that our own Universe is the one in which life is possible is simply due to a selection effect: only those universes that allow the creation of conscious life can ensure that their physics constants are appropriate for life. There is also an infinity of other universes, but these are inaccessible to our knowledge. This is known as the 'multiverse' hypothesis.

I think that both the watchmaker and multiverse are gratuitous speculations that cannot lead to any firm conclusion. The fine tuning idea is based on an assumption of a somewhat Pythagorean character: that the Universe is matter ordered according to mathematical laws, and their constants are produced on demand. This line of argument follows the same type of reasoning employed by those theoretical physicists who create mathematical models of the Universe by finding symmetries here and there, putting topologies of this or that type, of a toroidal or other shape, 32-dimensional universes, negative masses, time as an imaginary variable, or wormholes—in short, a whole zoo of fabulous animals created to speculate about our Universe and beyond it.

A mathematical model of physics is only a hypothetical representation of Nature; the more credible it is the more things it manages to explain the fewer alternative explanations there are. It has never been seen or proven that the constants of physics can take values different from those they are known to have in this Universe. To be able to speak of fine tuning, one has to assume that a parameter is adjustable within a range of values. It is like a radio, where one can tune a station within a spectrum of frequencies. But if the radio has a fixed and unchangeable dial, the probability of getting that frequency is 100%, because there is no other frequency. Are the constants of physics adjustable *à la carte*? Is the gravitational constant a variable that can be adjusted at the start of creation at one's whim? What, then, about the charge of the electron or Planck's constant? We know nothing about these questions. In this case, is not remaining silent better than talking about the low probability that a constant of physics may have a different value? In order to know a probability, one must know the statistical distribution of possible values of each constant. And what do we know about it? Nothing. Therefore, to say that a Universe with other constants could not harbour life is to say nothing of substance because such a Universe is only a mathematical game, not a representation of possible physical reality.

Besides, if there were such a fine tuning at work, it does not seem to be producing very fine results. If a god wanted to create a Universe habitable for life, why would he create one in which it is so extremely unlikely to find a planet like Earth that meets the conditions necessary for life? If we were to adjust and make universes *à la carte*, I think a more suitable creation could have been made for the purposes of harbouring life.

George F R Ellis (1939–), a theoretical cosmologist and philosopher of cosmology with religious leanings[2], classifies the causes of so-called fine tuning into six types (Ellis 2007 [section 9.1.6]): (1) mere chance; (2) necessity (it had to be this way and cannot be otherwise); (3) the constants of physics that we currently observe had

[2] George F R Ellis is a practising Quaker who has focused part of his scientific career on the more philosophical aspects of cosmology. In 2004 he won the Templeton Prize 'for Progress Toward Research or Discoveries about Spiritual Realities'. John D Barrow, previously mentioned in connection with the strong anthropic principle, was also a Templeton Prize winner in 2006 and was also a religious, a member of United Reformed Church. I suppose this interest of religious cosmologists in fine tuning and the anthropic principle is no coincidence.

a very high probability of taking the value they did; (4) all the possibilities occur in different multiverses; (5) there is a cosmological natural selection in which the universes, after successive collapse and expansion, survive if they have values of the constants that permit the generation of black holes that produce the collapse; (6) God is the cause. There is also a seventh answer by (Ellis 2007 [section 9.1.7]) with which I agree: that there are too many uncertainties in the matter to enable us to provide an answer to the question. The fact that there are answers to suit all tastes shows that anybody can speculate on the subject to fit his own preferences.

We should not confuse the fine tuning issue with the attempt to search for a theory that gives the values of the constants of physics, the latter being very praiseworthy effort that, when it comes to fruition, would be a step further in our understanding of the laws of Nature. We must first produce the theory that gives the values of the constants of physics or their probability distributions, and then, if there still remains something like fine tuning, we may speculate about the causes. But there is no such theory yet. Therefore, ruminating on the number of combinations of laws and physical constants capable of giving rise to the existence of life of similar characteristics to our own is very small is little more than mere speculation. There is no information on the number of possible combinations, and we do not know whether it is large or small or whether there is only one possible value.

In Newton's times one could still wonder what would happen to the Universe if, instead of a gravitational attraction inversely proportional to the distance squared, it were inversely proportional to the cubic distance or to a power with exponent greater than three. The result would be that no planet would be able to orbit around the stars: either it would fall into a spiral orbit towards the centre, or it would recede to an infinite distance, at which it would not be possible for life to appear on any planet. Could a physicist of the 17th or 18th century think that the laws of gravitation had been tuned[3] so as to give rise to life? A physicist of the late 19th century, however, would say, that the -2 exponent is explicable in terms of a gravitational vector field of constant total flux through a closed surface surrounding the source, as stated in Gauss' theorem, interpreting the field strength as a flux emitted by each mass. Since the spherical surface around a point mass is proportional to r^2, the intensity of the gravitational field must be proportional to r^{-2}. Gravitational theories have been proposed that do not comply with Gauss' theorem (or its equivalent in general relativity, the Birkhoff theorem), but this is not the topic of discussion here, The point here is to indicate that the value of a constant of nature that is not understood at a given epoch beyond the experimental datum might be understandable in terms of a later development of physics.

From a Pythagorean viewpoint, it is possible to think that physical theories are not distinguished from other kinds of mathematical models that are not found in Nature. Mathematically, one can build a Universe with any number of dimensions, and construct laws of attraction/repulsion between particles with any kind of

[3] In this case not so finely, since small variations of the exponent -2 can maintain orbits for an indefinite period of time between two distances from the focus, even though with an advance or recoil of the periaster; only with an exponent $\leqslant -3$ are the orbits unstable for any initial conditions.

dependence on position, velocity, time, or any other variable, but possible reality is more than just a mathematical model; it is the empirical reality we can ascertain as physicists (see section 8.1.1). We should therefore refrain from constructing imaginary universes and asking ourselves why such universes do not exist whereas ours does. Physics studies real nature and its laws, and we cannot see any fine tuning because, as far as we know, the constants of the Universe are constant—if the reader will forgive the redundancy—not variables to be freely tunable.

9.3 Binggeli's *Primum Mobile*

The Swiss astronomer Bruno Binggeli (1953–) in his book *Primum Mobile* (it is in German, but there is a short article in English that summarises its contents: Binggeli 2006, 2017) compares the standard model of modern cosmology with the cosmology in the Judaeo–Christiano–Gnostic beliefs of the Scholastic Middle Age, depicted in Dante Alighieri's (1265–1321) *Divine Comedy*. The author finds that there is a perfect correspondence in some essential points between both world views. The three basic tenets of *Primum Mobile* are present in the observable Universe of the Big Bang theory: (1) there is a maximum finite distance from us in the observable Universe, (2) the observable Universe is a sphere with us at the centre; and (3) it has a hierarchical structure. Comparison of figures 1.1 and 1.2 (in chapter 1 of present book Fundamental Ideas in Cosmology) illustrates this observation. The book also speculates about other correspondences: the photons and bosons of modern physics would play a role similar to the messenger angels in Dante's book; the black holes would occupy the role occupied by Hell; etc. His comparison of the three basic tenets seems to hit the nail on the head, although there is some degree of free speculation in this and other aspects of Binggeli's book.

One may wonder about the cause of these correspondences, and an answer is also provided by Binggeli (2006): There has to be a psychological mechanism dominating our visions of Nature. According to Binggeli, the result of our research is highly biased by the influence of the culture in which we are embedded (which has inherited the Scholastic cosmological view) and our own psychological patterns. Modern cosmology is a symbolic expression of the states of our mind. Binggeli argues that our view of external reality is indeed a reflection of our interior world, and that the way to understand modern science should go through psychological analysis.

Sigmund Freud (1856–1939) is not Binggeli's favourite psychologist, since Freud is compromised with a deterministic world-view proper to the 19th century. Binggeli prefers Carl Gustav Jung (1875–1961) because of the latter's greater affinity to quantum physics[4]. Rather than Freudian sexual drives, Binggeli analyses the behaviour of scientists in terms of Jungian religious necessity, as introspective self-examination, an unconscious wish to become like God. Large, expensive telescopes and particle accelerators, he affirms, are comparable to the cathedrals of the Middle Age as expressions of power and prestige for their symbolic value.

---

[4] Quantum mechanics is irrelevant for understanding the mind, and its flirting with subjectivity (John von Neumann (1903–1957), Eugene Paul Wigner (1902–1995), etc) is merely a speculation unrelated to physics and disconnected from the neurosciences (López-Corredoira 2009).

But Jung is too mystical and too mentalist. His suggestions that matter and mind are two aspects of the same thing sound too esoteric, very far from a naturalist view. Another exotic view is Binggeli's claim, 'Each star is a hot spot of the whole Cosmos, each individual consciousness is a hot spot of the divine Self', and the argument that science is a vehicle to reach a higher state of consciousness. These are beautiful and poetic statements, but they go beyond sociological or psychological explanations. Nonetheless, there are some merits in this point of view, some subtle intuitions, that may contain useful clues in the labyrinth of social and psychological causes affecting the construction of cosmological models. I agree with the consideration that cosmology nowadays should be analysed under the lens of expert sociologists and psychologists. I hazard the guess that, in the future, when the delusion of obtaining a complete cosmological theory fades away, humanists will spend a long time investigating these topics.

9.4 Politics and the democratisation of cosmology

9.4.1 Prestige and entertainment

> Politicians are directly interested in astronomy because it is something that can bring prestige to the country. (Davoust 1995)

A quite remarkable observation! Davoust (1995) also adds that politicians promote research in astronomy to give employment to the national industries in the building, for example, of large telescopes. Apart from such industrial considerations, the prestige associated with a science like astronomy for politicians is similar to that attaining to sports (López-Corredoira 2008), but not all branches of astronomy share the same degree of attraction, and people want colourful spectacle. In terms of mass media coverage cosmology occupies one of the prime positions, along with planetary science and the search for extraterrestrial life.

In western democracies, offering science as an entertainment through mass media is the most important part of the motivation to promote research in cases of pure science without direct applications to industry. But taxpayers expect some results and certain goals to be achieved, as most of them cannot understand why intellectuals should be eternally discussing ideas without reaching any conclusion. Cosmology is no exception since it offers just this kind of narrative. Popularised visions of the Universe also provide unconscious motivations that partly explain the success of some cosmological proposals (Ten Ros 2001).

It is claimed that our cosmological theories are very robust, with the intention of giving an impression of the certainty of our knowledge. Uncertainties are disquieting, and people prefer hard information, a secure world. They prefer to live in a world without surprises, with a guaranteed retirement salary, a world in which they can live a long life surrounded by children and grandchildren, and die peacefully in bed after that long life. A conservative science (see section 8.4.2) offers a view of calmness, in which everything is apparently under control, and the few discordant voices are those of outliers who do bad science.

The sociological reasons for favouring orthodox proposals might also be related to the preference of domesticity in our civilisation (López-Corredoira 2013 [chapter 5]). Finding a promising change of paradigm closer to the truth among tens of competing proposals is very difficult. In orthodoxy, although absolute truth is not guaranteed, at least a consensus version of the truth is offered that carries weight. And, as said by Luminet (2008), consensus in the scientific community is transformed into orthodoxy and ends up becoming counter-revolutionary. By means of it, if somebody is wrong then everybody is wrong and the fault is diluted among many people. Investment in science prefers security, it prefers domesticity and control, rather than a promising and challenging change of paradigm that is uncertain, with the attendant difficulty of guessing from which direction a new paradigm might come.

9.4.2 Cosmology influenced by the economy and social structure

The heterodox cosmologist Eric J Lerner (1947–) observes a relation between the ideas dominant in cosmology and those dominant in society due to economic circumstances (Lerner 1991 [chapter 4]). According to him, the Big Bang model was in eclipse from around 1957 to 1964, a period of expansion, postwar recovery, and confidence in progress. The golden age of the Big Bang in the seventies corresponds to a decade of pessimism, oil crises, and worldwide economic turmoil. 'Cosmology', says Lerner (1991 [chapter 4]), 'was scientific research on the cheap'. In contrast to expensive experimental science, theoretical cosmologists used only paper, pencil, and a little computer time. This, in the crisis-ridden seventies, was a reason for pursuing this kind of theoretical study. Lerner also thinks that Alan Guth (1947–) arrived at his idea of cosmic inflation just as the worst monetary inflation of the century was coming to a climax (Lerner 1991 [chapter 4]). This is a debatable interpretation of cause-and-effect connections in cosmology; indeed, one may think of many examples of important advances in cosmology that are not at all associated with any sort of social crisis, and vice versa: There are some social crises—the economical crisis in 2008 or the sanitary crisis of COVID-19 in 2020—that do not appear to be associated with any important advance in the standard model of cosmology. Anyway, in spite of the presence of much noise in the correlations, and the existence of many possible social or non-social causes, one may assume that such a social-science connection is not entirely misleading.

The standard model is not alone in being influenced by dominant ideas in society, the same has also occurred with competing theories and sceptical stances. George Gamow (1904–1968), one of the fathers of the Big Bang idea, argued against Steady State Theory by telling a story concerning one of its defenders: He told that the Soviet astronomer Boris Vorontsov-Velyaminov (1904–1994) was forced to accept Steady State by arguments deriving from dialectic materialism (Gamow 1952). However, Fred Hoyle (1915–2001) had occasion to verify in his visit to the Soviet Union in 1958 that Steady State Theory was not unanimously accepted by the Soviet authorities, since the application of physical theories to the whole Universe was viewed suspiciously as an anti-scientific extrapolation of local laws. Cosmology was considered in Soviet Union as a bourgeois science, a cancerous tumour that corroded proletarian science (Kragh 1996 [pp 260–2]). Whether in supporting heterodox models or none at all, it is clear the Soviet science distinguished itself from its western counterpart in the most speculative areas, such as cosmology. We cannot maintain that Soviet Union cosmology was embedded in an ideology whereas western science was not. All societies have ideologies, and all of them have some degree of influence in how their cosmology is produced. It is easier to recognise such influences in other societies while believing that here in the west we do pure science uncontaminated by ideology. Accusing the former Soviet Union of anti-scientific practices in such a way would might well be described as the pot calling the kettle black.

9.4.3 Doctors of a new faith

Cosmological theories are couched in technical language and mathematics that isolate it from the general public. More than being a mere inconvenience, such technicalities fit the role adopted by present-day cosmology, which has replaced the priests of old religions purporting to reveal the mysteries of the Universe. One of the reasons for the success of theoretical cosmologists during the 20th century among the masses was indeed the derivation of a mathematical formulation that few people could understand:

> The people were told that the true nature of the physical world could not be understood except by Einstein and a few other geniuses who were able to think in four dimensions. Science was something to believe in, not something which should be understood. Soon the best–sellers among the popular science books became those that presented scientific results as insults to common sense. (Hannes Alfvén (1908–1995), Nobel Prize of Physics in 1970, heterodox cosmologist with his proposal of Plasma Cosmology; at his article ‘How Should We Approach Cosmology?’ (Alfvén 1978))

9.4.4 Amateurs

One of the consequences of the popularisation of cosmology is the rise of hundreds or thousands of amateurs without specialised training in the topic, who engage in cosmological thinking. It is a purely sociological phenomenon that merits consideration.

Amateurs proposing new alternative cosmological models are often possessed by some sort of unappreciated genius complex, an ‘ego’ that drives them towards the creation of new alternative models. Their dream is to create a new paradigm in science that completely changes our view of the Universe. Many of them try to demonstrate that Einstein was wrong, perhaps because he is the symbol of genius and defeating his theory would mean that they are geniuses above Einstein. But most of their ideas are ill-founded, with little to be said in their support. Few of them merit serious consideration.

I often receive new theories from amateurs or semi-amateurs (some of them have some formal education in scientific disciplines, but they are not specialised researchers) who want to topple all the well-established tenets of physics to leave space for new physical or cosmological theories that fail in even the most basic aspects. This kind of work makes almost no reference to professional journals, and tends to cite popular books on science. Rather than studying a particular scientific problem, they talk in generalities. Their quixotic enterprises contain many weaknesses and inconsistencies because they have insufficient knowledge of the subject. And the result is that the hundreds or thousands of heterodox cosmologists in the shadows do not get to hear the voice of some possible genius who might be in their midst. As it is, autonomous research activities are never taken seriously; one must use official channels in order to be heard by professional researchers.

9.5 Anglo-Saxon cultural colonialism

In the first period of cosmology, important advances in observing techniques were produced in the US, where funds were available for building big telescopes. There were few theoretical cosmologists in the US, with the exception of Richard Tolman (1881–1948) and Howard P Robertson (1903–1961). This was in part due to the scepticism towards theoretical activities in that country, especially towards such topics as cosmology. In addition, modern theoretical cosmology required a level of mathematical accomplishment generally unavailable at that time in the US (Gale 1993). In the first half of the 20th century most theoretical cosmologists were therefore European, with an important part in the UK. Today, however, US cosmologists are predominant, together with the UK tradition, so we live in an era of Anglo-Saxon cosmology in terms of the production of both theories and observations.

Cosmology is not a pure science derived from observation, experiment, and mathematical modelling. It contains a world view and this is adapted to the mentality and culture where it was created, so the countries where it was developed influence the results, not only because of religious ideas, as we have seen. Had cosmological research originated in other countries (Germany, Russia (or the old Soviet Union), India, Japan, etc) the resulting cosmovision might have been different. It is certainly true that cosmology is now carried out in many other countries, as well as in the US and the UK, but the late-comers continue the lines of thought initiated by the pioneers.

The dogmas of standard cosmology are usually accepted when they come from a US researcher. Rashid Sunyaev (1943–), a cosmologist born in the former Soviet Union, in a private discussion with the Estonian cosmologist Jaan Einasto (1929–) in 1973, argued that nobody would take their results seriously until confirmed by an American astronomer (Einasto 2001). Einasto (2001) corroborated Sunyaev's impressions with his own observations:

> Indeed, our experience has confirmed Sunyaev prediction several times. Öpik's measurement of the discovery of the Andromeda galaxy was accepted only after Hubble (1925)'s discovery of Cepheids in M31 and other galaxies of the Local Group. Kuzmin's determination of the local mass density in the Universe remained almost unknown, even after the work by Gilmore *et al* (1989). Our work on the presence of dark matter around galaxies was noticed after Ostriker *et al* (1974) got similar results, and Vera Rubin confirmed flat rotation curves of galaxies; the void-filament structure of the Universe was accepted after the work by Geller and Huchra (1989). Is this a rule in scientific revolutions?

It is certainly the case, the important discoveries in cosmology announced with much fanfare are not homogeneously distributed around the planet, even by weighting the number of researchers per nation. There is a dominance of Anglo-Saxon cosmologists, not only in number of papers published, but also in the impact research exerts. This is part of the cultural colonisation of the most powerful countries after Second World War, not only English becoming the official language of science, but also ruling on how science may be pursued. Fields of science more sensitive to cultural contamination, such as cosmology, thus became embedded in British and North American philosophies.

9.5.1 Analytical philosophy

The philosophical tradition of English-speaking countries differs from those traditions of other European countries and differs even more from eastern traditional philosophies. As a matter of fact, on referring to western philosophy, the term 'continental philosophy' (Critchley 2001) was coined to for philosophical traditions from mainland Europe outside the analytic movement that characterises Anglo-Saxon philosophy. Continental philosophy includes German idealism, existentialism, psychoanalytic theory, critical theory, and many other elements that are almost absent in the Anglo-Saxon analytical tradition. The main distinguishing feature between continental and analytical philosophies is their opposing attitudes to objectivity and how we grasp reality. Indeed, it could be argued that the excessive use of the third person in English language academic texts enforces an ideology in the observer (who should properly use the first person to denote the subjectivity of approach) is to all intents and purposes effaced.

The analytical tradition is closer to an objective rational scientific method of analysis, whereas many branches of continental philosophy consider irrational elements in human nature. Both traditions include some epistemological features that cast doubt upon the validity of scientific arguments; both consider the mediations between reality and our representation of it filtered by human factors. However, continental epistemology also frequently includes metaphysical and irrational elements that are usually absent in the analytical tradition. Whatever metaphysics there is in the Anglo-Saxon tradition serves as a bridge between science/analysis/logic/reason and religion (anthropic principles, creationism, etc), but there is no truly metaphysical thinking detached from religion in the British and North American tradition. It resembles neither German nor ancient Greek philosophy.

Continental philosophy has more connection with literature; its practitioners have more character, more style. There is also concern about aesthetical topics, the search for beauty, etc, whereas the analytical and scientific traditions choose plain, formal, colourless prose, devoid of personality, eschewing aesthetic approaches, in an attempt to maintain neutrality with respect to subjective approaches, or at avoiding subjectivity. Analytical philosophy also lacks, or has very little, historical perspective in comparison with continental philosophy.

This attitude makes English-speaking philosophers much closer in their world view and tools of analysis to the scientific analytical mind than to the traditions of continental philosophy. Dialogue between analytical philosophy—which includes the philosophy of science—and science is possible. However, this analytical view usually forgets the human factors. In cases such as cosmology, which is closer to metaphysical speculation in some of its aspects, it also excludes other, richer world views beyond mathematical and empirical analysis.

Analytical philosophers also tend to specialise much more than their continental equivalents by emulating the behaviour of scientific colleagues. The result is that, in general and occasional exceptions, in the 20th and 21st centuries, philosophers born in or having migrated to the UK, the US, Canada, Australia, and other anglophone nations, abandon general views on the problems of philosophy and focus on particular problems, in the spirit of the analytical attitude: they break the Universe into multiple parts and study them separately in the hope that the whole will be the sum of the parts studied separately. This attitude has its advantages and disadvantages, and this is not the place to discuss whether the analytical or the continental perspective is better. What is clear is that different structures of thought give rise to different world views. Both philosophers and some researchers agree that the mode of thinking about a scientific problem greatly influences the result (e.g., Kurucz 1992). Particularly in the case of cosmology, whose aim is to understand existence in Nature, the different attitudes of philosophers arrive at totally different views.

9.5.2 Anglo-Saxon philosophy of cosmology

Philosophers more embedded in continental tradition—myself included—usually find misleading the contents of the many attempts to create a philosophy of cosmology linked exclusively to what scientists or analytical philosophers have to say about it. For instance, a conference proceedings volume with the title *The Philosophy of Cosmology* (Chamcham *et al* 2017) was published after a conference of the same name, celebrated in Tenerife (Canary islands, Spain; see figure 9.2)[5], although organised between Oxford and Cambridge Universities in the UK and in

[5] Although some of my works published before 2014 were very relevant to a conference on the combination of cosmology and philosophy, and I was living near the venue of the conference, I was not invited to give a talk, and the conference was not open to the submission of contributions either.

Figure 9.2. Poster announcing the conference on Philosophy of Cosmology celebrated in Tenerife (Canary islands, Spain) on 2014 September 12–16. [Reproduced from Public Domain https://philocosmology.files.wordpress.com/2014/08/schedule-and-abstracts-for-final-conference-12-16-sept-2014.pdf. University of Oxford and Templeton Foundation]

the US between Columbia, New York, Rutgers, UC Santa Cruz, and Yale Universities. It is a typical example of this kind of understanding of philosophy applied to cosmology: it was basically a book in which physicists or astronomers were invited to talk about cosmology and its speculations, and in which the very few philosophers who were invited focused only on the technical details of physics rather than talking about philosophy in a wider humanistic and historical sense. There were no philosophers addressing questions concerning what we call 'philosophy' in the continental tradition: nobody spoke about the cosmological approaches of German idealism by Immanuel Kant (1724–1804) or Friedrich Hegel (1770–1831); or the analysis of the structures of power associated with the discourse of science and reason, as in the tradition initiated by Friedrich Nietzsche (1844–1900) or Michel Foucault (1926–1984); or the psychoanalysis of culture as in Freud or Jung (as mentioned in section 9.3). There was almost no mention of classical philosophers, and many technicalities referring to inflation, quantum field theory and other aspects that are not philosophy (for a European philosopher, with the exception of the UK and followers of analytical philosophy); only physics was discussed. The few professional astronomers, such Bruno Binggeli, who have developed a truly original philosophical approach beyond a mere comment on a physics lesson, were not invited either. Also, no mention was made of any alternative to the standard model, not even MOND (Modified Newtonian Dynamics) (Merritt 2020 [Preface]), which is one of the most remarkable and with highest impact alternative ideas nowadays, received a mention. It seems that the purpose of this volume, like many other similar texts claiming to develop a philosophy of cosmology, was not to show the variety of possible philosophical approaches around the topic of cosmology, but to affirm that there is only one possible philosophy: acceptance of the standard cosmological model as the only possible truth about the Universe, with philosophy as a servant to this dogma.

While on the subject of MOND, David Merrit's (1955–) book *A Philosophical Approach to MOND* (Merritt 2020) is another example of how to do the philosophy of science applied to cosmology—in this case focusing on the MOND hypothesis, using a typically Anglo-Saxon approach. The author is not a philosopher, but purely an astrophysicist working with MOND. However, he adopts a philosophical approach that may be compared in many aspects to the work of many professional philosophers in Anglo-Saxon academia. His focus is on the ideas of Karl Popper (1902–1994) and Imre Lakatos (1922–1974), in which a hypothesis should be subjected to the test of falsifiability, tested against predictions with experiments or observations, before it can be accepted as a solid scientific theory. This is the main, and almost unique, philosophical idea in the book; the remainder of it being a lengthy dissertation in defence of the MOND hypothesis, trying to show that this is a better scientific hypothesis than that of dark matter; that is, it tries to be judge and party at the same time. In spite of the title, I think this is more a technical book of science with a mass of scientific details, the simple mention of Popper, Lakatos, and other philosophers of science serving more as a decorative background against which to expound a 'scientific' case in favour of MOND.

It is true that a hypothesis has to be verified with *a posteriori* experiments or observations, but that is a triviality that any scientist knows, and the ideas of Popper do not help scientists to be better in their research. In spite of its fame among scientists, I do not think Popper is a great revelation in philosophical spheres either, since many previous philosophers who have dedicated time to discussing the methodology of science since Galileo Galilei (1564–1642) or Francis Bacon (1561–1626) were clear about what the scientific method should be. Falsificationism is a new term for an old concept that had long been implicitly or explicitly present in any discussion of scientific methodology. I think that the expression 'Popper's falsificationism', together with William of Ockham's (c. 1280/1288–1349) 'razor', has become a mantra for invoking the spirit of philosophy every time scientists address technical and analytical issues with a nod to the Humanities.

Anglo-Saxon philosophy of cosmology is certainly a valid way to view the Universe: a few physical principles that can order the events of the Universe under the dogma that gravity rules the dynamics of matter–energy in space-time; and that we can understand the whole Universe in one century just by proffering few prejudiced ideas and adding ad hoc more elements every time some observation does not fit our preconceptions. Nonetheless, this is only one point of view among many. For a humanist trained in the history of ideas, discussing a single idea is not enough. Scepticism is indeed a philosophy of cosmology that is closer to the European tradition.

In my opinion, no true philosophy of cosmology has yet evolved. Philosophers in the wider continental tradition do not have much interest in the minutiae of scientific theories, and those who are deeply involved in the discipline of physical cosmology usually lack a wider historical, sociological, and metaphysical perspective to be able meaningfully to consider cosmology beyond the technical issues of science. The fact we can think about a philosophy of cosmology denotes a tendency towards an analytical position.

9.6 Cosmology as cultural expression

Throughout this chapter, I have examined different aspects of the influence of the ideas of a society in the cosmological pictures. There are religious, philosophical, political, and economic ideas among the social factors pushing cosmology in a given direction. Indeed, this happens with any science to some degree, but more so to cosmology than to others because of the high degree of speculation underlying its principles and the implications of that science for humanity, as factors that make cosmology particularly sensitive to social factors. Nonetheless, we must not forget that there is observational evidence in favour of the different cosmological theories. Nature is more than a social construct or any other kind of postmodern claim. Cosmology is a mixture of objective analysis and biased ideologies. Determining the ratio of natural and social components in this mix is the key to the question, for which we need a historical perspective that will only become available in the future, once we pass the present era of cosmological delusion.

References

Alfvén H 1978 How should we approach cosmology? *Problems of Physics and Evolution of the Universe* ed L V Mirzoyan (Yerevan: Publishing House of the Armenian Academy of Sciences) p 9

Barrow J D and Tipler F J 1986 *The Anthropic Cosmological Principle* (Oxford: Oxford University Press)

Ben Salem K 2005 The evolution of the universe: a new vision *Pac J. Sci. Technol* **6** 37

Binggeli B 2006 *Primum Mobile. Dantes Jenseitsreise und die moderne Kosmologie* (Zurich: Ammann Verlag & Co)

Binggeli B 2017 Primum mobile. Dante's journey and modern cosmology *Società Dante Alighieri, Rome, 16 Nov. 2017* (http://www.brunobinggeli.ch/pdf/DanteRoma-en.pdf)

Bube R H 1971 Man come of age: Bonhoeffer's response to the god-of-the-gaps *J. Evang. Theol. Soc.* **14** 203–20 (https://www.etsjets.org/files/JETS-PDFs/14/14-4/14-4-pp203-220_JETS.pdf)

Carmeli M 2000 The first six days of the universe (arXiv:astro-ph/0008040)

Chamcham K, Silk J, Barrow J D and Saunders S 2017 *The Philosophy of Cosmology* (Cambridge: Cambridge University Press)

Coulson C A 1955 *Science and Christian Belief* (Chapel Hill, NC: University of North Carolina Press)

Critchley S 2001 *Continental Philosophy: A Very Short Introduction* (Oxford: Oxford University Press)

Davies P C W 1983 *God and the New Physics* (New York: Simon and Schuster)

Davoust E 1995 The purpose of astronomy *Vistas Astron* **39** 315–22

Einasto J 2001 Dark matter and large scale structure *Historical Development of Modern Cosmology (ASP Conf. Ser. 252)* ed V J Martínez, V Trimble and M J Pons-Bordería (San Francisco, CA: Astronomical Society of the Pacific) pp 85–107

Ellis G R 2007 Issues in the philosophy of cosmology *Handbook in Philosophy of Physics* ed J Butterfield, J Earman, D M Gabbay, P Thagard and J Woods (Amsterdam: Elsevier) pp 1183–286

Gale G 1993 Philosophical aspects of the origin of modern cosmology *Encyclopedia of Cosmology* ed N S Hetherington (New York: Garland Publishing) pp 481–95

Gamow G 1952 *The Creation of the Universe* (New York: Viking Press)

Geller M J and Huchra J P 1989 Mapping the universe *Science* **246** 897–903

Gilmore G, Wyse R F G and Kuijken K 1989 Kinematics, chemistry, and structure of the galaxy *Annu. Rev. Astron. Astrophys.* **27** 555–627

Grünbaum A 1993 Creation in cosmology *Encyclopedia of Cosmology* ed N S Hetherington (New York: Garland Publishing) pp 126–36

Hu D 2004 Organized criticism of Einstein and relativity in China, 1949–1989 *Hist. Stud Phys. Biol. Sci.* **34** 311–38

Hubble E P 1925 NGC 6822, a remote stellar system *Astrophys. J.* **62** 409–33

Jaki S L 1988 *The Savior of Science* (Washington, DC: Gateway Editions)

Jastrow R 1978 *God and the Astronomers* (New York: Norton & Co)

John Paul II (Pope) 1981 Scientific research and man's spiritual heritage *Address of Pope John Paul II to the Pontifical Academy of Sciences (Oct. 3, 1981)* (Rome: Pontifical Academy of Sciences)

Kragh H 1996 Cosmology and Controversy *The Historical Development of Two Theories of the Universe* (Princeton: Princeton University Press)

Kragh H 2007a *Conceptions of Cosmos. From Myths to the Accelerating Universe: A History of Cosmology* (Oxford: Oxford University Press)

Kragh H 2007b The controversial universe: a historical perspective on the scientific status of cosmology *Phys. Phil.* **2007** 008 (https://www.semanticscholar.org/paper/The-Controversial-Universe%3A-A-Historical-on-the-c-Kragh/4177ebb95da4187407179acaddace1c52e0ad8bb)

Kurucz R L 1992 Gedanken astrophysics: the universe since recombination *Comments Astrophys.* **16** 1–15

Lemaître G 1931 The beginning of the world from the point of view of quantum theory *Nature* **127** 706

Lerner E J 1991 *The Big Bang Never Happened: A Startling Refutation of the Dominant Theory of the Origin of the Universe* (Toronto: Random House)

López-Corredoira M 2008 What is research? *Against the Tide. A Critical Review by Scientists of How Physics and Astronomy Get Done* ed M López-Corredoira and C Castro Perelman (Boca Raton, FL: Universal Publishers) pp 219–25

López-Corredoira M 2009 Quantum mechanics and free will: counter-arguments *NeuroQuantology* **7** 449–56

López-Corredoira M 2013 Peaks in the CMBR power spectrum II: Physical interpretation for any cosmological scenario *Int. J. Mod. Phys.* D **22** 1350032

López-Corredoira M 2015 Ajuste fino: Nueva versión del mito del dios-relojero para tapar agujeros en el conocimiento científico *Naturaleza y Libertad* **5** 83–94

Luminet J-P 2008 Is science nearing its limits? Summarizing dialogue *Is Science Nearing Its Limits?* ed G Steiner and E Rui Vilar (Manchester: Carcanet Press) pp 180–200

Merritt D 2020 A Philosophical Approach to MOND *Assessing the Milgromian Research Program in Cosmology* (Cambridge: Cambridge University Press)

Meyers R 2003 A Brief history of competing ideologies in cosmology and evidence for non-cosmological redshifts as a case for alternative theoretical interpretations in cosmology *PhD thesis* University of Western Sydney, Sydney

Narlikar J V 2001 The evolution of alternative cosmologies *Historical Development of Modern Cosmology (ASP Conf. Ser. 252)* ed V J Martínez, V Trimble and M J Pons-Bordería (San Francisco, CA: Astronomical Society of the Pacific) pp 175–99

Ostriker J P, Peebles P J E and Yahil A 1974 The size and mass of galaxies, and the mass of the universe *Astrophys. J. Lett.* **193** L1–4

Pius XII (Pope) 1952 Modern science and the existence of god *The Catholic Mind* **49** 182–92

Roberts R E 1924 *The Theology of Tertullian* (London: Epworth Press)

Sigmund P E 1963 *Nicholas of Cusa and Medieval Political Thought* (Cambridge, MA: Harvard University Press)

Singer D W 1950 *Giordano Bruno, His Life and Thought with Annotated Translation of His Work, on the Infinite Universe and Worlds* (New York: Abelard-Schuman)

Soler Gil F and López Corredoira M 2008 Dios o la materia *Un debate sobre cosmología, ciencia y religión* (Barcelona: Áltera)

Ten Ros A E 2001 Aspects of the popularization of cosmological ideas at the beginning of the XXth century *Historical Development of Modern Cosmology (ASP Conf. Ser. 252)* ed V J Martínez, V Trimble and M J Pons-Bordería (San Francisco, CA: Astronomical Society of the Pacific) pp 309–23

The Editors of Encyclopaedia Britannica 2013 Anthropic principle (accessed 2021, May 25th) *Encyclopaedia Britannica* (UK & US: Encyclopaedia Britannica, Inc.)

Wright R 1993 Science, god and man *Time* (4 Jan 1993)

Chapter 10

Epilogue

Final remarks, summing up, and discussion.

According to some authors, cosmology has been examined both as a science, just like any other field of research in the natural sciences, and as myth, a purely intellectual ideological construction derived from our culture. Possibly neither of these two exclusive approaches is correct, something intermediate being more appropriate. The science of astrophysics is a powerful tool for describing Nature, and many partial truths are possibly contained in it, but the Universe is very large and we are very small to be able to understand it as a whole with independence from our circumstances. Perhaps we should even avoid the words 'universe' and 'cosmologist', as Rutherford and Mike Disney have suggested. In any case, for those who decide to earn a salary in the business of cosmology some advice is offered in this chapter.

To summarise the contents of the book up to here, there are nowadays many cosmological theories—the standard Big Bang ΛCDM (chapter 1) and many others (chapter 2)—but only the standard one is seriously considered among most cosmologists, alternative ones being usually neglected except for a reduced group of professional cosmologists, with perhaps some exceptions in cases of alternative models including small variations with respect to the standard one, but not those that challenge the most fundamental ideas of modern cosmology.

Throughout this book, I have tried to argue that there are two main causes for the success of the standard model: physical observations and social factors. Many observations can be explained with the current version of the Big Bang theory, although many of these successes were obtained after a number of ad hoc corrections were made to the theory, a process somewhat comparable to the epicyclic constructions of Ptolemy. Moreover, many tenets of the standard cosmological theory that are being cast into doubt in a wide-ranging literature. While it is certainly

doi:10.1088/978-0-7503-3775-5ch10

true that many of the references that I have cited in the review of such problems in chapters 3–7[1] may contain wrong or uncertain arguments and results, and further research is needed to confirm or refute their statements, the presence of such a wide literature indicates at the very least that the model is not perfect yet, and that we have not reached the stage at which we can say that the model perfectly explains everything about stars, galaxies, background radiations, etc.

Many cosmologists prefer to ignore papers that talk about some controversy or problem with ΛCDM, simply consider them to be wrong, or assume that any apparent problems will soon be solved after a reanalysis that takes into account some ad hoc corrections that remain to be done. However, they do not level the same criticisms at papers supporting the standard model. Why are papers in disagreement with theoretical expectations suspected to be wrong and those supporting the standard model deemed to be definitively correct? It does not look like a perfect application of objective scientific method. There are also those who advocate the standard model because they believe it fits some types of observations. So, for example, when problems arise in large-scale structure, they consider these not to be serious because we know from the CMBR that ΛCDM works perfectly. And when problems or anomalies in the CMBR surface, they claim that they cannot be correct because we know that large-scale structure is perfectly reproduced with the standard model, and so on—not very objectively scientific.

In my opinion, the physical arguments that often adduced may confer some advantages on the standard model on many points, but not enough advantage to consider it an indisputable truth. Neither are there sufficiently strong arguments for many of the alternative proposals to be definitively rejected. Alternative models are certainly less developed than the standard model, because far few researchers have been working on them. Social factors are important in this competition. As argued in chapter 8, the 'Matthew effect' (in which resources get distributed to the most successful theory) acts as a potential in a field that attracts cosmologists, causing funds, research positions, prestige, telescope time, publication in top journals, citations, conferences, and other resources to be dedicated almost exclusively to the standard cosmology. Moreover, we have social factors that are external to science: religious, economic, and political ideologies of the society of which we are a part largely determine the contents of cosmological ideas, as explained in chapter 9. The imposition of Anglo-Saxon culture around the world is pushing cosmologies and other philosophical ideas towards a particular analytical view derived from the dominant culture.

In this chapter, I conclude with further short discussions on how close to a science or a myth cosmology should be considered. I also offer recommendations for new generations of cosmologists—older generations are unchangeable—and few final remarks.

---

[1] There are also other reviews with an even greater number of references than this book; for example, the magnificent recent piece 'Challenges for ΛCDM: An update' (Perivolaropoulos and Skara 2021).

10.1 Cosmology: science or myth?

10.1.1 Is a science of cosmology possible?

> And we would pretend to understand everything about cosmology, which concerns the whole Universe? We are not even ready to start to do that. All that we can do is to enter in the field of speculations. So far as I am concerned, I would not comment myself on any cosmological theory, on the so-called 'standard theory' less on many others. Actually, I would like to leave the door wide open. (Jean-Claude Pecker, in Narlikar *et al* (1997))

I agree with Jean-Claude Pecker (1923–2020), a classical heterodox and dissident cosmologist. Before wondering which is the true model of cosmology, we must wonder whether we have the conditions to create a theory on the creation (or non-creation) and evolution of the whole Universe, whether the psychological–sociological conditions of the cosmologists are or are not weightier factors than observations of Nature. Is present-day cosmology dominated by our culture or by Nature's objective truths?

Rather, I would say that, before understanding the Universe, we must understand the pieces of the puzzle separately (galaxies, their formation, their evolution, whether they are separate from each other, the origin of the elements, the origin of the background radiations, etc), instead of assembling all of them into a simple idea that could convert astrophysics into a speculative science. There are, however, many cosmologists, philosophers, and historians of science who think that cosmology became an empirical science beyond speculation after the discovery of the Cosmic Microwave Background Radiation. This is what Helge Kragh (1944–) (e.g., Kragh 2007), a remarkable historian of modern cosmology, says. As I have maintained throughout this book, I would not agree with the statement by Kragh (2007) that cosmology is a proper science like nuclear physics, hydrodynamics, etc. Even if there are aspects which are comparable with observations, these are just a few partial aspects of the whole reality, whereas cosmology stands for a science of the whole Universe and its entire history, something for which we do not have all the empirical or observational information that we need to fill in the many gaps in that history that are so far questions of pure speculation and hazardous extrapolations. Because of the limitations on inferring truth, cosmology should not be taken as a realistic approach (Hogg 2009)[2]. Cosmology has extremely little or no predictive power and is of a speculative nature (Hakim 1989). It employs principles that are unverifiable because of their metaphysical nature, especially when we talk about the primordial Universe.

[2] However, Hogg (2009) claims that we understand many aspects of cosmology, and that some facts are established beyond doubt, such as the expansion; and he says that 'it is difficult to imagine another theory truly replacing ΛCDM'.

Is cosmology perhaps comparable to palaeontology in its attempts to reconstruct the facts from fossils (Kragh 2001, Ćirković 2002, Ćirković and Perović 2018)? Kragh (2001) replies affirmatively since cosmologists do indeed look for fossils, traces from the past that can tell us about the different conditions of the Universe several billion years ago. Astronomy certainly does not follow the typical scientific method because it observes, rather than performs experiments on, the past of the Universe. However, it cannot be put on the same level of scientificity as palaeontology, because the objects of study in palaeontology are much more limited, and the geological and biological processes are known, whereas cosmologists play with elements of which there is no direct experience (dark matter, dark energy, new physics, etc), or they must make extrapolations and assumptions for which there is no evidence, such as the cosmological principle, that the laws of physics do not change over time, etc. This means that the process of choosing between standard and non-standard models in cosmology is less balanced, less based on evidence than in other scientific disciplines. In any case, there are also huge extrapolations involved in disciplines such as palaeogeography and palaeobiology. One may indeed doubt the different theories of different fields of science, but for different reasons.

Another argument considering cosmology as a science is made by Kragh (2007): That Nobel prizes have been given to a number of cosmologists. In my opinion, this does not necessarily mean anything if we consider that Nobel prizes are merely a part of social and cultural structures. Recognition does not necessarily endow a branch of knowledge or its creators with higher value, but simply heightens their status. Indeed, there are many social and economic interests at play in declaring cosmology to be a solid science; there is a lot of skin in the game, and this drives the claim that we can know the whole Universe and its history.

The physicist Ernest Rutherford (1871–1937) used to say, 'Don't let me hear anyone use the word "universe" in my department'. In the same style, the astrophysicist Mike Disney (1937–) dared to declare: 'The word 'cosmologist' should be expunged from the scientific dictionary and returned to the priesthood where it properly belongs' (Disney 2000). Naturally, the very word 'universe', meaning everything that exists or has existed, deserves some discussion, since we have access to only a small part of the observable cosmos that might not be generalisable to the whole, unless we work under the 'good luck' assumption that is part of the cosmologist's dogma (Disney 2000).

These are the words of an old-style scepticism. Nowadays, the young bloods of precision cosmology do not care for such statements and proudly claim that people in the past did not know what we now know today. Cosmologists show no trace of doubt and an amazing sense of security when dissertating on highly speculative topics. Science does indeed advance, and cosmology advances in terms of both the flood of new data and its epicycle-like patches to persuade theory to fit the data, but the great questions remain. Many wise men have deliberated on cosmology over many long centuries without ever reaching a definitive solution.

10.1.2 Is cosmology a pure myth?

Just as Galileo Galilei's (1564–1642) hopes were thwarted that the old authorities, who had spent decades of their careers studying Aristotelian physics and astronomy, would suddenly change their opinions about the geocentric world system, even after they were shown the moons of Jupiter or the craters of the Moon through Galileo's telescope, neither may we anticipate any seismic shift in mentality in cosmology today. 'Faith is the organ of knowledge, intellect is secondary. Your science without premise is a myth', says Naphta, a character in Thomas Mann's (1875–1955) *The Magic Mountain.*

While considering in general as being a myth is highly debatable, it is nevertheless quite reasonable to view cosmology as in a sense being a blend of scientific knowledge of Nature and sundry sociocultural ideologies. Is the standard model of cosmology—some authors wonder (Alfvén 1976, 1978, Madrid Casado 2018)—akin to, say, the Hindu myth of a cyclic Universe, the Chinese cosmic egg, the Biblical myth of the creation of the world in six days, the Ptolemaic cosmological myth, or does it somehow resemble a fairytale (Disney 2007)? For the heterodox cosmologist and Nobel Prize of Physics laureate Hannes Alfvén (1908–1995), it is a myth adorned with sophisticated mathematical formulae that make it more credible, but it is not necessarily true (Alfvén 1976, 1978). 'When the big bang arrives, with it again arrives metaphysics,' said the physicist Antonio E Ten Ros (1949–) when referring to cosmological ideas at the beginning of the 20th century (Ten Ros 2001). The philosopher Carlos M Madrid Casado (1980–) talks about the myth of cosmos, the belief in the existence of a cosmos that can be known thanks to science (Madrid Casado 2018 [chapter 23]). Madrid Casado, who advocates that our idea of the cosmos is just a myth, states that a science of cosmology is impossible because the whole reality of the Universe cannot be grasped by our science.

I disagree a 'pure' myth approach. Nothing is pure in human intellectual activities; there is not pure reason able to determine the contents of Nature independently of the mind of the observer, as one of the most important modern philosophers, Immanuel Kant (1724–1804), posited in his first *Critique*. However, neither is there a pure mind or pure society or culture in our scientific affairs either. Criticism is not dogmatism, and denying the existence of the Universe and the possibility of knowing some of its elements looks as dogmatic as entertaining no doubts on the Big Bang hypothesis as the reality of the Universe. There are limits to our knowledge of the Universe, but this does not mean that everything in present-day cosmology is a fable comparable to the superstitions of primitive societies. Or, as said by the philosopher Bjørn Ekeberg (1978–) with regard to cosmology, 'This is not to say scientific knowledge is equivalent to any kind of belief, only that our current world-picture is powered by this persistent historical form of faith'.

10.2 Desiderata for broadened perspectives in cosmology

10.2.1 Openmindedness

According to Hogg (2009), Loeb (2010), astrophysicists born before 1950 are much more likely to discuss an outrageous idea about the Universe than those born after 1970. We can also further distinguish a transition generation born between 1950 and 1970, and a hyperorthodox generation born after 1990, who think that dark energy is as real as the cutlery they use during lunch. The obstacle that a pluralistic approach to research in cosmology may encounter in the near future is that those ready to discuss ideas critically have now retired, or are close to retirement, and the new emerging generations, the millennials of science, seem to be absorbed exclusively in technology, acquiring funds to continue working in an area of science that has become extremely competitive, and contentedly pursuing an optimistic epistemology that allows them calmly to dedicate themselves to their hyperspecialized careers without falling prey to the unsettling philosophical approaches so typical of German *Kulturpessimismus*. Realistically, with the exception of a small coterie of individuals, the generations born between ~1970 and ~2010 are lost generations from the intellectual standpoint of open discussion within cosmology. But history shows us that cultural movements are cyclical, and that confidence in certain speculative ideas may be called into question every few generations. It is too much to hope that the present dominant generations working in cosmology will change their attitudes or views, but with luck future generations will more closely resemble those open-minded scientists born before 1950. This may indeed arrive within a few decades, if the avalanche of controversy worsens with time, perhaps in combination with the failure to identify dark matter candidates (perhaps even more so for the other three dark knights; see chapter 3).

Alternative theories are at present not as competitive as the standard model in cosmology. If they were to become more developed, they might possibly compete with some aspects of the ΛCDM theory. Perhaps new generations, dissatisfied with both Λ (dark energy or quintessence) solutions and Cold/Warm Dark Matter, Newtonian or relativistic gravity, etc, may move on to other fundamental ideas in cosmology. Those more embedded in philosophical thinking may also hook onto scepticism. Whatever that change in thinking might be, a more pluralistic and open-minded generation of cosmologists is both possible and highly to be recommended.

No specific dogma has been defended in this book, or comments made concerning the correctness for falsity of the standard cosmological model; neither has any particular stance been taken respecting constructivism or scientific realism (see, for example, Soler Gil 2012 for a discussion of these positions). A certain degree of scepticism has been expressed concerning the validity of the standard cosmology and a recommendation made to examine the problem from several perspectives. Readers may have different ideas about cosmology, and that is all to the good, provided no attempt is made to impose a specific line of thought. Free and tolerant exchange of different and opposing ideas ensures a healthy intellectual environment in which there may be frank and unfettered discussion on every subject, be it cosmology, philosophy, religion, politics, etc. It was such a plurality of ideas that gave rise to the Enlightenment, when science reached such dizzying heights. As pointed out by Bjørn Ekeberg, such an approach is sorely needed in cosmology:

> … despite all the reasons to be skeptical of the cosmological world–picture, and despite all the indications its paradigm may be creaking in its foundations,

> I can not claim that it is wrong, only that it is dubious. And I argue for a deeper questioning about the foundation of our sciences, open-eyed about the broader political context of power in which both the sciences and their critics operate. (Ekeberg 2019)

10.2.2 Modesty

In the 1990s, several leading cosmologists stated publicly that the true nature of dark matter would be discovered within five years (Turok 1996, Sanromà 2007). More than 25 years have passed since then and no dark matter non-baryonic particle has yet been found.

Nonetheless, quests to seek out answers to many of the lacunae in ΛCDM have not ceased, and an excessive epistemological optimism (see section 8.4.1) is still rampant. However confident we may inwardly be regarding the present-day cosmological enterprise, surely a little modesty might not be amiss. Perhaps dark matter will indeed be found in a few years' time, and the conflicts and problems of ΛCDM will be resolved. I am not denying such a possibility, but should things not turn out as optimistically expected, exercising a little caution when making claims would be prudent.

Taking into account that we have little direct information of the Universe as a whole, limiting our ambitions would seem to be a reasonable approach to adopt. There are limits to cosmology because we are finite human beings limited by our experiences and circumstances, not demigods able to read the mind of a creator playing mathematical games with the Universe, as some Pythagoreans may think. There is a lack of humility in Pythagoreanism, or in expressions such as 'precision Cosmology'. Orthodox cosmologists should count themselves lucky if they are able to argue in favour of the standard model as a 'rough' description of the Universe.

One of the most reputed physicists of the former Soviet Union, Lev Davidovich Landau (1908–1968), said: 'Cosmologists are often in error, but never in doubt'. Never truer than today, although many cosmologists are stung by it. A leading cosmologist more recently announced: 'I am certain that it is time to retire Landau's quote' (Turner 2001).

The great old masters, even the creators of the standard model, were cautious in their assertions. Edwin P Hubble (1889–1953) throughout his life doubted the reality of the expansion of the Universe that he (and others) associated with the redshifts of galaxies. Willem de Sitter (1872–1934) claimed: 'It should not be forgotten that all this talk about the Universe involves a tremendous extrapolation, which is a very dangerous operation' (de Sitter 1931). This scepticism is healthy since 'all caution is too little' (Spanish proverb). It is not a question of substituting one model for another, since it would be the 'same dog with a different collar' (another Spanish proverb), but of realising that cosmology is limited as a science.

We must also admit that, in terms of giving better explanations, alternative theories are not, in their present state, as competitive as the standard model of cosmology. If they were more developed, they might possibly compete in some aspects with the Big Bang theory, but every effort is made in today's scientific community to stifle their development (chapter 8). In any case, one advantage of alternative theories is that their study may lead to new useful cosmological tests, even if the theories themselves are not viable; for example, the redshift–angular size test (see section 4.5) in connection with studies of the Steady State theory proposed by Hoyle (1959). We need to stick to hard facts, free from interpretation, just as Lemaître or Hubble did when they discovered the expansion of the Universe, and good physicists to propose new theories in the tradition of Einstein, Gamow, and others.

The huge body of literature on cosmology makes it more and more difficult to propose a theory without discussing all previous studies. Heterodox cosmologists in their turn would therefore do better if they adopted an even more modest position than orthodox researchers. They must avoid adopting such unprofessional attitudes as claiming that the Big Bang idea is totally wrong, and that their own cosmological model would solve all of the issues that defy the standard theory.

10.2.3 Patience

It was only around 100 years ago when the origin of the spiral nebulae (Hetherington 1993) was hotly debated, and whether they were systems of stars outside our own Galaxy or part of it. We are now attempting to construct models of the Universe in its entirety. It is perhaps over-ambitious and precipitate to attempt such a huge undertaking. The astronomer Gérard de Vaucouleurs (1918–1995) pointed out in 1970 that perhaps our ideas about cosmology are quite premature (de Vaucouleurs 1970). Just because we have added some new patches to the theory and have more precise data about galaxies, light element abundances, and background radiation, the situation now is not radically different from the state of the art 50 years ago, so possibly we may also consider the development of whatever we try to build—be it science or metaphysical speculation—to be still far from a finished state.

In science to day there is constant pressure to publish, give talks at conferences, acquire the next postdoc or tenure, etc. There is little time to think, and this lack of time to devote to the most important things in science, the fundamental ideas, is substituted by a vast production of repetitive papers in a race to get 15 minutes of fleeting glory in a new press release announcing the next spectacular advance in cosmological knowledge, even though we are aware that we are playing the same repetitive game with minor variations. A leap forward in a field like cosmology can be produced only by a substantial advance in our understanding of the cosmos, but this may take very long time compared to the career time-span of individual scientists, who might well be barred from dominant positions of the business of cosmology, because of their low visibility and repetitious performances in conferences and press releases. In serious astrophysics, however, the way to be a real scientist rather than a businessman is by gradually improving one's ideas, not by participating in public debates but through distinctive work over many years of reading, thinking, and calculating before anything useful should see the light of day; in other words, by emulating the scientists of more than half a century ago.

Herbert Dingle (1890–1978), philosopher and scientist, gave this advice to the few cosmologists of 1950s. It still holds today:

> Let our younger cosmologists forget cosmology for the space of three years—the Universe is patient—, it can wait, and instead read the history of science—I mean the work of great scientists themselves. After asking themselves what meaning it has for the work of today, let them return to cosmology and give their attention again to the great problems into which they have prematurely rushed. (Dingle *et al.* 1953 [March 13])

This call to patience is of particular relevance to those few researchers who dare to create a new cosmology. The career of heterodox cosmologist is long and with many obstacles along the way. The researcher needs to publish some tens of papers, some of them on orthodox cosmological issues, before daring to present any innovative and revolutionary idea. The researcher must also advance step by step, focusing each paper on different astrophysical observations, rather than trying to create an all-encompassing theoretical idea that attempt to solve everything at once. There are no short cuts to a revolution in cosmology. We may repeat Euclid's (c. 325–c. 270 BC) reply to King Ptolemy I's request for an easier way of learning mathematics: 'There is no Royal Road[3] to geometry'.

10.2.4 Funding and professionalism

Some creators of alternative ideas in cosmology, having encountered difficulties in getting their ideas funded, have opted to follow lonely paths as solitary outsiders, away from the bustle of research centres and university departments, and have chosen to earn a salary from employment other than research in astrophysics in order to dedicate their spare time to free thinking in cosmology. Unfortunately, this is a sure way to be classed as dilettanti and be automatically shunned in any professional discussion of cosmological ideas.

A somewhat different approach, if you are a professional researcher in astrophysics interested in exploring heterodox ideas in cosmology, is to continue doing 'normal' research (without, say, trying to disprove the Big Bang is wrong, that Einstein was wrong, etc) for more than 50 percent of the research time in any field that provides funding, a position, and the possibility of publishing in major journals. In that way, one would be an insider, part of the system, which is the only way to get any attention (sometimes). Research is an endurance marathon, not a short sprint. There is, of course, a problem with this approach: many scientists who started with high ideals to do challenging research somehow lose themselves in mid-race, perhaps tempted by obtaining positions of power, or bogged down in endless bureaucratic

[3] The Royal Road was an ancient highway reorganised and rebuilt by the Persian king Darius the Great (Darius I) of the first (Achaemenid) Persian Empire in the 5th century BC, to facilitate rapid communication throughout his very large empire (Graf 1994).

tasks, quite forgetting pristine intentions. This endurance test also produces a selection effect, with only a few attaining their initial goals, and only a select few of these will ever achieve a true revolution in fundamental cosmological ideas.

10.2.5 Broad scientific and philosophical outlook

From a pessimistic point of view and epistemologically speaking (the opposite of the optimism described in section 8.4.1), cosmology is an impossible enterprise. It is true that there are many claims concerning the limits of cosmology (section 10.1.1), and there are even those who frown on the use of the word 'cosmology'. Nevertheless, physics and astronomy are great sciences able to throw light on many aspects of our Universe, at least partially. There can be little doubt that our sciences have harvested an impressive crop of knowledge about stars, galaxies, and astrophysical processes. The Universe is a vast laboratory enabling us to test physical laws and make predictions.

Rather than viewing cosmology as an occupation, we should instead speak about physics and astronomy, or combine both as in astrophysics. And rather than wasting our life within a very narrow specialisation of the standard cosmological model, either as a theoretician or as an observer, it would be better to carry out research in a wider sense, touching on many fields, learning many new techniques. This would make research less obsessively tied to a given cosmological model. It is of course understandable if, after having dedicated several decades of one's career to searching for non-baryonic dark matter, deriving cosmological parameters from the CMBR, or any similarly narrowly focused specialised task, that one would not take kindly to adopting a more broadly based research outlook; one might then be led to accept that the research of a lifetime has hitherto not been the best use of time. However, in a diversified career in many different areas of research, one may more easily come to admit that certain conclusions were wrong and others were correct. That is how research should work, betting on some ideas and abandoning them later when we change our mind, but for this we need to diversify our bet and not to put everything in the same portfolio (Loeb 2010). As a saying says, do not put all your eggs in one basket.

Science was once call the Philosophy of Nature, a branch within the great tree of intellectual thought. It is not recommended that professional scientists should also be professional philosophers, or vice versa; that is unfortunately too difficult a feat in today's cultural marketplace. However, any person dedicated to cultural activities should be aware of the general philosophical framework in which those activities take place. For cosmology in particular, or whatever else you choose to call it, a broad vision should be considered a plus. Cosmology cannot be pursued without taking into account human factors and biases in science. It cannot be blindly assumed that just looking at CMBR we are witnessing the beginning of the Universe. Cosmologists need to study their colleagues, observe how science is funded, and analyse cosmology as a multi-parametric problem—and not merely with reference to the cosmological parameters (H_0, σ_8, etc). Western science in general may possibly, but debatably, be absolute and universal, but, unlike other sciences, cosmology is greatly influenced by culture.

Scepticism is a good option, my favourite one, given the degree of uncertainties and the difficulties in separating the phenomena of Nature from cultural interaction. Nonetheless, other approaches are possible, either within some standard or alternative dogmatism or with an absolute postmodern relativism. Whatever option is chosen, scientific, philosophical, historical, and sociological analyses are helpful in understanding our place, if any, in the Universe.

10.3 Some closing personal remarks

Among the many messages throughout this book, I would like to emphasise one particular leitmotiv, indicated at the very beginning of the chapter 1: the Universe is very large and we are very small compared to it, not only in size. Creating a cosmology that attempts an explanation of the whole existing Universe is an obsession shared by all cultures. We think we now have the true cosmological model, like many civilisations that have gone before us. There are reasonable grounds for thinking that the total truth about the origin and evolution of the Universe is beyond our reach. Nonetheless, science is a powerful tool for describing Nature, and many partial truths are possibly contained in our astrophysical knowledge.

My hope is that the reader of this book may find it useful and inspiring in order to carry out new and challenging research as a scientist or as a humanist, rather than adopting the dull, conformist attitude so prevalent in present-day cosmology. I hope, after you have read this book or some parts of it, that you will seriously consider whether everything in the standard model is really quite so clear and well established, the only remaining mission left being the refinement of the measurement of certain parameters. I am convinced that the game is far from over. There are many pending tasks still to be addressed. We might indeed soon find what non-baryonic dark matter is—perhaps. Or someone may discover a definitive proof that the Big Bang theory is wrong—maybe. Or possibly, a new sociological, philosophical, or historical approach, or a combination of all of these, will shed new light on the psychological processes that have determined our present-day cosmovision. We cannot rashly declare that we are witnessing historical moments in cosmology. All moments are part of the history of the development of ideas. As western hegemony declines and new nations with different cultures come to the fore, the coming generations will infuse a welcome breath fresh air into the study of cosmology and humanity's view of the cosmos.

References

Alfvén H 1976 La cosmologie: Mythe ou science? *La Recherche* **69** 610–6

Alfvén H 1978 How should we approach cosmology? *Problems of Physics and Evolution of the Universe* ed Universe (Yerevan: Publishing House of the Armenian Academy of Sciences) p 9

Ćirković M M 2002 Laudatores temporis acti, or why cosmology is alive and well–a reply to Disney *Gen. Relativ. Gravit.* **34** 119–30

Ćirković M M and Perović S 2018 Stud. Hist. Phil. Mod. Phys. *Alternative explanations of the cosmic microwave background: a historical and epistemological perspective* **62** 1–18 de Sitter W 1931 Contributions to a British Association discussion on the evolution of the universe *Nature* **128** 706–9

de Vaucouleurs G 1970 The case for a hierarchical cosmology *Science* **167** 1203–13

Dingle H *et al* 1953 1953 March 13 meeting of the Royal Astronomical Society *Observatory* 73 41–9

Disney M J 2000 The case against cosmology *Gen. Relativ. Gravit.* **32** 1125–34

Disney M J 2007 Modern cosmology: science or folktale? *Am. Sci.* **95** 383–5

Ekeberg B 2019 *Metaphysical Experiments. Physics and the Invention of the Universe* (Minneapolis, MN: University of Minnesota Press)

Graf D F 1994 The Persian royal road system *Continuity & Change (Achaemenid History, vol 8)* ed A Kuhrt, M C Root and H Sancisi-Weerdenburg (Leiden: The Netherlands Institute for the Near East) pp 167–89

Hakim R 1989 The special status of cosmology in science *Foundations of Big Bang Cosmology* ed F Walter Meyerstein (Singapore: World Scientific) pp 85–139

Hetherington N S 1993 Great debate *Encyclopedia of Cosmology* ed N S Hetherington (New York: Garland Publishing) pp 260–2

Hogg D W 2009 Is cosmology just a plausibility argument? (arXiv:0910.3374)

Hoyle F 1959 The relation of radio astronomy to cosmology *Radio Astronomy (IAU Symp. 9)* ed R N Bracewell (Stanford: Stanford University Press) pp 529–33

Kragh H 2001 Nuclear archaeology and the early phase of physical cosmology *Historical Development of Modern Cosmology (ASP Conf. Ser. 252)* ed V J Martínez, V Trimble and M J Pons-Bordería (San Francisco, CA: Astronomical Society of the Pacific) pp 157–69

Kragh H 2007 The controversial universe: a historical perspective on the scientific status of cosmology *Phys. Phil.* **008** 14 (https://eldorado.tu-dortmund.de/bitstream/2003/24422/1/008.pdf)

Loeb A 2010 Taking "the road not taken": on the benefits of diversifying your academic portfolio (arXiv:1008.1586)

Madrid Casado C M 2018 *Filosofía de la Cosmología* (Oviedo: Pentalfa)

Narlikar J V, Padmanabhan T and Burbidge G *et al* 1997 Questions and comments that followed the panel discussion *J. Astrophys. Astron.* **18** 477–89

Perivolaropoulos L and Skara F 2021 Challenges for ΛCDM: an update (arXiv:2105.05208)

Sanromà M 2007 El big bang: història i sociologia d'una teoria científica *Master's thesis* (Barcelona: Universidad Oberta de Catalunya)

Soler Gil F J 2012 *Discovery or Construction? Astroparticle Physics and the Search for Physical Reality* (Frankfurt am Main: Peter Lang)

Ten Ros A E 2001 Aspects of the popularization of cosmological ideas at the beginning of the XXth century *Historical Development of Modern Cosmology (ASP Conf. Ser. 252)* (San Francisco, CA: Astronomical Society of the Pacific) pp 309–23

Turner M S 2001 Two theorists never in doubt *Phys. Today* **54** 10

Turok N 1996 *Critical Dialogues in Cosmology* (Singapore: World Scientific)

Appendix A

Notes on the content of the book

A.1 Reproduction of previous texts by the author

A minor part of the material in this book has already been published in two papers of Martín López-Corredoira: 2014 Non-standard models and the sociology of cosmology *Studies in History and Philosophy of Modern Physics* **46-A** 86–96 (special issue on *Philosophy of Cosmology*); 2017 Tests and problems of the standard model in Cosmology *Foundations of Physics* **47** 711–68. Permission of reproduction provided. Another paper is to be published during the production of this book: Martín López-Corredoira & Louis Marmet 2022 Alternative ideas in Cosmology *Int. J. Mod. Phys. D*, in press.

A.2 Bibliography

The total number of references is 969: 82 books, 3 PhD theses, 1 Master thesis, 735 articles published in journals, 93 proceedings of conferences or chapters of books, 49 preprints only posted at arXiv.org (43) or viXra.org (6), 6 unpublished manuscripts or oral presentations. A histogram of the distribution of years of publication of the references is given in figure A.1.

Of the 735 articles published in journals, the most cited journals are indicated in table A.1.

doi:10.1088/978-0-7503-3775-5ch11

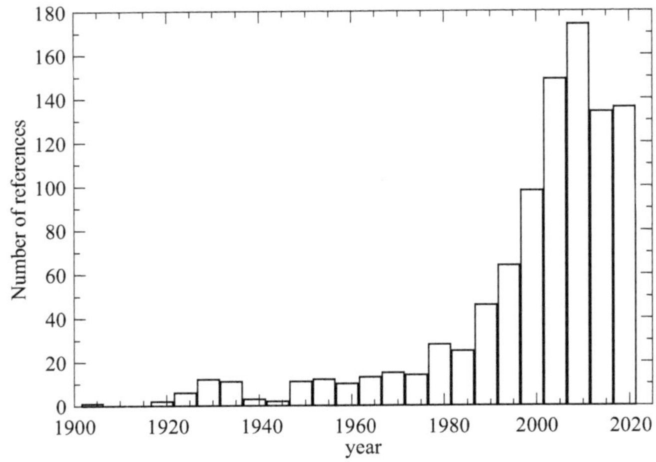

Figure A.1. Histogram of the distribution of years of publication for the references published in this book.

Table A.1. Journals of the bibliography, ordered in number of references or alphabetical order for the same number.

Journal	No. of cited papers
Astrophys. J.	141
Mon. Not. R. Astron. Soc.	99
Astron. Astrophys.	74
Nature	41
Astrophys. J. Lett.	28
Phys. Rev. D	21
Astrophys. Space Sci.	20
J. Cosmol. Astropart. Phys.	16
Annu. Rev. Astron. Astrophys.	9
Phys. Rev. Lett.	9
Science	9
Astron. J.	7
Int. J. Mod. Phys. D	7
Proc. Natl Acad. Sci. USA	7
Astrophys. J. Supp. Ser.	6
Universe	6
Class. Quantum Grav.	5
Gen. Relativ. Gravit.	5
J. Astrophys. Astron.	5
J. Cosmol.	5
Nat. Astron.	5
Phys. Dark Universe	5
Phys. Lett. B	5
Rev. Mod. Phys.	5
OTHERS	195

www.ingramcontent.com/pod-product-compliance
Ingram Content Group UK Ltd.
Pitfield, Milton Keynes, MK11 3LW, UK
UKHW051857100926
8556IPUK00005B/662

9 780750 337731